I. Harold Zimmerman

Computational Physics

Computational Physics

DAVID POTTER
Imperial College, London

A Wiley–Interscience Publication

JOHN WILEY & SONS
London · New York · Sydney · Toronto

Library of Congress catalog card number 72–8613

ISBN 0 471 69555 6

Set on Monophoto Filmsetter and printed by J. W. Arrowsmith Ltd., Bristol, England

Preface

My purpose in writing this book has been threefold. It originated in the first place from a course of lectures presented to final year undergraduate students at Imperial College London, and at the United Kingdom Atomic Energy Authority Laboratory at Culham, Berkshire. The book is firstly, therefore, an exposition of the subject of computational physics to introduce the uninitiated student to the field. Assuming no prior knowledge of the subject, the book introduces the reader to the fields of numerical analysis, numerical linear algebra and to the formulation of theoretical models on the computer.

Second, I have aimed to provide a brief and basic reference work for those working in the subject, which increasingly includes a wider and wider variety of physicists working in such diverse fields as fluid dynamics or solid-state physics. There has been a severe gap in the literature here, for while there have been many books on numerical analysis, and in particular one might mention the invaluable work by Richtmyer and Morton, *Difference Methods for Initial-Value Problems*, no book has presented a coherent description of the many finite models used in the simulation of physical problems. The physicist has had to search among the journals for the rapidly expanding literature on the subject. Again, many of the methods used in one field are appropriate in another and, due to the lack of communication between different branches of physics, there has been a great need to present a coherent view of the subject. It is my hope too that the book will be of value to other scientists and engineers, and to those developing computer models in meteorology, oceanography and aerodynamics.

But third, and most important, my purpose has been to gain a greater understanding of the future powerful role of the computer in theoretical physics. It is only twenty-five years since von Neumann envisaged the modern digital computer, and in that time the capacity of computers has multiplied many times. There has been an increasing proliferation of work on the simulation of physical phenomena, but there has been little time to appreciate the significance of what is being done, or to develop a well-tested methodology. Whereas experimental physics and theoretical analytical physics have had centuries to develop and standardize the method of approaching a problem, it is not yet clear what we can expect from computational physics, nor how we are to approach a particular physical problem.

The subject is so young and untried, that it is evident that there are many methods and ways of using the computer so far undreamt of, which can enormously enhance our understanding of the natural world. It is my hope that this book will be a small step along that intriguing path towards an understanding of the computer's role in physics.

I should like to thank the many people who have helped to make this book possible. In particular, I should like to thank Dr Harlow and his colleagues at Los Alamos, New Mexico, for providing figures from their magnificent work on the simulation of hydrodynamic phenomena. Similarly, I am indebted to Professor Roger Hockney of Reading University, Berkshire, Dr Frank Hohl of NASA, Langley, Virginia, Dr K. Bryan of the National Institute of Oceanography, Washington and Mr Jes Christiansen of the Culham Laboratory, Berkshire, for making available their original results which have been used in this book.

I should like to thank Dr Keith Roberts of the Culham Laboratory, Berkshire, for many illuminating discussions and insights over a long period and in supporting, with much enthusiasm, a considerable part of this work.

London — DAVID POTTER
May, 1972

Contents

Chapter IV Numerical Matrix Algebra

Chapter VIII Phase Fluids

Chapter IX Classical Fluid Dynamics

CHAPTER I

Introduction

1 The Nature of Computational Physics

Computation in physical theory is concerned with the formulation and numerical solution of large mechanical systems. The term 'mechanics' is used to mean the science that quantitatively describes the motion, or tendency to motion, of material objects or systems of objects in the natural world. The classical mechanics of Newton, which has been developed elegantly by the mathematical physicists of the eighteenth and nineteenth centuries—and in particular by D'Alembert, Lagrange, Hamilton and Jacobi—provides us with the laws of motion of particles and systems of particles which constitute much of the material world. Quantum mechanics provides us with the laws of motion of atomic, nuclear, or sub-nuclear particles. Above the nuclear level at least, the forces which operate between particles are well established; in particular the fundamental electromagnetic and classical gravitational forces are precisely known. Increasingly, therefore, physicists have been able to solve quantitatively for the dynamics of systems in the material world or at least they have been able to do so in principle. However, the systems of interest in the material world are assemblies of not a few but frequently a great number of elementary particles, whether they are stars in a galaxy, atoms in a classical fluid or ions and electrons in a plasma. Often it is the case that, although the principles and even the basic equations of interesting systems are well established, they are in effect unsolved, and little understood. We can solve one- or two-body problems analytically but three-body problems are generally beyond the wit of analysis. In all but the very simplest of examples, direct solutions of the few- or many-body problem by analysis proves impossible.

It is clear that the mathematical formulation of a set of physical principles is only useful in so far as such a formulation allows solutions. In many problems, which are concerned with the macroscopic or gross properties of very large assemblies of particles, either obeying quantum or classical laws, the methods of statistical mechanics have proved powerful and appropriate. Frequently, however, complex and 'nonlinear' equations arise on applying statistical mechanics, as occurs, for example, in the equations of hydrodynamics, in magnetohydrodynamics, and in the Vlasov equation describing collisionless phase fluids.

Computational mechanics provides a numerical method of solution for the nonlinear, many-variable problems which arise in describing mechanical systems. However, it is not possible on the computer to solve such systems by the direct application of the appropriate laws of motion, say Hamilton's equations, and the laws of force which apply to the system. The computer, now and in the foreseeable future, is finite in comparison to the number of stars in a galaxy, or the number of atoms in a drop of liquid and, furthermore, is incapable of directly describing the continuum, which occurs, for example, in the space of a molecular structure. Models of physical systems have to be developed for solution on the computer and, because the computer is finite, models that describe less than the maximal information of the system have to be used.

This book is concerned with articulating a coherent methodology, and the application, of computational mechanics. The subjects of data analysis and the accurate numerical expression of explicit solutions are of a different nature and are not included. The essence of the book is the formulation of finite numerical models of the few-, the many-, and the infinite-body problems. We are concerned with the wide range of problems which are 'unsolved' in many-body theory because they are implicitly formulated, and explicit solutions are not possible with conventional analysis. Descriptions are given of quantitative models for the few-particle direct-interaction system, for classical and quantum particles in self-consistent fields, for the 'infinite'-body problems which occur in phase fluids as in the Vlasov equation, for hydrodynamics and for applied hydrodynamics with long-range forces as in magnetohydrodynamics, or gravitational hydrodynamics.

2 The Computer in Physical Theory

High-speed electronic computers have been in general use only since the early fifties and the methodology of their application to physical theory is yet to be fully established. It is intuitively clear that we need them, but it is not at all clear in what manner we should apply them. The individual operations performed in a computer are in no sense different from arithmetic computations performed by hand, by a slide rule, by a book of logarithmic tables, or by a desk calculator. On the other hand, the number of computations, and the number of possible variables, is so considerable, and certainly on such a greater scale than available by previous computational devices, that, in effect, a qualitatively new computational medium is now available. It is necessary, therefore, to question the role that this new computational medium can provide in the development of a physical theory.

In the first place, to assert the nature of a physical theory, it is instructive to quote a definition due to Duhem (1954, p. 19):

> A physical theory is a system of mathematical propositions, deduced from a small number of principles, which aim to present as simply, as exactly, and as completely as possible, a set of experimental laws.

Essentially it is implied here that the set of principles on which the theory is based are abstractions devised by the imagination and that a theory is judged by how simply, how exactly and how completely it describes experimental laws.

We may now return to our central question and ask: What role, if any, does or can the computer have in the development of a physical theory? As we have seen there are three processes involved in the development of a physical theory. First, there is the proposal of a set of abstract principles; second, the deduction of a set of mathematical propositions; third, the resolution of those mathematical propositions to provide the experimental laws which the theory seeks to describe. Clearly, advancing a set of abstract principles is an act of imagination based on the discipline and experience of the physicist. The computer overtly has no significance here. Yet we see that the computer has an obvious relevance to that part of the development of a physical theory which deals with the resolution of our mathematical propositions to describe experimental laws. The computer can provide the means by which we relate mathematical propositions to experimental laws and the modern digital computer is a qualitatively new and unique way of doing this. Each arithmetic or Boolean operation performed by the computer is in itself no different from the operation of an abacus or sliderule. It is because the number of these operations which we are now capable of performing is so very great, and is certainly many orders of magnitude greater than we have been able to perform previously, that the computational medium the computer offers us is qualitatively new and is qualitatively different from anything which has previously been available.

Given this new mathematical medium wherein we may resolve mathematical propositions which we could not resolve before, more complete physical theories may possibly be developed. The imagination of the physicist can work in a considerably broader framework to provide new and perhaps more valuable physical formulations.

In the subject of computational physics it is necessary first to develop the technique of solving mathematical equations on the computer. Systems, and frequently very large systems, of mathematical equations must be solved as efficiently as possible. This, however, is only the technique of the subject. Of equal, if not of greater importance, is the formulation of those mathematical equations and the development of what may be called our intuition, based on the computational medium, so that the application of physics and the physical method may be extended to an increasing description of the natural world.

3 The Limitations of Mathematical Analysis

It is useful here to compare the scope and the limitations of mathematical analysis and computational physics and, furthermore, to consider the interrelation between mathematical theory, computational physics and experiment. By 'mathematical analysis' is meant the resolution, by conventional algebraic methods, of physical principles to describe experimentally observed phenomena.

Mathematical physics has been extremely successful when the theory is *linear*, when *symmetry* can be invoked and when only a few *variables* need be employed. For example, the linearity of electrodynamics has produced a formulation which is simple, beautiful and above all tractable. Similarly the central symmetry of the hydrogen atom has been usefully employed to obtain solutions for the electron energy levels, but in the hydrogen molecule or the helium atom the central symmetry is destroyed and solutions are not readily obtainable. Again, one- or two-body problems, which may be described by six or twelve variables respectively, have readily been solved by analysis, but frequently no analytic solutions can be obtained for three- or more body problems.

These restrictions in mathematical analysis are frequently severe, and limit the development of physics considerably. On the other hand, in computational physics, the mathematical properties of linearity, symmetry and a small number of variables are not demanded in obtaining solutions. The essence of the systems which may be described on the computer is that they be *finite* and *discrete*. Nevertheless, although a computational formulation cannot be infinite, *many-variable* systems (typically 10^6 or greater on present-day computers) can be described. Much of the physics which has been successfully described by mathematical analysis deals with continuous media, and the power of differential calculus, as a mathematical tool, has lead to the wide application in physics of the concept of a continuous medium. On the other hand, in computational physics, only discrete entities may be described and hence concepts such as particles, cells, or a set of waves are useful. If we wish to describe a continuous medium or a continuous field concepts which are inherently mathematical abstractions, the continuum must be divided into a set of many elements.

To summarize this comparison between mathematical analysis and computational physics, we may list the properties inherent in each approach. Mathematical analysis relies on linearity, on symmetry, on a small number of variables and is effective in describing continuous media. In computational physics, the physical systems described must be discrete and finite and the computational method is particularly effective in describing many-variable systems.

These approaches to physics are in no way exclusive, but rather the roles played by mathematical theory, computational physics and experiment are to be regarded as complementary. Each approach can contribute in part to our understanding of physical phenomena. It is true that, in some instances, only one approach proves temporarily possible. Until recently, this has been largely the case in the study of general relativity, where few experiments have been possible to corroborate, or otherwise, a sophisticated and well-developed theory. Similarly the application of computational physics can provide information on phenomena where experiment is difficult, if not impossible. Such is the case in the study of the structure of stars, where observation is limited to the surface of stars and any experimental probe to the core of a star would rapidly be destroyed by the extreme heat or gravitational pressure. Similarly in controlled thermonuclear fusion physics, probes introduced into a hot plasma are quickly destroyed, while computer simulation can provide the fullest information. However, if experiment fails to contribute, progress in physics is limited, for the models employed by theory, whether analytical or computational, are only models and must continually be compared with the natural world through experiment. It is in this all-important sense that mathematical theory, computational physics and experiment are complementary.

4 The Discrete Nature of the Computer

To appreciate the significance of the physical models and the manner in which they are formulated for solution on the computer, it is necessary to consider the essential structure of digital computers. We need not concern ourselves with the details of, or variations between, different computers. It is sufficient to appreciate that all computers have as their kernel two features: a logical unit, frequently called the central processor, in which simple, individual computational operations are performed and a central memory bank where information may be stored.

Information in the central memory, or core store, is stored in units called bits, where each bit has two modes: say, yes or no, or zero or one. However, the memory is conventionally organized by grouping a number of bits (typically 18, 32, 36, 48, or 60 bits) into a 'word' of information. In practice, information is dealt with in words. A word may be assigned to represent an integer number, a real number, a name (that is, a set of alphabetic characters), a vector, the variables describing one particle, etc. Typically, on large, present-day computers, the central memory contains of the order of 100,000 words or variable locations and, although it is obviously capable of storing a very large number of variables, it is, nevertheless, still finite. It is true that the storage capacity of computers can be expanded very greatly by the use of external memories such as magnetic drums, disks, or tapes but the access time associated with these additional memories is considerably longer than that of the central memory, and in many problems we are frequently limited to the main central memory.

In operation, words of information are transported from the central memory and reproduced in the central processor, a procedure called *fetching*. Simple arithmetic or Boolean logical operations are then performed in the central processor and the new words of information, thus created, are *stored* back in the central memory. Clearly these operations of fetching and storing and the operations of the central processor occur on a finite timescale, which on present-day computers (say the IBM 360/91 or the CDC 6600) is of the order of 10^{-6} to 10^{-7} seconds. Because both the memory of the computer has a finite capacity and the number of operations which may be performed per unit time is finite, the physical systems that are described by computational physics must be represented by discrete finite mathematical models.*

As a pertinent example of the discrete mathematical representations which are employed we shall consider the case of the description of galactic structures. In many galaxies, astronomy has revealed the apparent existence of spiral arms, the analysis of which can provide considerable information on the rotational properties of the galaxy, or the history of the evolution of particular galaxies (see Chapter 6). This problem can be studied in two obvious ways computationally; by the methods of *particles* or *fluids*. In the first method, the galaxy may be thought of as a finite set of interacting stars or 'particles' in which each star moves under the self-consistent gravitational field arising from all the other stars in the assembly. However, the number of stars in a galaxy (typically 10^{10}) is very considerably greater than the number which may be represented in the computer.

An alternative approach is to employ the methods of statistical mechanics, where it is assumed that the galaxy is composed of so large a number of stars that, effectively, it may be described as a continuous fluid. Hence equations which define the evolution in time of the distribution, or density, of the galactic fluid may be derived to describe the system. However, to satisfy the finite requirements of the computer, the fluid is divided into a set of elementary cells to provide difference equations which can again describe the motion and structure of the galaxy.

Another simple example to illustrate the discrete requirements of computational mechanics occurs in the vibrations of ion lattices. Using Debye theory, the solid may be described as a continuum, but the resulting dif-

* For a discussion of the structure of present-day computers the reader is referred to Philips and Taylor (1969).

ferential equations must be reformulated as difference equations and consequently the properties of elements or cells of the solid are described. Alternatively the particle method could again be employed by following directly the motions of a limited number of interacting ions on the lattice.

These two themes—of particles, which lead to large sets of ordinary differential equations coupled frequently by a field, and of fluids, which lead to partial differential equations in the continuum or partial difference equations in the discrete representation—recur throughout this book. The word 'particle' as applied in this book, is not to be taken literally as meaning one star belonging to a galaxy, or one electron belonging to a plasma, but rather it applies to a mathematical representation describing Lagrangian motion.

In conclusion, it is to be appreciated that, at the present time, although the discrete and finite methods which are employed in computational mechanics are consistent with the problems at hand, they have not necessarily been justified rigorously. We are usually describing less than the maximal information within the system and it is not at all clear under what conditions this is justifiable. For example, the use of consistent difference equations on a discrete mesh to describe, say, the equations of hydrodynamics, is certainly valid for long-wavelength phenomena and can be shown to be valid in the linear case. But in the nonlinear case, with which we are usually interested, and in the absence of diffusion, the abandonment of fine-scale phenomena is a procedure which is not at all clear. Possibly there is some practical justification, since the results of computational mechanics are frequently in agreement with experiment. In the future development of the subject, consideration will have to be given to the general principles which are employed when a reduced set of variables in the computational model is used to describe an infinite- or a many-variable physical system.

5 Summary of the Content

Because of the simplicity and solubility of the differential calculus, a considerable range of physics is formulated in terms of a continuum and is quantitatively described by systems of differential equations. We have seen the necessity on the computer of employing discrete and finite models, and the subject matter of computational physics must consequently be formulated in terms of the difference calculus rather than the differential calculus. In Chapters 2, 3 and 4, the basic methods of numerical algebra and the difference calculus are introduced and discussed.

Chapter 2 introduces the concept of a space-time mesh or lattice, which in finite methods replaces the continuum. The requirements of a difference solution to the initial-value problem are considered and the basic techniques of integration in time are introduced. In partial differential equations, discussed in Chapter 3, the application of Fourier techniques is instructive and for partial difference equations the methods of Fourier analysis are widely applied to examine difference properties. The point of view adopted in discussing such difference techniques is that of the physicist rather than that of the mathematician. Although the discussion is not concerned with the exactitude of the mathematical techniques, or the properties of pathological examples, it is concerned with the essential concepts involved, such as stability, accuracy and dispersion. It is still more concerned with the pertinence and wide applicability of the finite methods to physical problems.

The difference calculus transforms the physical equations of interest into matrix equations, which it is essential to solve on the computer with as great an efficiency as possible. In Chapter 4, effective and applicable methods of matrix algebra and, in particular, methods for the solution of important classes of matrix equations and the determination of eigenvalues and eigenvectors, are discussed. Again the approach here is to maintain the relation to physical problems.

With the mathematical techniques of finite methods established, Chapters 5 to 10 are concerned with the formulation of the many-body problem in computational mechanics. In Chapter 5 the 'exact' action-at-a-distance N-body problem is discussed. Here the forces on any one particle are evaluated explicitly as the sum of the interactions of the $(N - 1)$ other particles. The term 'exact' is used here not in the sense that numerical errors do not occur through the use of finite methods, but rather in the sense that the formulation is a direct application of Newton's laws of motion without recall to statistical mechanics or physical approximations. Since there are approximately N^2 interactions, only relatively few particles in such an assembly can be described, but nevertheless the models have considerable use both in studying statistical mechanics in an 'experimental' manner, and in applications to particles with short-range forces which occur in classical molecular fluids.

By defining an average self-consistent field in the Hamiltonian of an assembly, models with considerably more particles can be developed both in classical and quantum systems and are discussed in the classical case in Chapter 6 and in the quantum case in Chapter 7. These models give rise to 'particle-in-cell' methods and have wide applicability. The so-called collisionless particle-in-cell method may be applied to the description of galaxies and plasmas, while in the collision-dominated case hydrodynamic systems are described. The methods and difficulties in the classical problem are closely tied to the Hartree–Fock method of self-consistent fields in the quantum problem and, in both instances, the essential difficulty is the nonlinear coupling between the Hamiltonian and the particle distribution.

In the limit of an infinite number of particles in a gravitational or plasma assembly, the Vlasov equation, formulated in the continuum of phase space, is obtained. Chapter 8 discusses the formulation of the Vlasov equation and the description of a continuous distribution in phase space by computational mechanics. Both Lagrangian and Eulerian methods may be applied in phase space and the evolution of the distribution may be followed directly in phase space.

For collision-dominated infinite-body problems, the equations of hydrodynamics are obtained. In Chapter 9, classical hydrodynamics for the incompressible and compressible cases is formulated within the framework of computational physics. There are two main approaches here. Either Eulerian difference methods may be applied, or the application of Lagrangian techniques can lead to discrete 'particle' methods, as in the collision-dominated particle-in-cell model, or in the use of 'vortex particles'. Water waves and fluid surface problems are also considered.

The inclusion of long-range body-forces in hydrodynamics leads to the wide extension of the hydrodynamic equations to magnetohydrodynamics and gravitational hydrodynamics. Simulations in astrophysics and in thermonuclear fusion physics are considered in Chapter 10.

It is clear that the range of physical phenomena considered is extremely broad. However, the point of view adopted is not to consider particular solutions or the details of a particular problem, but rather to establish the methods of computational mechanics. Examples of a general nature which are instructive, or intuitively helpful, are used to illustrate the approach.

CHAPTER II

Elements of the Difference Method

1 Introduction: Finite Elements in Physics

Because the arithmetic dealt with in the computer is both discrete and finite, rather than continuous and infinite, it is essential in computational physics to formulate the quantitative equations of interest in a discrete form. The most basic and widely used method is the method of differencing, whereby the properties of small elements of a continuous physical system are examined. Discretization of both time-like variables, denoted by t, and space-like variables, denoted by x, is necessary and in this chapter the essential concepts of a time-mesh and a space-mesh are introduced. Unlike the case for space variables, the creation of a time-mesh is frequently achieved by an integration procedure and derivatives in time and space are therefore usually treated differently.

The technique of differencing has wide application and certainly many concepts in physics arise from considering small elements of a continuum. For example, in analysing the dynamics of a stretched string, equations of motion are applied to the displacement of a small element of the string and, in the limit of a set of infinitely small elements, the differential wave equation is obtained. If, instead of taking the limit where an element becomes infinitely small, we were to write down the finite set of equations describing the finite set of elements of the stretched string, we would arrive at a system of difference equations. Similarly much of electromagnetic theory has been obtained by considering discrete finite elements. The concept of a finite set of charged particles leads to the concept of a charge distribution $\rho(x, t)$ defined in the space-time continuum. Again in magnetostatics, Ampère expressed the magnetic force as the force between two current elements. It is only in the limit of infinitely small elements in electromagnetic theory that the differential equations of Maxwell are obtained in the continuum.

Because of the power of the differential calculus as a mathematical tool, a wide range of physical phenomena have been formulated by the application of the abstract concept of a continuous medium. Consequently there is left to us a legacy of differential equations, many of which have not been solved and which are only understood in the most rudimentary way. To solve these equations and to interpret them, it is necessary, in a sense, to perform the reverse process of their originators and obtain algebraic difference equations from the differential equations.

2 Discrete Representation of a Continuous Variable

We consider the independent continuous variable x, which lies in the domain,

$$X = (X_1, X_2)$$

$$X_1 \leqslant x \leqslant X_2$$

The continuum is replaced by a *mesh*, or *lattice* of points, by dividing the domain X into a set of $J - 1$ elements, Δx_j. A vector $\{x_j\}$, of finite dimension J, may then be constructed by defining the continuous variable x only at the points j (see Figure 2.1),

$$1 \leqslant j \leqslant J \tag{2.1}$$

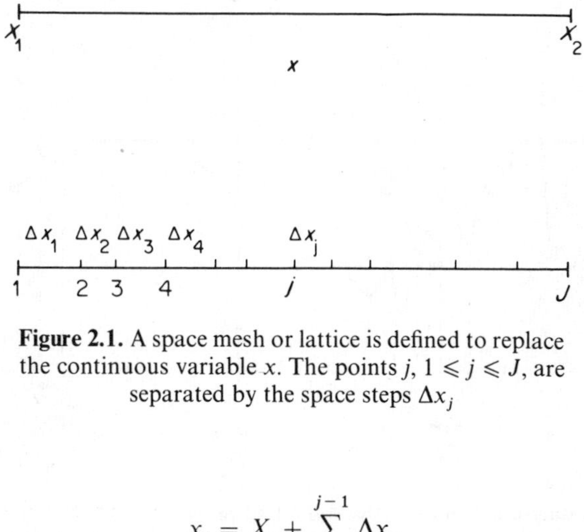

Figure 2.1. A space mesh or lattice is defined to replace the continuous variable x. The points j, $1 \leqslant j \leqslant J$, are separated by the space steps Δx_j

Then,

$$x_j = X_1 + \sum_{\nu=1}^{j-1} \Delta x_\nu \tag{2.2}$$

Thus, given a dependent function $f(x)$ in the continuum, the function f can be approximated in some sense by defining a corresponding vector $\{f_j\}$, on the independent variable mesh $\{x_j\}$:

$$f_j = f(x_j) \tag{2.3}$$

Since the dependent function f is defined everywhere on the continuous variable x, the representation $\{f_j\}$ is an incomplete description of $f(x)$.

Nevertheless the function f may be approximated from $\{f_j\}$ at any point x',

$$x_j \leqslant x' \leqslant x_{j+1} \tag{2.4}$$

by interpolation from the vector components f_j and f_{j+1} between adjacent points. If

$$\epsilon = \frac{x' - x_j}{x_{j+1} - x_j} \tag{2.5}$$

then, with first-order interpolation,

$$f^* = \epsilon f_{j+1} + (1 - \epsilon) f_j \tag{2.6}$$

and f^* approximates f. The nature of this approximation is that we are only describing 'long-wavelength' properties of the function f in the continuum. Clearly if f changes rapidly over the element Δx_j, then f^* is in some sense a 'poor' approximation to f. And clearly we cannot describe wavelengths less than Δx_j (Figure 2.2). Thus the essence of the difference method is its applicability to long-wavelength phenomena and, equally, the more points that are included on the domain X of the dependent variable (the larger J), the better is the representation $\{f_j\}$ of f.

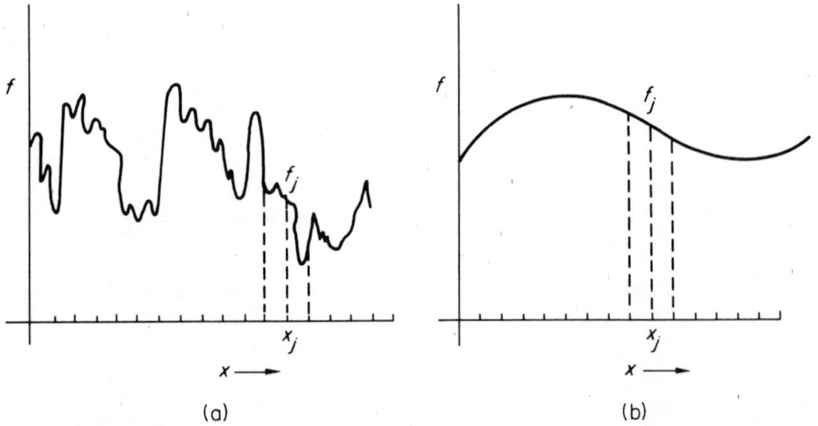

Figure 2.2. A function f of the independent variable x is represented as a vector of components f_j on the mesh points j: (a) the mesh representation is a poor approximation to a rapidly varying function; (b) a good approximation for a slowly varying (long-wavelength) function

To quantify these ideas, of a poor or a good approximation, when applying the difference method, it is useful to employ the techniques of Fourier analysis, whereby functions may be written as a sequence of Fourier modes or waves.* For simplicity it is assumed that the function f is periodic over

* For example, see Courant and Hilbert (1953).

the domain X: that is, outside X there exists a periodic repetition of the function f. Then, providing the dependent function satisfies simple conditions (Dirichlet conditions), it may be expanded as an infinite Fourier series:

$$f(x) = \sum_{k=-\infty}^{\infty} \hat{g}_k \, e^{i2\pi kx/X} \tag{2.7}$$

where

$$\hat{g}_k = \frac{1}{X} \int_X f(x) \, e^{-i2\pi kx/X} \, dx \tag{2.8}$$

and i is the square root of minus one. This expression is a statement that any reasonably behaved dependent function f, which is defined everywhere on the continuous independent variable x, can be analysed into an *infinite* set of Fourier modes, where $\hat{g}_k$ is the amplitude of the mode with wavelength X/k. The usefulness of this procedure is clear, since the discrete approximation $\{f_j\}$ for the continuous function f may be expanded in a totally analogous manner, and we may compare the representation of f with the equivalent representation of $\{f_j\}$. Since the components f_j form a vector of finite dimension J, it follows that the discrete representation $\{f_j\}$ may only be written as the sum of a *finite* set of (J) orthogonal functions.

It is assumed that each mesh element Δx_j, for all j, is constant, say Δ and, again, the function f and correspondingly the discrete representation f_j are periodic. Then f_j may be written as a finite Fourier series:

$$f_j = \sum_{k=1}^{J} \hat{g}_k \, e^{i2\pi kj/J} \tag{2.9}$$

with the amplitude $\hat{g}_k$ of each mode specified by the summation,

$$\hat{g}_k = \frac{1}{J} \sum_{j=1}^{J} f_j \, e^{-i2\pi kj/J} \tag{2.10}$$

This expansion is clearly consistent with the infinite Fourier series (equations 2.7, 2.8) for the case of the function $f(x)$ defined in the continuum, since $J\Delta$ has replaced X and $j\Delta$ has replaced x. As no wavelength smaller than Δ can be defined on the mesh, the infinite series becomes a finite series. More formally, however, we shall prove that the amplitudes $\hat{g}_k$ in the finite Fourier series (equation 2.9) are defined by the given relations (equation 2.10). First, by determining the summation over a pair of modes, we show that the Fourier modes form an orthogonal set,

$$\begin{aligned} S &= \sum_{j=1}^{J} e^{i2\pi jk/J} \, e^{-i2\pi jk'/J} \\ &= \sum_{j=1}^{J} e^{i2\pi(k-k')j/J} \end{aligned} \tag{2.11}$$

We may re-express the summation S as:

$$S = e^{i2\pi(k-k')/J} \sum_{j=0}^{J-1} \{e^{i2\pi(k-k')/J}\}^j \tag{2.12}$$

The summation on the right-hand side of equation 2.12 has the form of a geometric progression, for which the sum is:

$$\sum_{n=0}^{N} x^n = \frac{1 - x^{N+1}}{1 - x} \tag{2.13}$$

and it follows that we obtain for the summation S (equation 2.12),

$$S = e^{i2\pi(k-k')/J} \left\{ \frac{1 - e^{i2\pi(k-k')}}{1 - e^{i2\pi(k-k')/J}} \right\} \tag{2.14}$$

After factorizing the expressions on the top and bottom of the curly brackets, we obtain:

$$S = e^{i\pi(J+1)(k-k')/J} \left\{ \frac{\sin[\pi(k - k')]}{\sin[\pi(k - k')/J]} \right\} \tag{2.15}$$

$$S = \begin{cases} 0 \text{ if } k \neq k' \\ J \text{ if } k = k' \end{cases} \tag{2.16}$$

since k and k' are by definition non-negative integers smaller than or equal to J and we have used $\sin \pi m = 0$ if m is an integer. The result $S = J$ for $k = k'$ can most simply be seen by expanding the two sine functions (equation 2.15) for small arguments.

To summarize, therefore, it has been shown that the Fourier modes form an orthogonal set:

$$\sum_{j=1}^{J} e^{i2\pi kj/J} e^{-i2\pi k'j/J} = J\delta_{kk'} \tag{2.17}$$

This result may now be used to verify the relations (equations 2.9 and 2.10), which demonstrate that the discrete representation $\{f_j\}$ may be expanded as a finite Fourier series. From the expansion (equation 2.9), we prove that the amplitudes must satisfy the relations in equation (2.10). From the components f_j of the vector, we form the sum,

$$\begin{aligned} \sum_{j=1}^{J} f_j e^{-i2\pi k'j/J} &= \sum_{j=1}^{J} \sum_{k=1}^{J} \hat{g}_k e^{i2\pi kj/J} e^{-i2\pi k'j/J} \\ &= \sum_{k=1}^{J} \hat{g}_k \sum_{j=1}^{J} e^{i2\pi kj/J} e^{-i2\pi k'j/J} \end{aligned} \tag{2.18}$$

and, using the orthogonality relations (equation 2.17),

$$\sum_{j=1}^{J} f_j \, \mathrm{e}^{-\mathrm{i}2\pi k' j/J} = \sum_{k=1}^{J} \hat{g}_k J \delta_{kk'}$$

$$\hat{g}_{k'} = \frac{1}{J} \sum_{j=1}^{J} f_j \, \mathrm{e}^{-\mathrm{i}2\pi k' j/J}$$

which is the result given in equation 2.10.

Thus we see that the representation $\{f_j\}$ on a discrete mesh, of the function f in the continuum, is finite in that only a finite set of wavelengths are described on the mesh and, more particularly, there is a cut-off wavelength Δ, below which no phenomena are described. The representation $\{f_j\}$ is a long-wavelength approximation to the function $f(x)$ in the continuum.

It is extremely useful and instructive, when comparing difference operators with differential operators, to determine the extent of the approximation with respect to a single Fourier mode, on the basis of the above analysis. By this technique, the severity of the difference approximation on different wavelength phenomena may be evaluated and the technique will be used extensively in subsequent chapters.

3 Difference Derivatives in Space

Having considered the representation of a continuous function on a discrete mesh, we shall now analyse approximations to the differential derivatives of functions. A derivative of a function yields information about the local variation of the function in space and consequently a difference derivative couples neighbouring points on the mesh. The obvious approximation to the first derivative $\mathrm{d}f/\mathrm{d}x$ at the point j on the mesh, $1 < j < J$, is:

$$\Delta'_x f_j = \frac{f_{j+1} - f_{j-1}}{2\Delta} \tag{2.19}$$

where Δ is the mesh-step length as before. Again this is clearly a good approximation to the derivative $\mathrm{d}f/\mathrm{d}x$, if f does not change very rapidly over Δ (Figure 2.3). In order to estimate the extent of the approximation Δ'_x for $\mathrm{d}/\mathrm{d}x$ (equation 2.19), it is useful to compare the difference derivative and the differential derivative of a Fourier mode on the basis of the analysis in Section 2.2. For a Fourier mode,

$$u = g\,\mathrm{e}^{\mathrm{i}kx} \tag{2.20}$$

the operation of the first differential derivative yields:

$$\begin{aligned} \frac{\mathrm{d}u}{\mathrm{d}x} &= \mathrm{i}kg\,\mathrm{e}^{\mathrm{i}kx} \\ &= \mathrm{i}ku \end{aligned} \tag{2.21}$$

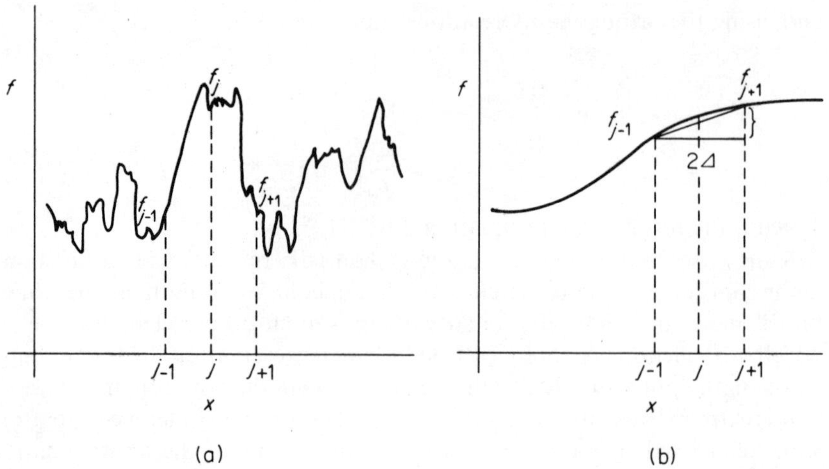

Figure 2.3. The difference approximation to the first derivative of a function: (a) a poor approximation; (b) a good approximation

Operating the difference derivative (equation 2.19) on the same Fourier mode u leads to the expression,

$$\Delta'_x u = \frac{g\,\mathrm{e}^{ikx_{j+1}} - g\,\mathrm{e}^{ikx_{j-1}}}{2\varDelta} \tag{2.22}$$

Now $x_{j+1} = x_j + \varDelta$, and $x_{j-1} = x_j - \varDelta$, so

$$\Delta'_x u = \frac{g}{2\varDelta}(\mathrm{e}^{ik(x_j+\varDelta)} - \mathrm{e}^{ik(x_j-\varDelta)})$$

$$= \frac{g}{\varDelta}\mathrm{e}^{ikx_j}\frac{1}{2}(\mathrm{e}^{ik\varDelta} - \mathrm{e}^{-ik\varDelta}) \tag{2.23}$$

$$\Delta'_x u = \frac{\mathrm{i}u}{\varDelta}\sin k\varDelta \tag{2.24}$$

Since, for small $k\varDelta$, $\sin k\varDelta$ approximates $k\varDelta$, the right-hand side of equation 2.24 approximates the result for the differential derivative (equation 2.21). Hence the difference formula (equation 2.19) is a good approximation for the first derivative when the wave number k is small (the wavelength $2\pi/k$ is large). The longer the wavelength, the better is the approximation. To estimate more exactly the approximation Δ'_x for $\mathrm{d}/\mathrm{d}x$, we may expand the result for the difference derivative (equation 2.24) for small $k\Delta$:

$$\Delta'_x u = \frac{iu}{\Delta}\left(k\Delta - \frac{(k\Delta)^3}{6} + O(k^5\Delta^5)\right)$$

$$= iuk\left(1 - \frac{k^2\Delta^2}{6} + O(k^4\Delta^4)\right) \tag{2.25}$$

$$\Delta'_x \equiv \left(1 - \frac{k^2\Delta^2}{6} + O(k^4\Delta^4)\right)\frac{d}{dx} \tag{2.26}$$

We say the difference operator is second-order accurate in $k\Delta$, or 'space-centred'.

For the second derivative, clearly a good and simple approximation to d^2f/dx^2 on the mesh $1 < j < J$ is (Figure 2.4),

$$\Delta''_x f_j = \frac{f_{j+1} - 2f_j + f_{j-1}}{\Delta^2} \tag{2.27}$$

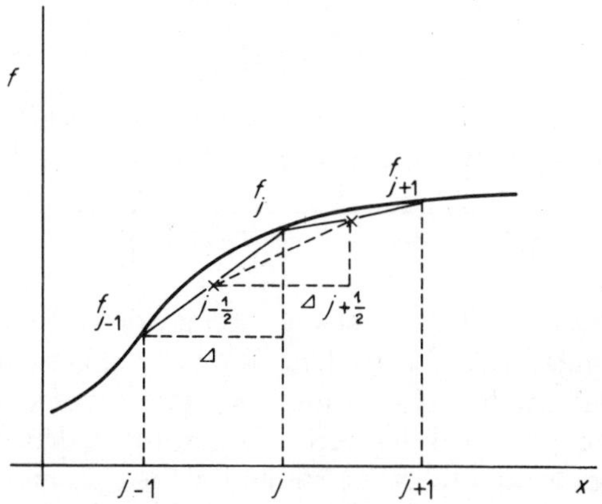

Figure 2.4. The difference approximation to the second derivative of a function

Again, the accuracy of this approximation can be established by considering the effect of Δ''_x operating on a Fourier mode of wavelength $2\pi/k$. For the differential operator d^2/dx^2,

$$\frac{d^2u}{dx^2} = \frac{d^2}{dx^2}(g\, e^{ikx})$$

$$= -gk^2\, e^{ikx}$$

$$= -k^2 u \tag{2.28}$$

On the other hand, for the second-difference operator, using the algorithm (equation 2.27),

$$\Delta_x'' u_j = \frac{g}{\Delta^2}(e^{ik(x_j+\Delta)} - 2\,e^{ikx_j} + e^{ik(x_j-\Delta)}) \tag{2.29}$$

$$= \frac{2g\,e^{ikx_j}}{\Delta^2}(\tfrac{1}{2}e^{ik\Delta} + \tfrac{1}{2}e^{-ik\Delta} - 1)$$

$$= \frac{2u}{\Delta^2}(\cos k\Delta - 1) \tag{2.30}$$

If we compare this result with the result for the second differential derivative, it is clear that the difference operator only approximates the differential operator for small $k\Delta$ where the wavelength of interest is much larger than the mesh step length. For this case, we expand the cosine of $k\Delta$ for small $k\Delta$:

$$\Delta_x'' u = \frac{2u}{\Delta^2}\left(1 - \frac{k^2\Delta^2}{2} + \frac{k^4\Delta^4}{24} - 1 + O(k^6\Delta^6)\right)$$

$$= -\,k^2 u\left(1 - \frac{k^2\Delta^2}{12} + O(k^4\Delta^4)\right) \tag{2.31}$$

Thus the second-difference operator is an approximation to second-order in the mesh step length over the wavelength:

$$\Delta_x'' \equiv \left(1 - \frac{k^2\Delta^2}{12} + O(k^4\Delta^4)\right)\frac{d^2}{dx^2} \tag{2.32}$$

These approximations to the first and second derivatives are commonly employed in simulating differential equations of interest in physics. More complex formulae can be used for these cases and more complex formulae are required when higher-order derivatives must be defined, but precisely the same philosophy and the same method of analysis may be employed in examining these more complex operators.

4 General Formulation of the Initial-Value Problem

We have formulated the essential ideas inherent to the difference technique where functions defined in the continuum are represented by a mesh vector and, at least for space-like variables, differential operators are readily defined on the mesh. For time-like problems we must evolve the solution in the real time of the computer and time derivatives may not be defined so readily, since the solutions at the new time-point are not known. Rather, in time-dependent problems, an integration procedure on a time-mesh must be specified. 'Time-like' is used to describe the class of problems which have one-point boundary conditions and generally this is a broader class of problems, than the class which arises from specifying a solution in real time. Nevertheless, problems in real time are particularly interesting and important, and the notation generally used is adopted from these problems.

The initial-value problem, occurring in every branch of physics, is of fundamental importance, involving as it does the idea of prediction and it is of prime interest too in computational physics. In particular it occurs when defining the equations of motion of a particle or a system of particles, whereby an ordinary differential equation or a system of ordinary differential equations is obtained. Equally it arises when describing the evolution of a continuous system, in which partial differential equations define the state quantitatively. In both cases the philosophy behind the formulation of the problem is the same and the general situation will be described here without distinguishing between the ordinary and partial differential cases. In the later sections of this chapter, with a view to initial simplicity, particular integration procedures will be introduced by linking the discussion to equations of the ordinary type.

Given a system defined by the state vector $\mathbf{u}(\mathbf{r}, t)$, in the space domain $R = R(\mathbf{r})$, then if $\mathbf{u} = \mathbf{u}^0$ is defined at time $t = 0$, and if $\mathbf{u}$ is defined on the surface S of R for all time t, we wish to determine $\mathbf{u}$ for all time t in R. The state of the system may be obtained for all time t as solutions to the initial-value equation,

$$\frac{\mathrm{d}\mathbf{u}}{\mathrm{d}t} = \mathrm{L}\mathbf{u} \tag{2.33}$$

In general, L is a nonlinear operator which is algebraic for ordinary differential equations. For partial differential equations L is a spatial differential operator. As a simple example to illustrate the formalism, we may take the case of a one-dimensional simple harmonic oscillator of mass m, force constant a, position variable x, and velocity v. Then the state vector $\mathbf{u}$ of the system is,

$$\mathbf{u} = (x, v) \tag{2.34}$$

and in this case the operator is a linear matrix operator,

$$\mathbf{L} = \begin{pmatrix} 0 & 1 \\ -a/m & 0 \end{pmatrix} \tag{2.35}$$

since the equations of motion of the oscillator satisfy

$$\frac{\mathrm{d}x}{\mathrm{d}t} = v \tag{2.36}$$

$$\frac{\mathrm{d}v}{\mathrm{d}t} = -\frac{a}{m}x$$

As a second simple example, we may formulate the problem of heat conduction in a one-dimensional rod, where the state vector of the system **u** is now a scalar, $u(x, t) = T$, the temperature, and the operator L is a differential operator in space,

$$\mathrm{L} = \frac{\partial}{\partial x}\kappa\frac{\partial}{\partial x} \tag{2.37}$$

by virtue of the diffusion equation with the coefficient of conductivity κ,

$$\frac{\partial T}{\partial t} - \frac{\partial}{\partial x}\kappa\frac{\partial T}{\partial x} = 0 \tag{2.38}$$

In such an example, where the state of the system is defined in the space-continuum and L is a differential operator in space, the spatial dimensions must be discretized by defining the state vector **u** on a space-mesh. The operator L becomes a difference operator of possibly complex form, which may be defined by the procedures outlined in section 2.3.

Returning to the general initial-value problem (equation 2.33), we shall integrate the equation in time over small steps, in parallel with the real time of the computer. The time dimension is divided into finite small intervals which separate points in time (Figure 2.5),

$$t^n = \sum_{\nu=1}^{n} \Delta t_\nu \tag{2.39}$$

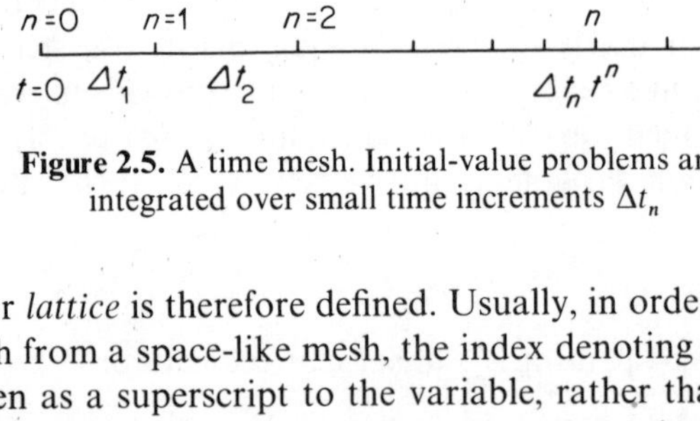

Figure 2.5. A time mesh. Initial-value problems are integrated over small time increments Δt_n

A *time mesh* or *lattice* is therefore defined. Usually, in order to distinguish a time-like mesh from a space-like mesh, the index denoting a particular time point is written as a superscript to the variable, rather than as a subscript as in the space-mesh case. By integrating the time-dependent equations (2.33) over a small time step, the state vectors $\mathbf{u}^{n+1}$ and $\mathbf{u}^n$ of the system at adjacent time points t^{n+1} and t^n are related:

$$\mathbf{u}^{n+1} = \mathbf{u}^n + \int_{t^n}^{t^{n+1}} \mathrm{L}\mathbf{u}\,\mathrm{d}t' \tag{2.40}$$

It is clear that the integral on the right-hand side of equation 2.40 cannot be evaluated exactly, since the vector $\mathbf{u}(t')$ is not known for all time t' in the interval, $t^n \leqslant t' \leqslant t^{n+1}$. Hence the essential finite difference approximation is introduced by assuming that for small time steps, $\Delta t = t^{n+1} - t^n$, the integrand in equation 2.40 may be approximated by a finite Taylor series expansion,

$$\mathbf{u}^{n+1} = \mathbf{u}^n + \int_{t^n}^{t^{n+1}} \left\{ \sum_{r=0}^{p-1} \left[\frac{\mathrm{d}^r}{\mathrm{d}t^r}(\mathrm{L}\mathbf{u}) \right]_{t^n} \frac{t'^r}{r!} + O(t'^p) \right\} \mathrm{d}t' \tag{2.41}$$

and after integrating the right-hand side:

$$\mathbf{u}^{n+1} = \mathbf{u}^n + \sum_{r=1}^{p} \left[\frac{\mathrm{d}^{r-1}}{\mathrm{d}t^{r-1}}(\mathrm{L}\mathbf{u}) \right]_{t^n} \frac{(\Delta t)^r}{r!} + O(\Delta t^{p+1}) \tag{2.42}$$

p is the order of accuracy in the time step Δt to which the integration scheme is carried out, but usually, and particularly for partial differential equations, higher-order terms than the second are not included,

$$\mathbf{u}^{n+1} = \mathbf{u}^n + \mathrm{L}\mathbf{u}^n\,\Delta t + \left[\frac{\mathrm{d}}{\mathrm{d}t}(\mathrm{L}\mathbf{u}) \right]_{t^n} \frac{(\Delta t)^2}{2} \tag{2.43}$$

The question arises of how we may define the time derivative in the last term of equation 2.43. In practice we can achieve second-order accuracy by using the variables at previous time levels (say t^{n-1}), or by defining intermediate time levels, or again we may use the unknown variables at the time step ahead t^{n+1} and particular algorithms will be introduced in the following sections of this chapter. There is the particular situation, however, where variables at the time step ahead t^{n+1} are used and the equation for the integration scheme takes the form,

$$\mathbf{u}^{n+1} = \mathbf{u}^n + \mathrm{L}\mathbf{u}^n(1 - \epsilon)\,\Delta t + \mathrm{L}\mathbf{u}^{n+1}\epsilon\,\Delta t \tag{2.44}$$

Here ϵ is an interpolation parameter, $0 \leqslant \epsilon \leqslant 1$, and second-order accuracy is only maintained when $\epsilon = \frac{1}{2}$. In the special case when $\epsilon = 0$, the new state $\mathbf{u}^{n+1}$ is defined explicitly by the known state $\mathbf{u}^n$ at the previous time step,

$$\mathbf{u}^{n+1} = (\mathrm{I} + \Delta t\mathrm{L})\mathbf{u}^n \tag{2.45}$$

In this event the method is said to be *explicit*, while on the other hand if $\epsilon \neq 0$ the method is said to be *implicit*,

$$(\mathrm{I} - \epsilon\,\Delta t\mathrm{L})\mathbf{u}^{n+1} = (\mathrm{I} + (1 - \epsilon)\,\Delta t\mathrm{L})\mathbf{u}^n \tag{2.46}$$

Assuming the operator on the left-hand side is non-singular, it is necessary at each time step to solve equation (2.46) for the new state $\mathbf{u}^{n+1}$:

$$\mathbf{u}^{n+1} = (\mathrm{I} - \epsilon\Delta t\mathrm{L})^{-1}(\mathrm{I} + (1 - \epsilon)\,\Delta t\mathrm{L})\mathbf{u}^n$$

$$\mathbf{u}^{n+1} = \mathrm{T}(\Delta t, \Delta)\mathbf{u}^n \tag{2.47}$$

The operator T is the difference operator which relates the states of the system over subsequent points on the time mesh. For partial differential equations, where L is a differential operator in space, the operator T is a difference operator which couples the dependent variables on the space mesh, of space step Δ. The formulation of equation 2.47 is the difference analogue of the initial-value problem in differential form (equation 2.33). This is the general problem and we are concerned to determine the circumstances in which such an integration in time is meaningful and the errors associated with such an integration.

5 Requirements of a Difference Solution to the Initial-Value Problem

By integrating over small time steps, we have reduced the initial-value problem to the problem of obtaining a sequence of solutions defined at discrete time points, where the state at a given time point is related to the state at the previous time point (equation 2.47) by an integration operator, $T(\Delta t, \Delta)$. This formulation applies equally to the example of a system of ordinary differential equations and to the more general situation of a system of partial differential equations defined in the continuum. In the latter case we replace the space continuum by a cell or mesh structure, but the nature of the time coupling is essentially the same.

The integration operator T, which relates time levels, is not unique and depends upon the choice of the particular integration scheme in time, and of the difference scheme in space, and we must question what criteria must be satisfied in choosing a particular difference scheme, or integration operator T. Here, we shall introduce the essential concepts behind such criteria, but we shall develop the criteria more rigorously for partial difference equations in the next chapter. The important properties, with which we shall be concerned, may be summarized under the headings: *consistency*, *accuracy*, *stability*, *efficiency*.

(*a*) *Consistency of a Difference Approximation*. Clearly the first property to be demanded is that in some manner the difference system approximates the differential system, or more specifically we demand that the difference system is consistent with the differential system. Essentially it is required, that in the limit of a small time step and a small space step, the difference system becomes identical with the differential system. Formally the requirement of *consistency* may be specified as:

$$\lim_{\substack{\Delta t \to 0 \\ \frac{\Delta t}{\Delta} \to \beta}} \lim_{\Delta \to 0} \frac{T(\Delta t, \Delta) - I}{\Delta t} = L \tag{2.48}$$

where L is the differential operator as in equation 2.33 and β is finite. Clearly, if this condition were not to be satisfied, the difference scheme would in no manner simulate the initial-value problem of interest. We may view this requirement as the most fundamental, but providing it is satisfied it is necessary to examine the difference scheme in more detail, since we use finite time steps and space steps which produce errors and make the solution deviate from the differential solution.

(*b*) *Accuracy of a Difference Approximation.* Accuracy of a numerical solution, as an approximation to the solution of a differential system, is impaired by the occurrence of two sources of errors. The first of these is termed the *truncation error* and is caused by the approximation involved in simulating the differential equations by the difference equations. The essence of this approximation has been pointed out previously (Sections 2.2, 2.3, 2.4), where we have seen that it arises from representing a continuous independent variable by a set of discrete points. Such errors are dependent on the magnitude of the mesh intervals in time and space, Δt, Δ, and we may readily determine the magnitude of the error. Patently in our choice of a difference scheme, a fundamental requirement is the minimization of the truncation error.

However, a second source of errors occurs (*round-off* errors) due to the accuracy with which a particular variable is described in the memory of the computer. Usually, real arithmetic is performed on the computer, in which numbers are stored in terms of an exponent and a decimal fraction and arithmetic of this type on the computer is not perfect. Clearly if the calculation is pursued with i decimal places, it will be less accurate than if it were executed with $i + 1$ decimal places. The error depends on how a number is 'rounded-off' in the lowest 'bits' of a 'word' in the computer and, to determine the total round-off error, a statistical analysis would have to be carried out. With modern digital machines, however, many bits are used in a word of information and correspondingly many decimal figures are, or can be, used. It follows that the cumulative round-off error is usually not serious.

Nevertheless it is important to note, at least in principle, that the arithmetic in the computer is not exact. As the result of arithmetic operations, errors in the eigenvectors belonging to a difference system are continually generated in the lowest bits of the words representing dependent variables, even though frequently such errors arise with small amplitude. This consideration leads us to question whether the difference solution is bounded, or more specifically, whether any possible eigenvector of the difference operator can grow without bound, a property termed the stability of the difference scheme.

(*c*) *Stability of a Difference Scheme.* Clearly, if a difference scheme can produce a solution which is not bounded, the consequence will be catastrophic and we say such a scheme is numerically unstable. If any error is amplified from time step to time step, the error will quickly swamp the solution and the result will be of no value. The property of stability may be stated as: *a numerical method is stable if a small error at any stage produces a smaller cumulative error*. Every useful numerical method for the initial-value problem must, at least under some conditions, be a stable numerical method.

To formulate the concept of stability quantitatively, we shall first consider a single independent ordinary differential equation. For this simple case the dependent variable is a scalar quantity, and it is easy to define an error, say ϵ^n, which occurs at step n. We are interested in the amplification of this error at step $n + 1$, ϵ^{n+1}:

$$\epsilon^{n+1} = g\epsilon^n \tag{2.49}$$

g is termed the amplification factor and it is related to the integration operator T and hence to the particular integration scheme which is employed. The amplification factor may therefore be related to the truncation errors of the integration scheme, but it does not take into account additional round-off errors which may occur at the new time step. Now, for the stability of the numerical method as expressed above, we require,

$$|\epsilon^{n+1}| \leqslant |\epsilon^n| \tag{2.50}$$

and using the definition of the amplification factor (equation 2.49),

$$|g\epsilon^n| \leqslant |\epsilon^n| \tag{2.51}$$

Hence numerical stability will be achieved if the condition,

$$|g| \leqslant 1 \tag{2.52}$$

is satisfied. It is necessary, however, to qualify this requirement for the class of problems which permit a growing solution. Certainly, in some physical problems, an exponential growth of the solution in time can occur and it follows that it is permissible for an error to grow without bound providing it remains smaller than the growing solution (see Section 3.2). This is not the general case, however, and we restrict ourselves by excluding the class of problems with growing solutions.

It is not difficult to extend the application of the concept of stability to a system of differential equations and to a system of partial differential equations (Chapter 3).* For a system of N first-order ordinary differential equations, it is necessary to define an error vector $\boldsymbol{\epsilon}^n$, each component of which is the error associated with each dependent variable in the system. Over a time step, the error is now amplified by an *amplification matrix* $\mathbf{G}$ to the error vector at the new time step $\boldsymbol{\epsilon}^{n+1}$:

$$\boldsymbol{\epsilon}^{n+1} = \mathbf{G}\boldsymbol{\epsilon}^n \tag{2.53}$$

* For a rigorous treatment and discussion, the reader is referred to Richtmyer and Morton (1967).

The amplification matrix is related to the integration operator $\mathrm{T}(\Delta t)$ (equation 2.44), which couples solutions at successive time steps, for if there exists an error $\boldsymbol{\epsilon}^n$ in the solution $\mathbf{u}^n$ at step n, then by the integration equation 2.47,

$$\mathbf{u}^{n+1} + \boldsymbol{\epsilon}^{n+1} = \mathrm{T}(\mathbf{u}^n + \boldsymbol{\epsilon}^n) \tag{2.54}$$

In the case where T is a linear operator, it is clear that the amplification matrix **G** is equal to the integration operator T. In the more general nonlinear case, however, we employ the assumption that the error vector is of small amplitude and it follows that equation 2.54 may be *linearized* by a Taylor expansion about the small error, namely,

$$\mathrm{T}(\mathbf{u}^n + \boldsymbol{\epsilon}^n) = \mathbf{Tu}^n + \left\{\frac{\partial}{\partial \mathbf{u}}(\mathbf{Tu})\right\}_n \boldsymbol{\epsilon}^n \tag{2.55}$$

where only the first two terms in the Taylor expansion have been included. Inserting this expression on the right-hand side of equation 2.54 and subtracting the 'exact' solution (equation 2.47),

$$\mathbf{u}^{n+1} = \mathbf{Tu}^n$$

the linear equation which relates error vectors over the two time levels is obtained,

$$\boldsymbol{\epsilon}^{n+1} = \left\{\frac{\partial}{\partial \mathbf{u}}(\mathbf{Tu})\right\}_n \boldsymbol{\epsilon}^n \tag{2.56}$$

The operator on the right-hand side of this equation is a linear matrix operator and the equation has precisely the form of the equation which defines the amplification matrix (equation 2.53). Hence, by comparing these equations, we deduce that each component of the amplification matrix, $G_{\mu\nu}$, is defined as:

$$G_{\mu\nu} = \frac{\partial u_\mu^{n+1}}{\partial u_\nu^n} \tag{2.57}$$

Accordingly we may determine the amplification matrix by simply differentiating the difference equations 2.47, in the manner defined by this algorithm.

Given the amplification matrix corresponding to the integration scheme of a system of ordinary differential equations, the error vector at the new time step is related to the error vector at the previous step (equation 2.53) and we may question how the concept of stability may be defined for an error vector rather than a scalar error. If the amplification equation is diagonalized, then the amplitudes of each error eigenvector, say $\boldsymbol{\epsilon}_\mu$, are related by the corresponding eigenvalues g_μ of the amplification matrix:

$$\boldsymbol{\epsilon}_\mu^{n+1} = g_\mu \boldsymbol{\epsilon}_\mu^n \tag{2.58}$$

The condition for stability must now be applied separately to the amplitude of each error eigenvector,

$$|\epsilon_\mu^{n+1}| \leqslant |\epsilon_\mu^n| \quad (\text{all } \mu),$$

or,

$$|g_\mu| \leqslant 1 \quad (\text{all } \mu) \tag{2.59}$$

The condition for stability, therefore, reduces to the requirement that the magnitude of each and every eigenvalue g_μ of the amplification matrix **G** must be smaller than or equal to unity. In general the eigenvalues of the amplification matrix may be complex, in which event we mean by the magnitude of the eigenvalue its amplitude in the complex plane, namely:

$$|g_\mu| = \sqrt{(g_\mu g_\mu^*)} \tag{2.60}$$

where g_μ^* is the complex conjugate of g_μ.

(*d*) *Efficiency of a Difference Scheme.* The fourth essential property of a particular numerical algorithm for the initial-value problem which must be considered is the efficiency of the method. Since any computer is finite both in operation time and memory, an arbitrarily complex numerical integration scheme may not in general be used. The efficiency of a particular difference scheme may be defined as the total number of arithmetic, logical and storage operations performed by the central processor of the computer, to obtain a solution over a unit time-length of the problem. On the one hand the efficiency decreases with greater complexity of the particular difference method being applied, while on the other hand, the accuracy of the solution may be increased with increasing complexity. A compromise must be reached in order to obtain a viable method which is both accurate and efficient. For a simple problem with a small number of variables, little computer time is required for a solution and a sophisticated scheme of high-order accuracy may be used, while for a system of many variables, accuracy has to be compromised in the interests of efficiency.

6 Integration of Ordinary Differential Equations

We shall here develop a number of important integration schemes in time, which are frequently used to obtain difference solutions to the initial-value problem. The methods will be developed for a single ordinary differential equation, or a pair of coupled ordinary differential equations. Although the methods will be introduced by considering these simple examples, they extend themselves to more complex situations and, in particular, to systems of partial differential equations. In addition, the schemes may be applied to large systems of ordinary differential equations.

The set of methods developed below is not a complete set. Rather, we start with the simplest possible scheme and develop more sophisticated methods, but very complex methods are avoided as their applicability is limited.

The ordinary differential equation is considered,

$$\frac{\mathrm{d}u}{\mathrm{d}t} + f(u, t) = 0 \tag{2.61}$$

where $u = u(t)$ and the initial conditions are defined,

$$u(t^0) = u^0 \tag{2.62}$$

This is a restricted example of the initial-value problem defined in Section 2.4 where, in equation 2.61, a functional rather than an operator formalism is used. In general, the equation is to include the case where the dependent variable u and the function f are complex, in which event a pair of coupled equations are represented.

As in Section 2.4, equation 2.61 may be integrated on a time mesh between the time points t^n and t^{n+1} over the time interval Δt,

$$u^{n+1} = u^n - \int_{t^n}^{t^{n+1}} f(u, t)\,\mathrm{d}t \tag{2.63}$$

$$\Delta t = t^{n+1} - t^n \tag{2.64}$$

Approximations to the integral in time, on the right-hand side of equation 2.63 define a number of methods.

(*a*) *The Euler First-order Method.* The simplest approximation to the integral in equation 2.63 is to approximate the integrand f, for all time $t, t^n \leqslant t \leqslant t^{n+1}$, by the function f at time t^n (Figure 2.6), and hence the

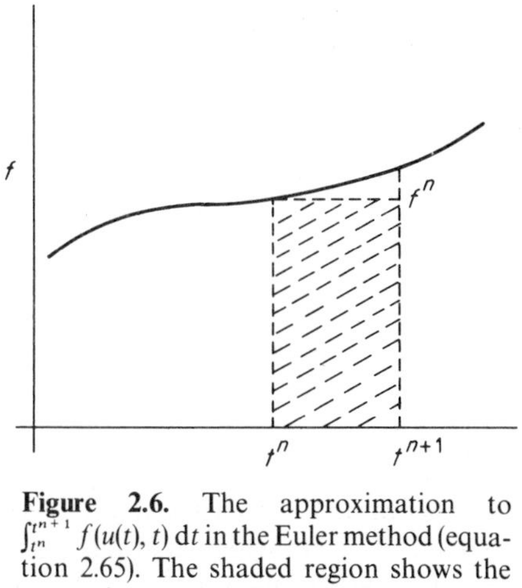

Figure 2.6. The approximation to $\int_{t^n}^{t^{n+1}} f(u(t), t)\,\mathrm{d}t$ in the Euler method (equation 2.65). The shaded region shows the approximation to the integral under the curve

algorithm for the Euler method is obtained:

$$u^{n+1} = u^n - f(u^n, t^n)\,\Delta t \tag{2.65}$$

The Euler method is explicit and accurate only to the first order in the time step Δt. It is clearly extremely simple and efficient, but we question under what conditions, if any, it is stable.

The analysis described in Section 2.5 is applied by assuming that there exists an error ϵ^n in the dependent variable u^n, at time t^n. We then determine the error which is produced in consequence in the dependent variable u^{n+1}. From equation 2.65,

$$u^{n+1} + \epsilon^{n+1} = u^n + \epsilon^n - f((u^n + \epsilon^n), t^n)\,\Delta t \tag{2.66}$$

If ϵ^n is small, then the function f at $u = u^n + \epsilon^n$ may be expanded about u^n, by a Taylor expansion:

$$f(u^n + \epsilon^n, t^n) = f(u^n, t^n) + \left.\frac{\partial f}{\partial u}\right|_n \epsilon^n + o(\epsilon^n), \tag{2.67}$$

and, using the original equation 2.65, an equation for the amplification of a small error is obtained:

$$\epsilon^{n+1} = \epsilon^n - \left.\frac{\partial f}{\partial u}\right|_n \Delta t\,\epsilon^n + o(\epsilon^n) \tag{2.68}$$

Hence, for the Euler scheme the amplification factor g is:

$$g = 1 - \left.\frac{\partial f}{\partial u}\right|_n \Delta t \tag{2.69}$$

We distinguish between three classes of problems: 'decay' type equations where $\partial f/\partial u > 0$; 'growth' type equations where $\partial f/\partial u < 0$; and 'oscillatory' type equations, where the dependent variable u is complex and a pair of coupled equations are described, in which $\partial f/\partial u$ is imaginary. The condition for the stability of the scheme (equation 2.52) must be applied. In the first case of decay type equations, where $\partial f/\partial u > 0$, we require for stability:

$$\left.\frac{\partial f}{\partial u}\right|_n \Delta t \leqslant 2 \tag{2.70}$$

$$\Delta t \leqslant \frac{2}{\partial f/\partial u|_n} \tag{2.71}$$

This condition therefore defines an upper limit on the allowable time step if a stable solution is to be obtained. If the time step, Δt, is chosen to be larger than this quantity, errors will occur which will increase without limit and the method will be 'unstable'. It is to be noted that the method by which we arrived at this condition is a very common one in physics when dealing with nonlinear equations. We have 'linearized' the analysis, by considering a small displacement ϵ^n from the known parameter u^n, and hence a linear equation relating ϵ^{n+1} and ϵ^n has been obtained.

As a simple example to illustrate the method, the equation,

$$\frac{\mathrm{d}u}{\mathrm{d}t} + \frac{u}{\tau} = 0 \tag{2.72}$$

with the initial condition

$$u(0) = 1 \tag{2.73}$$

where τ is a constant, may be examined. This equation arises physically, for example, in the decay of an electric current in an inductive (L) and resistive (R) circuit when $\tau = L/R$. It arises in radioactive decay problems where u is the neutron flux from a radioactive source, and τ is the half-life of the radioactive material. Again it arises in describing the decay of the average electron momentum due to collisions in the conduction band of an ion lattice. We know the solution analytically:

$$u = \mathrm{e}^{-t/\tau} \tag{2.74}$$

To illustrate the Euler method, we shall obtain the solution to the decay equation by the explicit first-order difference method (equation 2.65). To ensure stability of the solutions, however, we must choose time steps which satisfy condition 2.71. For this example,

$$\left|\frac{\partial f}{\partial u}\right|_n = \frac{1}{\tau}$$

$$\Delta t \leqslant 2\tau \tag{2.75}$$

The problem is written in dimensionless form by measuring the time variable t in units of τ, the decay time (hence $\tau' = 1{\cdot}0$), and the results of three solutions to the difference equation 2.65 are summarized in Table 2.1 and are illustrated in Figure 2.7. In the case where the stability criterion is well satisfied ($\Delta t = 0{\cdot}1\tau$), the solution is regular with each time step and is clearly a good approximation to the exact solution (equation 2.74). If a larger time step is taken, the accuracy of the solution is diminished though it remains regular in following the exact solution. When the time step is chosen to satisfy the equality in the stability criterion ($\Delta t = 2{\cdot}0\tau$), the solution bears no relation to the exact solution, but at the same time, u^n does not grow without bound as n increases, so that the solution is neutrally stable. However, in the third example, when the stability criterion is violated ($\Delta t > 2{\cdot}0\tau$), within a few time levels the solution grows catastrophically and bears no relation to the exact solution. It is clear that the stability condition is a fundamental property of a difference method and care must always be taken to ensure the satisfaction of any stability criterion.

Table 2.1 The decay equation solution by $u^{n+1} = u^n - \Delta t u^n$

Exact solution	Stable solution $\Delta t = 0{\cdot}1\,(\tau)$		Neutrally stable solution $\Delta t = 2{\cdot}0\,(\tau)$		Unstable solution $\Delta t = 10{\cdot}0\,(\tau)$	
$u(0{\cdot}0) = 1{\cdot}000$	$t = 0{\cdot}0$	$u^0 = 1{\cdot}000$	$t = 0{\cdot}0$	$u^0 = 1{\cdot}0$	$t = 0{\cdot}0$	$u^0 = 1{\cdot}0$
$u(0{\cdot}1) = 0{\cdot}9048$	$0{\cdot}1$	$u^1 = 0{\cdot}9000$	$2{\cdot}0$	$u^1 = -1{\cdot}0$	$10{\cdot}0$	$u^1 = -9{\cdot}0$
$u(0{\cdot}2) = 0{\cdot}8187$	$0{\cdot}2$	$u^2 = 0{\cdot}8100$	$4{\cdot}0$	$u^2 = 1{\cdot}0$	$20{\cdot}0$	$u^2 = 81{\cdot}0$
$u(0{\cdot}3) = 0{\cdot}7408$	$0{\cdot}3$	$u^3 = 0{\cdot}7290$	$6{\cdot}0$	$u^3 = -1{\cdot}0$	$30{\cdot}0$	$u^3 = -729{\cdot}0$
$u(0{\cdot}4) = 0{\cdot}6703$	$0{\cdot}4$	$u^4 = 0{\cdot}6561$				
$u(0{\cdot}5) = 0{\cdot}6065$	$0{\cdot}5$	$u^5 = 0{\cdot}5895$				
$u(0{\cdot}6) = 0{\cdot}5488$	$0{\cdot}6$	$u^6 = 0{\cdot}5306$				
$u(0{\cdot}7) = 0{\cdot}4966$	$0{\cdot}7$	$u^7 = 0{\cdot}4775$				
$u(0{\cdot}8) = 0{\cdot}4493$	$0{\cdot}8$	$u^8 = 0{\cdot}4298$				
$u(0{\cdot}9) = 0{\cdot}4066$	$0{\cdot}9$	$u^9 = 0{\cdot}3865$				
$u(1{\cdot}0) = 0{\cdot}3679$	$1{\cdot}0$	$u^{10} = 0{\cdot}3481$				

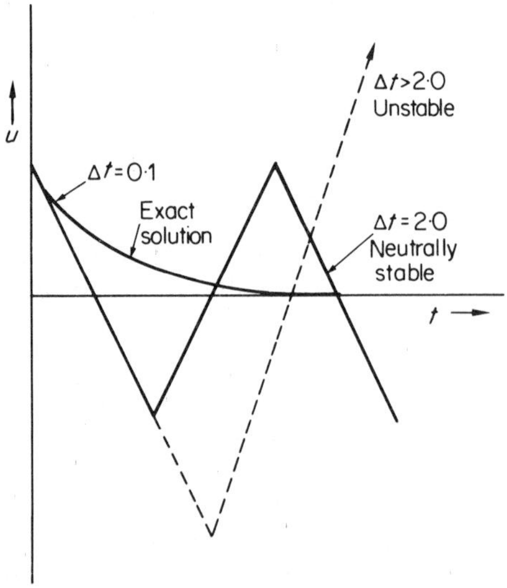

Figure 2.7. Solutions to the decay equation (2.72) obtained by the Euler method. The three cases of a numerically stable ($\Delta t = 0{\cdot}1\tau$), neutrally stable ($\Delta t = 2{\cdot}0\tau$) and unstable ($\Delta t > 2{\cdot}0\tau$) solution are illustrated (cf. Table 2.1)

It is informative at this stage to consider the alternative example of 'oscillatory' type equations where, in the equation for the amplification factor (equation 2.69) of the Euler method, the parameter $\partial f/\partial u$ is imaginary. As the simplest example of such a situation, the equation for a simple harmonic oscillator, of frequency ω, is taken,

$$\frac{d^2x}{dt^2} + \omega^2 x = 0 \tag{2.76}$$

where typically the dependent variable x represents the displacement from an equilibrium position. A normalized velocity v is defined, so that the second-order equation may be rewritten as two first-order equations:

$$\frac{dx}{dt} - \omega v = 0$$

$$\frac{dv}{dt} + \omega x = 0 \tag{2.77}$$

We shall use a complex variable notation,

$$\frac{du}{dt} + i\omega u = 0 \tag{2.78}$$

with $u = x + iv$.

In this example the amplification factor is complex,

$$g = 1 - i\omega\,\Delta t \tag{2.79}$$

Here and in the general case, where the parameter $\partial f/\partial u$ is imaginary and consequently where the amplification factor g is complex, to examine the stability of the method we must consider the magnitude of the amplification factor in the complex plane (equation 2.60):

$$|g|^2 = gg^* = 1 + \left|\frac{\partial f}{\partial u}\right|_n^2 \Delta t^2 \tag{2.80}$$

The magnitude of the amplification factor is always larger than unity for all values of the time step Δt and, for purely oscillatory type equations, the Euler method is inappropriate and unconditionally unstable.

In more sophisticated examples, where the problem is nonlinear, the derivative $\partial f/\partial u$ is a function of u. Hence at each time level, a new time step must be determined to satisfy, if possible, the stability criterion. For an optimized solution, the magnitude of the time step is a function of time.

At least for decay-type equations, we have seen that we must choose a small time step if a stable solution for the Euler method is to be obtained. In addition, in the worked example, it is clear that a more accurate solution is obtained when a smaller time step is used. The Euler method is accurate only to first order in the time step Δt and the error arises through approximating the integral in equation 2.63 by evaluating the integrand f at time t^n only. Effectively, only the first term in a Taylor expansion of the integrand has been used (equation 2.42) and the solution can be improved if the next term in the Taylor expansion is included, when we say the integrand is 'time-centred'.

(*b*) *The Leapfrog Method.* One common method to time-centre the integrand in equation 2.63 and, consequently, to obtain second-order accuracy, is to use a 'two-level' formula (Figure 2.8). We determine the

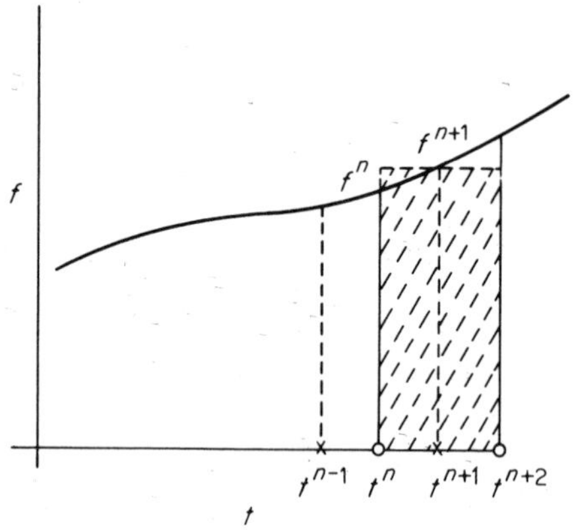

Figure 2.8. The approximation to $\int_{t^n}^{t^{n+2}} f(u(t), t)\, dt$ in the leapfrog method (equations 2.81)

derivative in time across a double timestep and we use the intermediate timestep to define the integral of f,

$$\begin{aligned} &\text{at } n+1 \qquad u^{n+1} = u^{n-1} - f(u^n, t^n) 2\,\Delta t \\ &\text{at } n+2 \qquad u^{n+2} = u^n - f(u^{n+1}, t^{n+1}) 2\,\Delta t \end{aligned} \tag{2.81}$$

u^{n+1} and u^{n+2} may be used in turn to determine u^{n+3} and for obvious reasons the method is most conveniently termed the leapfrog method. It is clearly very simple and second-order accuracy has been obtained.

There is a difficulty in this method, however, for the proper boundary conditions define only $u^0 = u(0)$, and to proceed we require $u^1 = u(\Delta t)$. Obviously we can determine u^1 by, say, the Euler method and then proceed for all subsequent time levels by the leapfrog method. It turns out, however, that the overall accuracy of the solution by the leapfrog method is a very sensitive function of the accuracy of u^1. In order to minimize the total errors,

u^1 is frequently determined with greater accuracy either by using sub-intervals in the first time interval Δt, or by using a high-order power series expansion.

Another difficulty arises in the leapfrog scheme with a nonlinear problem when, in order to minimize the errors, the time step is varied. The method is no longer time-centred and difficulties can be expected.

To investigate the stability of the leapfrog scheme, we notice that the two formulae in the algorithm (equations 2.81) are interchangeable. The same method of stability analysis is applied as before, but now errors at three time levels, ϵ^{n-1}, ϵ^n, ϵ^{n+1}, are related,

$$\epsilon^{n+1} = \epsilon^{n-1} - \left.\frac{\partial f}{\partial u}\right|_n 2\,\Delta t \epsilon^n \tag{2.82}$$

thus,

$$g^2 = 1 - \left.\frac{\partial f}{\partial u}\right|_n 2\,\Delta t g \tag{2.83}$$

$$g = -\alpha \pm \sqrt{(\alpha^2 + 1)} \tag{2.84}$$

where $\alpha = \partial f/\partial u|_n\,\Delta t$. There are two roots for the amplification factor g due to the use of a method of second-order accuracy and, for non-oscillatory equations, where the parameter α is not imaginary, the magnitude of the amplification factor for one of the roots is always greater than unity. Hence the leapfrog scheme is not usually appropriate for equations of the decay or growth type. On the other hand, for equations describing oscillations (for example equation 2.78) where α is purely imaginary, the amplification factor g is complex and, providing α is small, the magnitude of the amplification factor is identically equal to one. If $\alpha = \mathrm{i}\beta$, where β is real,

$$g = -\mathrm{i}\beta \pm \sqrt{(-\beta^2 + 1)}$$

$$gg^* = 1 \quad (\text{for } \beta \leqslant 1) \tag{2.85}$$

For the particular case of the pair of coupled harmonic equations (equation 2.78), the stability requirement is, therefore,

$$\beta = \omega\,\Delta t \leqslant 1$$

$$\Delta t \leqslant \frac{1{\cdot}0}{\omega} \tag{2.86}$$

Thus again, for stability, the time step must be smaller than the characteristic time described by the equations, which in this case is the period through one radian of the oscillation.

We shall illustrate the particular difficulties and properties of the leapfrog scheme, which have been suggested above, by referring to the decay equation (2.72). We notice from the two-level formulae (equations 2.81) that the variables defined on the even mesh (time levels $2n$) are weakly coupled to the variables defined on the odd mesh (time levels $2n + 1$) and vice versa. We shall define ξ as the variable u on the even mesh and ζ as the variable u on the odd mesh. Then, with no loss of generality, the leapfrog formulae (equations 2.81) for the decay equation (equation 2.72) may be written as:

$$\xi^{2n} = \xi^{2n-2} - \zeta^{2n-1}\frac{2\,\Delta t}{\tau}$$

$$\zeta^{2n+1} = \zeta^{2n-1} - \xi^{2n}\frac{2\,\Delta t}{\tau} \tag{2.87}$$

These equations are equivalent to the coupled first-order differential equations:

$$\frac{d\xi}{dt} + \frac{\zeta}{\tau} = 0$$

$$\frac{d\zeta}{dt} + \frac{\xi}{\tau} = 0 \tag{2.88}$$

If we add and subtract these two equations, we obtain the normal modes of the system,

$$\frac{d}{dt}(\xi + \zeta) + \frac{(\xi + \zeta)}{\tau} = 0 \tag{2.89}$$

$$\frac{d}{dt}(\xi - \zeta) - \frac{(\xi - \zeta)}{\tau} = 0 \tag{2.90}$$

The first of these modes is the solution which we seek and which satisfies the original first-order differential equation. However, the second mode is not consistent with the original equation and consequently, in the leapfrog scheme, an additional extraneous 'numerical mode' can occur. If this latter mode is permitted to exist, it will lead to large errors in the solution and we can expect a large oscillation to develop between the variables on the even mesh and those on the odd mesh. To minimize the effect of this mode, we must determine the dependent variable u^1, at the first step, as accurately as possible since the magnitude of each of the modes is defined by the 'initial conditions' u^0 and u^1.

To summarize, we have seen that, in the leapfrog scheme, second-order accuracy can be obtained with an extremely simple difference formula. This is obtained, however, first with a dependence on a stability criterion and second at the risk of introducing, particularly for nonoscillatory equations and in the nonlinear case, an extraneous computational mode which can seriously corrupt the solution.

(*c*) *The Explicit Two-step Method*. An extremely useful method with wide application is to 'time-centre' the integral $\int_n^{n+1} f\,dt$ in equation 2.63 by

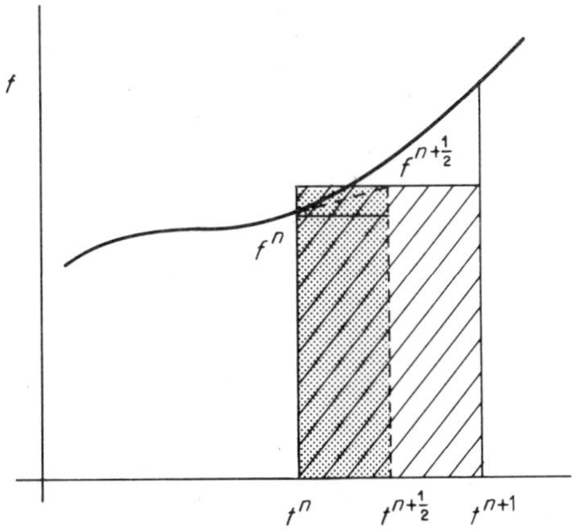

Figure 2.9. The approximation to $\int_{t^n}^{t^{n+1}} f(u(t), t)\, dt$ in the two-step method (equations 2.91, 2.92). The first step uses the Euler method to 'time-centre' the integral at $t^{n+\frac{1}{2}}$ (dotted region). The second step evaluates the integral to second-order accuracy (shaded region)

a two-step procedure (Figure 2.9). The two-step method applies the Euler explicit method as a first stage, in evaluating the dependent variable u at the intermediate time $t^{n+1/2}$ (the auxiliary calculation):

auxiliary: $$u^{n+1/2} = u^n - f(u^n, t^n)\frac{\Delta t}{2} \tag{2.91}$$

main: $$u^{n+1} = u^n - f(u^{n+1/2}, t^{n+1/2})\,\Delta t \tag{2.92}$$

The values $u^{n+1/2}$ are intermediate and are not retained after the time level t^{n+1}. Applying the stability analysis as before, the errors at the two time-levels are related:

$$\begin{aligned} \epsilon^{n+1} &= \epsilon^n - \left.\frac{\partial f}{\partial u}\right|_n \Delta t \left\{1 - \left.\frac{\partial f}{\partial u}\right|_n \frac{\Delta t}{2}\right\} \epsilon^n \\ g &= 1 - \alpha + \tfrac{1}{2}\alpha^2 \end{aligned} \tag{2.93}$$

where $\alpha = \partial f/\partial u|_n\, \Delta t$. Again for the case $\partial f/\partial u > 0$, stability is achieved for a sufficiently small time step,

$$\Delta t \leqslant \frac{2{\cdot}0}{\partial f/\partial u|_n} \tag{2.94}$$

It is noteworthy that the two-step formulae may be obtained directly from a Taylor expansion of u^{n+1} in the time step Δt, about u^n, though it is most conveniently applied in the form presented above. The method is in effect the special second-level case of the Runge–Kutta methods (Fox and Mayers 1968).

(*d*) *Implicit Second-order Method.* In the three methods which have been described above, the solution for the dependent variable at each time step is obtained in explicit form and a stability criterion must always be satisfied. In simple ordinary differential equations, it is frequently useful to use an implicit method, where the integral in equation 2.63 is determined with second-order accuracy in the time step Δt, by the use of a time average for the integrand f between the time levels t^n and t^{n+1} (Figure 2.10):

$$u^{n+1} = u^n - \{f(u^n, t^n) + f(u^{n+1}, t^{n+1})\}\frac{\Delta t}{2} \tag{2.95}$$

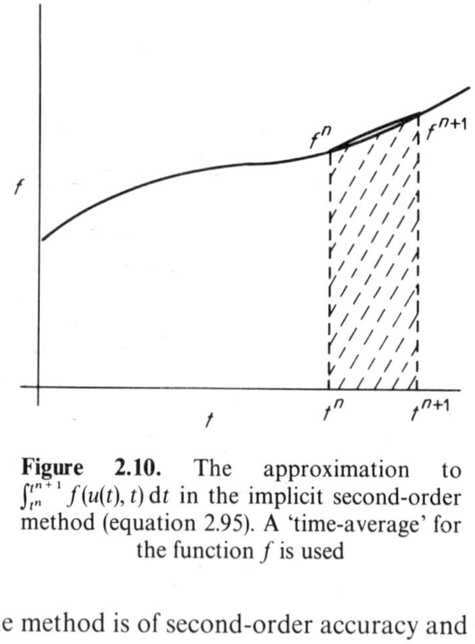

Figure 2.10. The approximation to $\int_{t^n}^{t^{n+1}} f(u(t), t)\,\mathrm{d}t$ in the implicit second-order method (equation 2.95). A 'time-average' for the function f is used

It is clear that the method is of second-order accuracy and we shall examine its stability:

$$g = 1 - \left.\frac{\partial f}{\partial u}\right|_n \frac{\Delta t}{2} - \left.\frac{\partial f}{\partial u}\right|_{n+1} \frac{\Delta t}{2} g \tag{2.96}$$

$$g = \frac{1 - \left.\dfrac{\partial f}{\partial u}\right|_n \dfrac{\Delta t}{2}}{1 + \left.\dfrac{\partial f}{\partial u}\right|_{n+1} \dfrac{\Delta t}{2}} \tag{2.97}$$

For decay equations ($\partial f/\partial u > 0$), the modulus of the amplification factor g is always smaller than unity and equally, for oscillatory equations where $\partial f/\partial u$ is imaginary, the magnitude of the complex amplification factor is equal to unity. For these general cases, therefore, the method is *unconditionally stable.*

Clearly a method which is stable, irrespective of the choice of time step, is of great advantage, particularly in complex and nonlinear problems. This desirable property is achieved, however, at the expense of some algebraic complexity, for in the difference formulation (equation 2.95) the new dependent variable u^{n+1} at each time step is obtained implicity. A possibly difficult algebraic equation remains to be solved at each time step. If the function f is not a complicated function of the dependent variable, a simple method for the algebraic solution of u^{n+1} may exist, but if this is not the case, we must devise an iterative formula to obtain a consistent solution.

In this section we have developed a number of important methods for integrating in time and it is the simplicity of these methods which permits their application to more complex problems. More sophisticated methods exist, but their applicability to partial differential equations is limited, and our concern has been to introduce broadly useful schemes and the techniques required to analyse them. A summary of integration methods for ordinary differential equations is included in Table 2.2.

Table 2.2 Methods for the integration of time-dependent ordinary differential equations:

$$\frac{du}{dt} + f(u, t) = 0$$

where $f^n = f(u^n, t^n)$, and $\alpha = \left.\frac{\partial f}{\partial u}\right|_n \Delta t$

Method	Algorithm	Amplification factor
1. Euler	$u^{n+1} = u^n - f^n \Delta t$ Stable for α real, if $\Delta t \leqslant 2 \Big/ \left.\frac{\partial f}{\partial u}\right\|_n$ Unstable for α imaginary $\epsilon = O(\Delta t)$	$g = 1 - \alpha$
2. Leapfrog	$u^{n+1} = u^{n-1} - f^n 2\Delta t$ Unstable for α real Stable for α imaginary, if $\Delta t \leqslant 1 \Big/ \left.\frac{\partial f}{\partial u}\right\|_n$ $\epsilon = O(\Delta t^2)$	$g = -\alpha \pm \sqrt{(\alpha^2 + 1)}$

Table 2.2 (*continued*)

Method	Algorithm	Amplification factor
3. Two-step	$u^{n+\frac{1}{2}} = u^n - f^n \Delta t/2$ $u^{n+1} = u^n - f^{n+\frac{1}{2}} \Delta t$ Stable for α real, if $\Delta t \leqslant 2 \Big/ \left.\frac{\partial f}{\partial u}\right\|_n$ Marginally unstable for α imaginary. $\epsilon = O(\Delta t^2)$	$g = 1 - \alpha + \frac{1}{2}\alpha^2$
4. Implicit	$u^{n+1} = u^n - \frac{1}{2}(f^n + f^{n+1})\Delta t$ Stable for α imaginary, for all Δt Stable for α real, for all Δt $\epsilon = O(\Delta t^2)$	$g = (1 - \alpha)/(1 + \alpha)$
5. Adams–Bashforth	$u^{n+1} = u^n - \frac{1}{2}(3f^n - f^{n-1})\Delta t$ Stable for α real, if $\Delta t < 1 \Big/ \left.\frac{\partial f}{\partial u}\right\|_n$ Marginally unstable for α imaginary. $\epsilon = O(\Delta t^2)$	$g = \frac{1}{2} - \frac{3}{4}\alpha \pm \frac{1}{2}\sqrt{(\frac{9}{4}\alpha^2 - \alpha + 1)}$

7 High-Order Ordinary Differential Equations

In computational physics we are seldom interested in solving merely one first-order ordinary differential equation and, if this were the case, we would not be interested in the detailed significance and analysis of various methods, since one first-order, ordinary differential equation can be solved in a negligible amount of time on the computer. But the methods described above carry over precisely to systems, and very large systems, of ordinary differential equations and indeed to partial differential equations (see Chapter 3).

For more complex differential equations than those described in the previous section, it is to be noted that a differential equation of order n may always be reduced to n first-order differential equations. For example, we consider a particle of mass m moving in an electric field $E_x(x, t)$, where x is the one-dimensional position of the particle. Then,

$$m\frac{d^2x}{dt^2} - eE_x(x, t) = 0 \tag{2.98}$$

and, if v is the velocity of the particle, this second-order differential equation can be written conveniently as two first-order equations:

$$m\frac{dv}{dt} - eE_x(x, t) = 0$$
$$\frac{dx}{dt} - v = 0 \tag{2.99}$$

The general problem may now be formulated. We are required to solve for a set of K variables $\mathbf{u}$,

$$\mathbf{u} = (u_1, u_2, \ldots, u_K)$$

which satisfy K, first-order, ordinary, nonlinear differential equations,

$$\frac{d\mathbf{u}}{dt} + \mathbf{F}(\mathbf{u}, t) = \mathbf{0} \tag{2.100}$$

with the initial conditions,

$$\mathbf{u}(t_0) = \mathbf{u}^0 \tag{2.101}$$

We are frequently interested in physical problems where the dimension K is a very large number (for example $K \approx 100{,}000$ on present-day machines) and it is obviously essential to solve these equations with optimized efficiency and accuracy. It follows that we should analyse particular difference methods with some care.

As an example where K can be large, we may be interested in a many-body Newtonian gravitational system, where the particles move in one dimension. The exact classical equations for such a system are:
for particles $\mu = 1$ *to* K

$$\frac{dx_\mu}{dt} - w_\mu = 0 \tag{2.102}$$

$$m_\mu \frac{dw_\mu}{dt} - \sum_{\substack{\nu=1 \\ \nu \neq \mu}}^{K} \frac{Gm_\nu m_\mu}{(x_\nu - x_\mu)^2} = 0 \tag{2.103}$$

where, x_μ and w_μ are the positions and velocities of the μth particle, G is the gravitational constant, m_μ is the mass of the μth particle and the sum is over all other particles acting on the μth particle. Here, we could use one of the explicit methods described above to follow the whole system of K particles for all time, $t > 0$. Many-body problems of this form are of great interest in physics and will be discussed in Chapter 5.

For such a system, we should examine the stability of any difference method employed. The stability must now be analysed in terms of the amplification matrix (equation 2.53), where we require that the modulus of the eigenvalues of the amplification matrix each satisfy the stability condition (equation 2.59).

CHAPTER III

Partial Difference Equations for Continuous Media

1 The Occurrence and Some Properties of Partial Differential Equations in Physics

The essential principles of the difference method have been introduced in the previous chapter, where we have seen that functions which are continuous in space and time may be replaced by finite vectors, the components of which are defined only at discrete points in space and time. The properties of a difference solution have been enumerated (Section 2.5). For ordinary differential equations in the initial-value problem, the properties of accuracy and stability have been related to the characteristic times associated with the equation (for example the decay time, or oscillation time). Similarly in partial differential equations the accuracy and stability of a numerical solution depend on the characteristic times described by the equations. Generally, therefore, and before applying the difference method to partial differential equations, it is important to establish some of the essential properties of such equations in physics.

In this section, we consider how partial differential equations arise in physics and we relate widely understood physical processes to the mathematical nature of the equations.

(*a*) *Principles of Conservation Applied to Continuous Media.* The abstract concepts of continuous media and continuous fields have very wide application in physics: in classical electrodynamics, Maxwell's equations are formulated by defining continuous source functions; a solid is frequently treated for simplicity as a continuum; a great variety of fluids (liquids, gases, plasmas, the galactic fluid) may be treated with most ease as a continuum. Other examples are phase fluids and 'continuous fields of force' in classical and quantum mechanics. When using such concepts, continuous functions in space and time are defined which describe the properties of the medium and, on applying the quantitative principles of physics, partial differential equations are obtained which couple the properties of the medium in space and time.

Although it is clear that the partial differential equations of physics arise in a great variety of ways and from very different problems, nevertheless such equations, and systems of equations, repeatedly take the same or similar forms. This is essentially because a great deal of the philosophy of physics and, in particular of classical physics, has been formulated in terms of principles of conservation. A few examples of great importance illustrate the point: mass cannot be created or destroyed; momentum is conserved; total electric charge is an invariant. The partial differential equations, which result from the application of such ideas, are themselves said to be conservative. We develop some particular examples to stress the approach and to illustrate the method.

The diffusion equation occurs frequently when describing the transport of particles, of momentum, or of energy. As a particular example it occurs when describing the temperature distribution in a solid, where classically the energy is transported by conduction.

Since the solid is a rigid stationary body, the variable energy density in the solid is related only to the thermal energy or temperature. Hence the principle of *conservation of energy* is invoked and must be satisfied when considering the energy in a finite volume V of the solid of surface S (Figure 3.1). By the principle of conservation of energy, the rate of change of energy

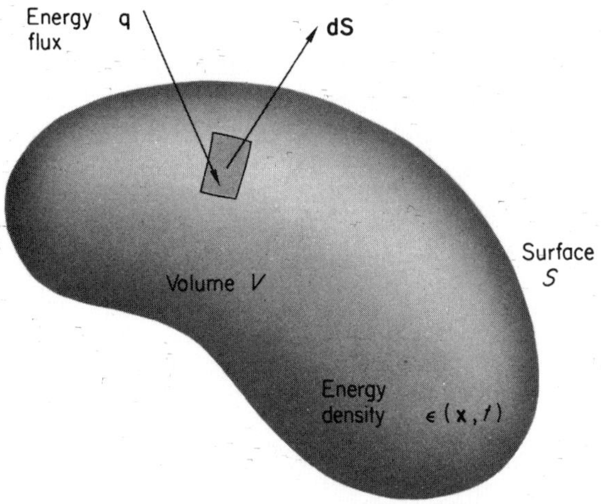

Figure 3.1. To obtain the heat-conduction equation in a solid, the principle of the conservation of energy is applied to an arbitrary volume V of the solid. Accordingly, the total energy in V may only alter by the occurrence of a flux of energy $\mathbf{q}$ across the surface S of V

in V must be equal to the flux of energy $\mathbf{q}$ across the surface S of V:

$$\text{energy in volume } V = \iiint \epsilon(\mathbf{x}, t)\, \mathrm{d}\tau$$

$$\text{flux across } S = -\oiint_S \mathbf{q} \,.\, \mathrm{d}\mathbf{S}$$

$$\frac{\partial}{\partial t}\iiint_V \epsilon(\mathbf{x}, t)\, \mathrm{d}\tau = -\oiint_S \mathbf{q} \,.\, \mathrm{d}\mathbf{S} \tag{3.1}$$

Applying the divergence theorem to the second term and for a constant volume V,

$$\iiint_V \frac{\partial \epsilon}{\partial t} + \nabla \,.\, \mathbf{q}\, \mathrm{d}\tau = 0. \tag{3.2}$$

Now ϵ, the energy density, is proportional to the temperature T and experimentally it is found that the heat flux $\mathbf{q}$ depends on the gradient of the temperature. Hence, defining a proportionality constant, the conductivity κ,

$$\frac{\partial T}{\partial t} - \boldsymbol{\nabla} \,.\, \kappa \boldsymbol{\nabla} T = 0 \tag{3.3}$$

the diffusion equation is obtained. The essential principle described by the diffusion equation in this case is then the principle of the conservation of energy.

In electromagnetic theory the partial differential equations of Maxwell describe conservation laws, as for example in the application of the principle of the conservation of *electric charge*. Again therefore in a volume V, of surface S, the rate of change of charge in V must be equal to the flux of charge (the current $\mathbf{j}$) across the surface. If ρ is the charge density $\rho(\mathbf{x}, t)$ then,

$$\frac{\partial}{\partial t}\iiint_V \rho\, \mathrm{d}\tau = -\oiint_S \mathbf{j} \,.\, \mathrm{d}\mathbf{S} \tag{3.4}$$

and, applying the divergence theorem,

$$\frac{\partial \rho}{\partial t} + \boldsymbol{\nabla} \,.\, \mathbf{j} = 0 \tag{3.5}$$

This is, of course, consistent with Maxwell's equations, since, using Gauss' law ($\boldsymbol{\nabla} \,.\, \mathbf{E} = 4\pi\rho$, where $\mathbf{D}$ is the electric field),

$$\boldsymbol{\nabla} \,.\, \left(\frac{\partial \mathbf{E}}{\partial t} + 4\pi \mathbf{j}\right) = 0 \tag{3.6}$$

and, on integrating and noting that $\nabla \cdot (\nabla \wedge \mathbf{X}) = 0$,

$$\frac{\partial \mathbf{E}}{\partial t} + 4\pi \mathbf{j} = \nabla \wedge \mathbf{X} \tag{3.7}$$

where the unknown vector **X** is identified as proportional to the magnetic field **B** by Ampère's law.

Similarly, Faraday's law is an expression of the *conservation* of *magnetic flux*: magnetic flux cannot be created or destroyed, since the rate of change of the total flux through a surface S is related only to the electric field around the boundary l of S (Figure 3.2):

$$\frac{\partial}{\partial t}\iint_S \mathbf{B} \cdot \mathbf{dS} = -\oint_l c\mathbf{E} \cdot \mathbf{dl} \tag{3.8}$$

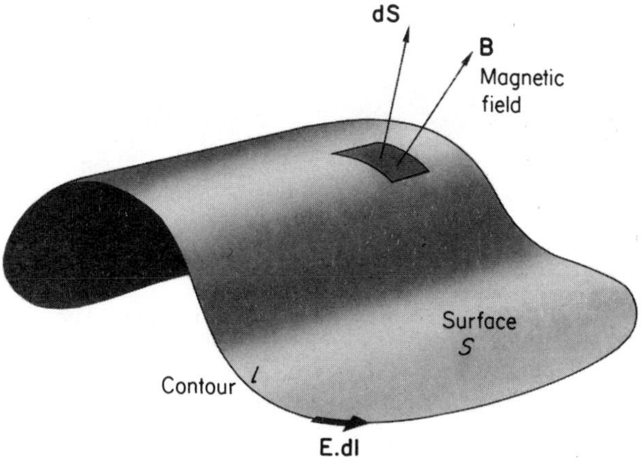

Figure 3.2. The diagram illustrates the conservation of magnetic flux according to Faraday's law (equation 3.9). The magnetic flux within the surface S may only alter by the existence of an electric field around the contour l of S

and, applying Stoke's theorem,

$$\frac{1}{c}\frac{\partial \mathbf{B}}{\partial t} + \nabla \wedge \mathbf{E} = \mathbf{0} \tag{3.9}$$

For a fluid, the basic classical principles of *conservation* of *mass*, *conservation* of *momentum* (Newton's third law of motion) and *conservation* of *energy*, are used to derive equations to describe the dynamics of a fluid. Defining a variable $\rho(\mathbf{x}, t)$ as the density of the fluid, we invoke the first principle to state that the rate of change of mass in the volume V must equal the mass flux crossing the surface S of V(Figure 3.3). The mass flux through any element of surface $\mathbf{dS}$ is just $-\rho\mathbf{v}\,.\,\mathbf{dS}$. Hence,

$$\frac{\partial}{\partial t}\iiint_V \rho\,\mathrm{d}\tau = -\oiint_S \rho\mathbf{v}\,.\,\mathbf{dS} \tag{3.10}$$

and, using the divergence theorem, a differential equation for the conservation of mass is obtained,

$$\frac{\partial\rho}{\partial t} + \nabla\,.\,\rho\mathbf{v} = \mathbf{0} \tag{3.11}$$

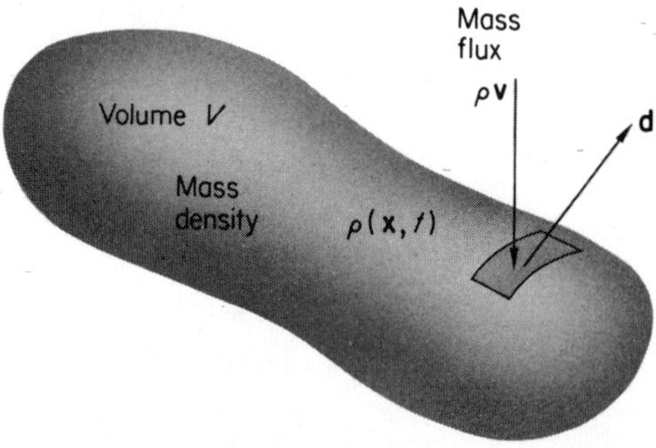

Figure 3.3. The conservation of mass in a fluid. The mass within the volume V may only alter by the existence of a net mass flux $\rho\mathbf{v}\,.\,\mathbf{dS}$ across the surface S of V. The continuity equation (3.11) is obtained

Similarly, by demanding that momentum be conserved, an equation of motion in the fluid is obtained. We shall consider the conservation of momentum in the X-direction (Figure 3.4). The total X-momentum in the volume V is:

$$\iiint_V \rho v_X\,\mathrm{d}\tau$$

The X-component of momentum of the fluid in the volume V is increased in time, by the convection of momentum and the effect of pressure (p) in the X-direction ($\hat{\mathbf{e}}_X$ is the unit vector in the X-direction).

$$-\oiint_S (\rho v_X\mathbf{v} + p\hat{\mathbf{e}}_X)\,.\,\mathbf{dS}$$

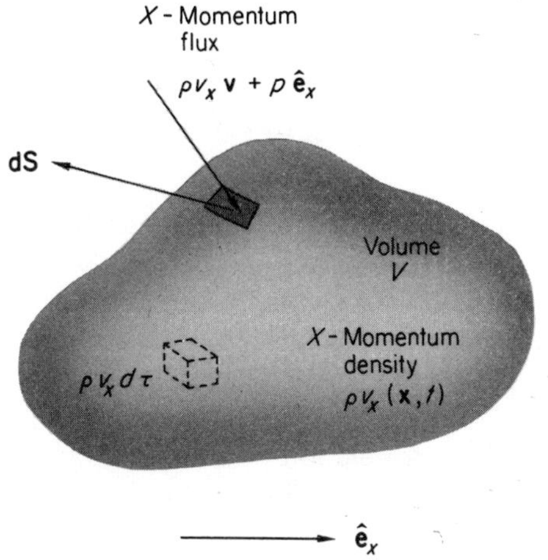

Figure 3.4. The conservation of momentum in a fluid. The momentum flux across the surface S includes the advected momentum and pressure flux

Hence the conservation of X-momentum yields the equation:

$$\frac{\partial}{\partial t}\iiint_V \rho v_X \, \mathrm{d}\tau = -\oiint_S (\rho v_X \mathbf{v} + p\hat{\mathbf{e}}_X) \, . \, \mathbf{dS} \tag{3.12}$$

Using the divergence theorem, we obtain,

$$\frac{\partial}{\partial t}\iiint_V \rho v_X \, \mathrm{d}\tau = -\iiint_V \boldsymbol{\nabla} \, . \, (\rho v_X \mathbf{v} + p\hat{\mathbf{e}}_X) \, \mathrm{d}\tau \tag{3.13}$$

$$\frac{\partial \rho v_X}{\partial t} + \boldsymbol{\nabla} \, . \, (\rho v_X \mathbf{v} + p\hat{\mathbf{e}}_X) = 0 \tag{3.14}$$

Similarly the equations of motion are obtained for the Y and Z directions. We summarize these three equations as:

$$\frac{\partial \rho \mathbf{v}}{\partial t} + \boldsymbol{\nabla} \, . \, (\rho \mathbf{v}\mathbf{v} + p\mathbf{I}) = 0 \tag{3.15}$$

where $\mathbf{I}$ is the unit tensor. These equations are the hydrodynamic equations (Chapter 9) describing the motion of a compressible fluid.

In summary, we have demonstrated how partial differential non-linear equations governing a variety of physical systems are obtained from simple fundamental principles of conservation. The differential equations are said to be conservative and take the particular form,

$$\frac{\partial \mathbf{u}}{\partial t} + \nabla . \mathbf{f} = 0 \tag{3.16}$$

where $\mathbf{u}$ is a vector of the dependent variables and the 'fluxes' are $\mathbf{f} = \mathbf{f}(\mathbf{u})$.

(*b*) *Physical Processes and the Dispersion Relation.* Partial differential equations couple points in space and time and the simple linear properties of a partial differential equation, or a system of partial differential equations, may be described by the behaviour of a wave in space and time. Given a dependent variable $u(x, t)$, a function of the space variable x, and the time variable t, and, if u satisfies a partial differential equation, we consider the effect of the partial differential equation on a single wave or Fourier mode in space and time;

$$u(x, t) = \hat{u}\mathrm{e}^{\mathrm{i}(\omega t - kx)} \tag{3.17}$$

ω is the frequency of the wave, and k is the wave number related to the wavelength λ, $k = 2\pi/\lambda$. After the insertion of such a mode in the partial differential equation of interest, a *dispersion relation* is obtained:

$$\omega = \omega(k) \tag{3.18}$$

The dispersion relation relates the frequency, and correspondingly the characteristic timescale, to a particular wavelength for the physical phenomena described by the partial differential equation. The frequency ω may be real, when oscillations or wave phenomena are described, or it may be imaginary when the growth or decay of the mode is described.

In a difference solution to the initial-value problem, we are particularly concerned with the timescales of the problem and the dependence of the timescales on the wavelength for different physical processes. The dispersion relation provides this information. We shall consider four types of processes which may be described by rather simple partial differential equations, but which, in more complex forms, keep recurring in interesting physical problems.

(*c*) *Waves and the Wave Equation.* The phenomenon of waves and wave motions occurs so frequently that it is not necessary to enumerate some examples. We consider the particular case of a wave on a stretched string, where the displacement $\xi(x, t)$ of the string is described by a wave equation (Figure 3.5).

$$\frac{\partial^2 \xi}{\partial t^2} - V_s^2 \frac{\partial^2 \xi}{\partial x^2} = 0 \tag{3.19}$$

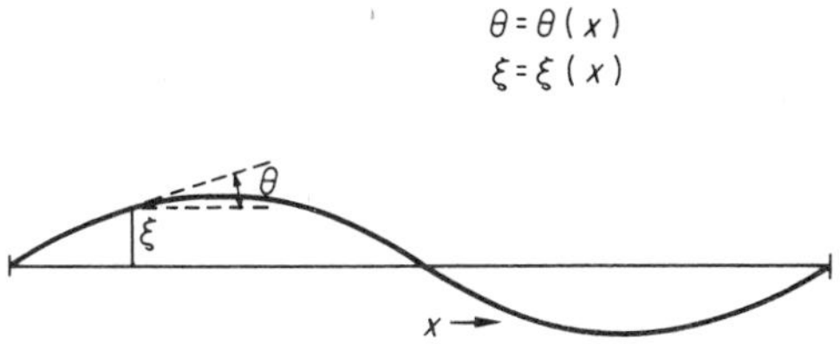

Figure 3.5. Wave on a string. $\xi(x, t)$ is the displacement from equilibrium while $\theta(x, t)$ is the angular displacement

The parameter V_s is, in this case, defined by the tension T in the string and the mass per unit length m of the string, $V_s = \sqrt{T/m}$. If L is a characteristic length along the string, then we can define a characteristic time τ, as the time for a travelling wave to propagate over the length L,

$$\tau \sim \frac{L}{V_s} \tag{3.20}$$

Using a more sophisticated approach, we shall consider a Fourier mode on the string:

$$\xi(x, t) = \hat{\xi}\, e^{i(\omega t - kx)}$$

Inserting the mode in the wave equation, for a given wave number k the frequency ω must satisfy:

$$-\omega^2 + k^2 V_s^2 = 0 \tag{3.21}$$

It follows that a characteristic timescale may be associated with the wave,

$$\tau = \frac{2\pi}{\omega} = \frac{2\pi}{V_s k} = \frac{\lambda}{V_s} \tag{3.22}$$

Equation 3.21 is the dispersion relation of the partial differential equation.

It is to be noted too, that if a velocity $v = \partial\xi/\partial t$ of the displacement and an angular displacement $\theta = \partial\xi/\partial x$ are defined, the second-order wave equation may be written as two coupled first-order equations,

$$\begin{aligned} \frac{\partial v}{\partial t} - V_s \frac{\partial \theta}{\partial x} &= 0 \\ \frac{\partial \theta}{\partial t} - V_s \frac{\partial v}{\partial x} &= 0 \end{aligned} \tag{3.23}$$

(*d*) *The Advective Equation.* The advective equation is related to the wave equation and arises when properties of a fluid are advected (or convected) by the fluid. We have already seen the term arise in the equations of hydrodynamics (3.11 and 3.15). The conservation of fluid mass may be written as,

$$\frac{\partial \rho}{\partial t} + \mathbf{v} \,.\, \nabla \rho + \rho \nabla \,.\, \mathbf{v} = 0$$

or,

$$\frac{\mathrm{d}\rho}{\mathrm{d}t} + \rho \nabla \,.\, \mathbf{v} = 0$$

where $\mathrm{d}/\mathrm{d}t = \partial/\partial t + \mathbf{v} \,.\, \nabla$ is the total time derivative or Lagrangian derivative. The property of fluid density is a property which may be regarded as localized to a fluid element and as the element moves in the fluid it follows that its density is advected with it. In the incompressible case (though with a variable density fluid) the conservation of mass equation is:

$$\frac{\mathrm{d}\rho}{\mathrm{d}t} = \frac{\partial \rho}{\partial t} + \mathbf{v} \,.\, \nabla \rho = 0 \tag{3.24}$$

namely the advective equation. It is clear that, when obtaining equations for any intensive property of any fluid in the initial-value problem, the advective terms will arise in the equation describing that property.

Again it is important to assign a time scale to the process of advection: obviously the time scale of interest here is simply the time for a point in the fluid to move over the characteristic distance L,

$$\tau = \frac{L}{|\mathbf{v}|} \tag{3.25}$$

$\mathbf{v}$ is now the centre of mass velocity and not a phase velocity. More precisely we obtain the dispersion relation for the one-dimensional advective equation for a Fourier mode,

$$\rho = \hat{\rho}\, \mathrm{e}^{\mathrm{i}(\omega t - kx)}$$

$$\omega = kv \tag{3.26}$$

and,

$$\tau = \frac{2\pi}{\omega} = \frac{2\pi}{vk} = \frac{\lambda}{|v|} \tag{3.27}$$

(*e*) *The Diffusion Equation.* The diffusion equation is very familiar and arises in a large number of problems in physics. In one dimension and in the simplest case the equation takes the form:

$$\frac{\partial u}{\partial t} - \frac{\partial}{\partial x} \kappa \frac{\partial u}{\partial x} = 0 \tag{3.28}$$

where $u(x, t)$ is a dependent variable and κ is a diffusion coefficient. In more complex forms the equation can include inhomogeneous or source terms on the right-hand side and it becomes nonlinear when the conductivity or diffusion coefficient is a function of the dependent variable, $\kappa = \kappa(u)$. The time scale of interest here is the diffusion time,

$$\tau = \frac{L^2}{\kappa} \tag{3.29}$$

More specifically we may again consider the effect of the diffusion equation on a Fourier mode,

$$u = \hat{u}\, e^{i(\omega t - kx)}$$

and consequently for a constant conductivity κ,

$$i\omega + \kappa k^2 = 0$$

$$\omega = i\kappa k^2 \tag{3.30}$$

This equation is the dispersion relation for the simple diffusion equation. The angular frequency ω is now imaginary and hence the mode decays in time. The time scale for this decay is,

$$\tau = \frac{2\pi}{\omega} = \frac{2\pi}{\kappa k^2} = \frac{\lambda^2}{2\pi\kappa} \tag{3.31}$$

(*f*) *The Elliptic Equation.* Finally one might include the elliptic equation, arising from boundary-value problems, as a fourth example. Again such equations are common and familiar in physics: Laplace's equation and Poisson's equation are examples,

$$\nabla^2\Phi = 0 \tag{3.32}$$

$$\nabla^2\Phi = -\rho \tag{3.33}$$

The dependent variable Φ might be an electrostatic or gravitational potential, while the inhomogeneous term or known 'source function', ρ, might be a charge or mass density. These equations result from considering static solutions, or alternatively from systems where it has been assumed that information is transported instantaneously. If the analogy with the previous three processes of wave propagation, advection, and diffusion is maintained, the frequency of a Fourier mode is effectively infinite,

$$\omega \to \infty$$

$$\tau \to 0$$

and the time scale for information to propagate over a scale length L is effectively zero.

(*g*) *Classification of Partial Differential Equations.* Four processes of importance in continuous media have been enumerated above where each process is associated with a commonly recurring partial differential equation. In their linear forms, three of these equations are each particular examples of the general second-order two-dimensional equation:

$$a\frac{\partial^2\phi}{\partial x} + b\frac{\partial^2\phi}{\partial x\partial y} + c\frac{\partial^2\phi}{\partial y^2} + d\frac{\partial\phi}{\partial x} + e\frac{\partial\phi}{\partial y} + f\phi + g = 0 \tag{3.34}$$

a, b, c, d, e, f and g may be functions of the independent variables x, y and possibly of the dependent variable ϕ, in which event the equation is nonlinear. Formally, partial differential equations are classified as,

hyperbolic: when $b^2 - 4ac > 0$
parabolic: when $b^2 - 4ac = 0$
elliptic: when $b^2 - 4ac < 0$

The difference formulation, though not the solution of boundary-value problems (elliptic equations), has been expressed in Chapter 2 and from now on this chapter will consider only partial differential equations from the initial-value problem.

2 The Stability of Difference Schemes for Partial Differential Equations

The important properties of a difference solution to the initial-value problem have been discussed in Chapter 2, and the nature of the difference method as a long-wavelength approximation has been discussed. For ordinary differential equations the important property of stability in the explicit case depends primarily on the magnitude of the time step in comparison to the physical times of interest in the problem and by evaluating the dispersion relations of particular partial differential equations we have discussed some important time scales of interest which occur in physical problems. We may now turn to the central task in this chapter and synthesize these ideas in evaluating the properties of a numerical solution to a system of partial differential equations.

In Chapter 2, the initial-value problem was formulated generally (equation 2.33), where the state of the system of interest in time $\mathbf{u}(\mathbf{x}, t)$ satisfies the equations,

$$\frac{d\mathbf{u}}{dt} = \mathrm{L}\mathbf{u} \tag{3.35}$$

where, for partial differential equations, the operator L is a spatial differential operator. In difference form, the initial-value problem was transformed as a sequence of solutions at points in time t^n (equation 2.47),

$$\mathbf{u}_j^{n+1} = \mathrm{T}(\Delta t, \mathit{\Delta})\mathbf{u}_j^n \tag{3.36}$$

the solutions at adjacent time levels being related by an integration operator T. We include the subscript j to stress that, for partial difference equations, the operator $\mathrm{T}(\Delta t, \Delta)$ couples both the dependent variables which comprise the vector state of the system and different points on the space mesh. To determine the stability as before, or to obtain the dispersion relation of the difference scheme, we apply the same techniques of analysis (see Section 2.5), but the problem may be simplified here *ab initio*, by decoupling points on the space mesh. This may be achieved by investigating the Fourier modes on the mesh separately and by demanding that the scheme be stable for each Fourier mode separately.

If we assume that the integration operator $\mathrm{T}(\Delta t, \Delta)$ is a constant, and otherwise a linear approximation is applied (see Section 2.5c), the Fourier modes on the difference mesh are independent. For the purpose of stability analysis, therefore, a Fourier mode in the dependent variables is applied,

$$\mathbf{u}_j^n = \hat{\mathbf{u}}^n \, \mathrm{e}^{ikx_j} \tag{3.37}$$

and from the difference equation 3.36, the amplitudes of the Fourier mode are related,

$$\begin{aligned} \hat{\mathbf{u}}^{n+1} \, \mathrm{e}^{ikx_j} &= \mathrm{T}(\Delta t, \Delta)\hat{\mathbf{u}}^n \, \mathrm{e}^{ikx_{j'}} \\ \hat{\mathbf{u}}^{n+1} &= \mathrm{e}^{-ikx_j}\mathrm{T}(\Delta t, \Delta) \, \mathrm{e}^{ikx_{j'}}\hat{\mathbf{u}}^n \end{aligned} \tag{3.38}$$

This equation is of the form

$$\hat{\mathbf{u}}^{n+1} = \mathbf{G}(\Delta t, \Delta, k)\hat{\mathbf{u}}^n \tag{3.39}$$

where $\mathbf{G}$ is the amplification matrix of the difference scheme, for the particular Fourier mode of wavenumber k. The formulation is essentially similar to that outlined in Section 2.5 and now for stability we demand that the amplitude of a Fourier mode be bounded. *For stability, if the amplitude of a Fourier mode is finite at time $t = 0$, then it must remain finite for all time steps n.*

To apply this requirement, we may write the vector amplitude of the Fourier mode of interest in terms of the eigenvectors $\mathbf{s}^{(\mu)}$ of the amplification matrix $\mathbf{G}$. If $\hat{u}_\mu^0$ is the amplitude of the vector Fourier mode along the eigenvector $\mathbf{s}^{(\mu)}$ at step zero,

$$\hat{\mathbf{u}}^0 = \sum_\mu \hat{u}_\mu^0 \mathbf{s}^{(\mu)} \tag{3.40}$$

Then, applying the amplification matrix equation (3.39), the amplitude of the vector Fourier mode at time step n is,

$$\begin{aligned} \hat{\mathbf{u}}^n &= \mathbf{G}^n \hat{\mathbf{u}}^0 \\ &= \mathbf{G}^n \sum_\mu \hat{u}_\mu^0 \mathbf{s}^{(\mu)} \end{aligned} \tag{3.41}$$

Since $\mathbf{s}^{(\mu)}$ is the eigenvector of the amplification matrix $\mathbf{G}$ with eigenvalue g_μ,

$$\mathbf{G}\mathbf{s}^{(\mu)} = g_\mu \mathbf{s}^{(\mu)} \tag{3.42}$$

then,

$$\hat{\mathbf{u}}^n = \sum_\mu \hat{u}_\mu^0 g_\mu^n \mathbf{s}^{(\mu)} \tag{3.43}$$

Thus, to satisfy the requirement for stability (as above),

$$|\hat{u}_\mu^0 g_\mu^n| < K|\hat{u}_\mu^0| \tag{3.44}$$

where K is a positive finite number, and this condition must be satisfied for each Fourier mode k and along each eigenvector (all μ) of $\mathbf{G}$,

$$|g_\mu^n| < K$$

or,

$$|g_\mu| < K^{1/n} \tag{3.45}$$

This condition imposes a requirement on each of the eigenvalues g_μ for all time steps n and, for large n, $K^{1/n} \to 1$. We therefore make plausible the condition for stability that:

$$|g_\mu| \leqslant 1 \tag{3.46}$$

Von Neumann has taken into account the possible occurrence of a growing local term in such a system of partial differential equations, and he has shown that a necessary and sufficient condition for stability is (Richtmyer and Morton 1967):

$$|g_\mu| \leqslant 1 + O(\Delta t) \quad \text{(all modes } k\text{, all eigenvalues } \mu) \tag{3.47}$$

Again, the eigenvalues g_μ can in fact be complex, and $|g_\mu|$ is defined as the amplitude of the eigenvalue in the complex plane,

$$|g_\mu| = \sqrt{(g_\mu^* g_\mu)} \tag{3.48}$$

where g_μ^* is the complex conjugate of g_μ.

The procedure outlined here for analysing the stability of a difference scheme for the initial-value problem in the case of partial difference equations, is essentially the same as the procedure outlined in Section 2.5c. However, we have explicitly demonstrated the manner by which the complex coupling between both dependent variables and different mesh points in space may be resolved for the purpose of the stability analysis of partial differential equations. Decoupling is achieved, first, from the space mesh points, by considering Fourier modes in space separately, and second by diagonalizing the resultant amplification matrix which now relates Fourier modes over subsequent time steps.

In the partial differential equations of interest, the integration operator T (equation 3.36) may possibly be nonlinear and consequently, after linearization, the Fourier mode amplification matrix **G** (equation 3.39) may not be constant. Instead, **G** may vary over the space and time lattice. The stability conditions (equations 3.47) then reduce to 'local conditions': that is, they must be satisfied everywhere on the space mesh, and they must be satisfied at all time steps n.

3 The Diffusion Equation: First-Order Explicit Integration

The simplest way of solving the diffusion equation in time is to use a first-order explicit method, analogous to the Euler method for ordinary differential equations. As before, at time $t = 0$, the initial conditions define the dependent variable u on the space mesh x_j (Section 2.2). We wish to integrate the equation,

$$\frac{\partial u}{\partial t} - \kappa \frac{\partial^2 u}{\partial x^2} = 0$$

over time steps Δt. The spatial operator $\partial^2/\partial x^2$ is defined as before (Section 2.3). Then $u_j^{n+1} = u(t^{n+1}, x_j)$ is defined by the difference equation,

$$u_j^{n+1} = u_j^n + \frac{\kappa \Delta t}{\Delta^2}(u_{j+1}^n - 2u_j^n + u_{j-1}^n) \tag{3.49}$$

We represent this integration scheme on the two-dimensional lattice in Figure 3.6.

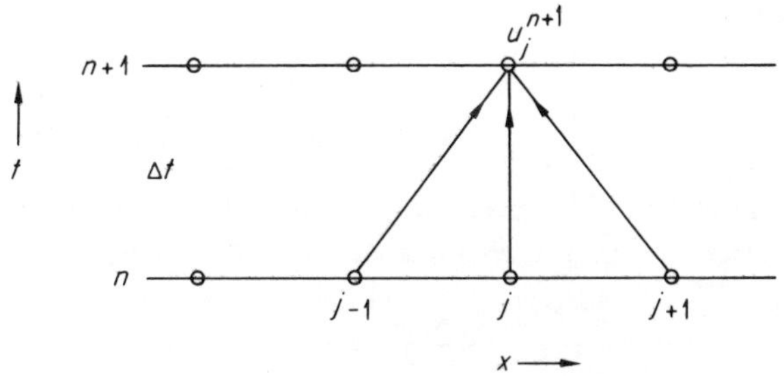

Figure 3.6. The diagram illustrates a space-time mesh on which the diffusion equation may be integrated by the first-order explicit method. The dependent variable is only defined at points in the $(x - t)$ plane. The second derivative in space in the diffusion equation is evaluated using the three points $(n, j - 1)$, (n, j), and $(n, j + 1)$. The time derivative is evaluated across the points $(n + 1, j)$ and (n, j)

As has been discussed previously, there are clearly errors associated with the time step Δt and the space step Δ, and we enquire into the accuracy, stability and efficiency of this method. The general approach (Section 2) in considering the stability of the scheme is to derive the amplification factor (or in more complex situations the amplification matrix) for a Fourier mode in space,

$$u = \hat{u}(t)\,e^{ikx}$$

Applying this functional form to the difference algorithm (equation 3.49),

$$\hat{u}^{n+1}\,e^{ikx_j} = \hat{u}^n\,e^{ikx_j} + \frac{\kappa\Delta t}{\Delta^2}\hat{u}^n(e^{ikx_{j+1}} - 2e^{ikx_j} + e^{ikx_{j-1}})$$

$$\hat{u}^{n+1} = \hat{u}^n\{1 + \frac{2\kappa\,\Delta t}{\Delta^2}(\tfrac{1}{2}\,e^{ik\Delta} + \tfrac{1}{2}\,e^{-ik\Delta} - 1)\} \tag{3.50}$$

Now $\frac{1}{2}\,e^{ik\Delta} + \frac{1}{2}\,e^{-ik\Delta} = \cos(k\Delta)$ and the equation relates the amplitude of a Fourier mode over consecutive time levels. Since the problem in this example concerns only one dependent variable and only one partial differential equation, the Fourier mode is coupled by an amplification factor g, and not an amplification matrix,

$$g(\Delta t, \Delta, k) = 1 + 2\frac{\kappa\,\Delta t}{\Delta^2}\{\cos(k\Delta) - 1\}$$

$$= 1 - \frac{4\kappa\,\Delta t}{\Delta^2}\sin^2\left(\frac{k\Delta}{2}\right) \tag{3.51}$$

Therefore, if we are to satisfy the von Neumann criterion for stability (equation 3.47),

$$\left|1 - \frac{4\kappa\,\Delta t}{\Delta^2}\sin^2\left(\frac{k\Delta}{2}\right)\right| \leqslant 1 \tag{3.52}$$

This condition must hold for every wavenumber k. So, taking the maximum value of the sine function, the requirement for stability is:

$$-4\frac{\kappa\,\Delta t}{\Delta^2} \geqslant -2$$

$$\Delta t \leqslant 0{\cdot}5\frac{\Delta^2}{\kappa} \tag{3.53}$$

Hence, to obtain stable numerical solutions, we must choose a time step smaller than a maximum time step (equation 3.53). This result can be interpreted physically and, indeed, is not surprising. The maximum permissible time step Δt_m is just the diffusion time associated with a scale length Δ, the mesh step length (cf. equation 3.29). Δt_{m} is the time for information to travel over the mesh length Δ. Since the method is explicit, information at time step $n + 1$ at point j is obtained only from the surrounding points. Information, therefore, only 'travels' on the mesh with a speed $\Delta/\Delta t$ in the explicit case and, if this speed is too small as a result of a large time step, catastrophic results can be expected. $\Delta/\Delta t$ is the lattice speed.

To determine the accuracy of the method, we may insert a Fourier mode in time and space and obtain the dispersion relation for the difference scheme. This can then be compared with the dispersion relation of the differential equation. The scheme is an approximation to the differential equation if, in the limit of zero Δt and large wavelength, the same dispersion relations are obtained. It is sufficient here to say that in this case the accuracy is of the order of Δt in the time step and Δ^2 in the space step ($O(\Delta t) + O(\Delta^2)$).

4 The Advective Equation: First-Order Explicit Integration

We consider the simple first-order explicit integration of the advective equation 3.24,

$$\frac{\partial u}{\partial t} + v\frac{\partial u}{\partial x} = 0$$

Defining the spatial derivative as in Section 2.3 and integrating only to the first order in the time step Δt (as for the Euler method),

$$u_j^{n+1} = u_j^n - \frac{v\,\Delta t}{2\Delta}(u_{j+1}^n - u_{j-1}^n) \tag{3.54}$$

The scheme is represented on the two-dimensional lattice in Figure 3.7. Again, we examine the amplification of a Fourier mode,

$$\hat{u}^{n+1}\,\mathrm{e}^{ikx_j} = \hat{u}^n\,\mathrm{e}^{ikx_j} - \frac{v\,\Delta t}{2\Delta}\hat{u}^n(\mathrm{e}^{ikx_{j+1}} - \mathrm{e}^{ikx_{j-1}})$$

or,

$$\hat{u}^{n+1} = \left\{1 - \mathrm{i}\frac{v\,\Delta t}{\Delta}\sin(k\Delta)\right\}\hat{u}^n \tag{3.55}$$

Thus the amplification factor $g(\Delta t, \Delta, k)$ for the scheme is complex,

$$g = 1 - \mathrm{i}\alpha \tag{3.56}$$

where,

$$\alpha = \frac{v\,\Delta t}{\Delta}\sin(k\Delta)$$

and consequently we must take into account the magnitude of the amplification factor in the complex plane,

$$|g|^2 = gg^* = 1 + \alpha^2 \tag{3.57}$$

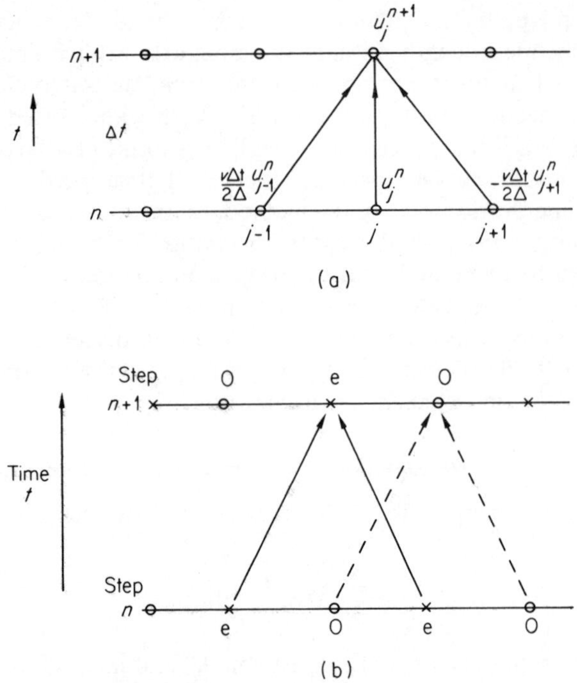

Figure 3.7. An unstable first-order scheme (equation 3.54) for the advective equation: (a) the time-derivative is evaluated between the points $(n + 1, j)$ and (n, j) while the space-derivative is evaluated between the points $(n, j + 1)$ and $(n, j - 1)$; (b) the scheme is unstable since there exist two interlocking uncoupled meshes, in the sense that advection on the mesh, e, occurs independently of advection on the mesh, o

For this scheme, therefore, the magnitude of the amplification factor is always greater than unity and the von Neumann stability condition (equation 3.47) cannot be satisfied for any nonzero value of α and for any time step. The difference scheme (equation 3.54) is inherently *unstable* for all time steps Δt and illustrates that not all consistent and obvious schemes are useful.

We may indeed find a stable, useful and first-order scheme for the advective equation by replacing u_j^n in the explicit algorithm (equation 3.54) by a spatial average,

$$u_j^{n+1} = \tfrac{1}{2}(u_{j+1}^n + u_{j-1}^n) - \frac{v\,\Delta t}{2\Delta}(u_{j+1}^n - u_{j-1}^n) \tag{3.58}$$

This algorithm, applicable to hyperbolic equations in general, is known as the Lax scheme (Lax 1954) and has wide applicability. Again, we shall analyse its stability by determining the amplification of a particular Fourier mode,

$$\hat{u}^{n+1} = \left\{\cos(k\Delta) - i\frac{v\,\Delta t}{\Delta}\sin(k\Delta)\right\}\hat{u}^n \tag{3.59}$$

which has the form of the amplification equation, and it follows that the amplification factor is

$$g = \cos(k\Delta) - i\frac{v\,\Delta t}{\Delta}\sin(k\Delta) \tag{3.60}$$

The magnitude of the amplification factor in the complex plane is:

$$\begin{aligned} gg^* &= \cos^2(k\Delta) + \left(\frac{v\,\Delta t}{\Delta}\right)^2 \sin^2(k\Delta) \\ &= 1 - \sin^2(k\Delta)\left\{1 - \left(\frac{v\,\Delta t}{\Delta}\right)^2\right\} \end{aligned} \tag{3.61}$$

It follows that the von Neumann stability condition (equation 3.47) will be satisfied for all wavenumbers k, if,

$$\frac{|v\,\Delta t|}{\Delta} \leqslant 1$$

$$\Delta t \leqslant \frac{\Delta}{|v|} \tag{3.62}$$

This condition for stability on the time step is an example of the *Courant–Friedrichs–Lewy* condition (Courant *et al.* 1928) applicable to hyperbolic equations. Again the result is not surprising. For an explicit method, we are required to choose a time step smaller than the smallest characteristic physical time in the problem, which for the advective equation is just the time for the velocity v to lead to a flow over the distance Δ (Section 3.1d). The Courant–Friedrichs—Lewy condition ensures that the physical velocity v is less than the lattice speed $\Delta/\Delta t$.

5 Dispersion and Diffusion on a Difference Mesh

It is informative to enquire further why, in the direct first-order explicit integration of the advective equation (equation 3.54), an unstable scheme is obtained while in the Lax method (equation 3.58) a stable scheme may be obtained. The inherent instability in the first-order integration of hyperbolic equations is caused by the failure to time-centre the equations. For example, we may rewrite the algorithm for the Lax method, without further approximation from equation 3.58, as:

$$\tfrac{1}{2}(u_j^{n+1} - u_j^{n-1}) + \tfrac{1}{2}(u_j^{n+1} - 2u_j^n + u_j^{n-1}) = \tfrac{1}{2}(u_{j+1}^n - 2u_j^n + u_{j-1}^n) - \frac{v\,\Delta t}{2\Delta}(u_{j+1}^n - u_{j-1}^n) \quad (3.63)$$

This equation, now to second-order accuracy, is equivalent to the differential equation,

$$\frac{\partial u}{\partial t} + \frac{\Delta t}{2}\frac{\partial^2 u}{\partial t^2} - \frac{\Delta^2}{2\,\Delta t}\frac{\partial^2 u}{\partial x^2} + v\frac{\partial u}{\partial x} = 0 \quad (3.64)$$

By using the Lax method, therefore, we are attempting to simulate the advective equation by equation 3.64. The second term arises, as in the unstable method, from attempting a first-order integration in the time step and from not time-centring the integration scheme. It is this term which gives rise to instability or growth in first-order methods and it is only by introducing the third term in the Lax method, which arises from spatially averaging the dependent variable at the previous time step, that stability is achieved. Thus, in the Lax method, we can only ensure stability if the third term, a diffusion term, is made larger than the second term. For a Fourier mode of wavenumber k and frequency $\omega \sim vk$, this is achieved by ensuring a small time step,

$$\frac{\Delta^2 k^2}{2\,\Delta t} \geqslant \frac{\Delta t}{2}\omega^2$$

or,

$$\Delta t \leqslant \frac{\Delta}{|v|}$$

This is the condition for stability which was obtained previously (equation 3.62).

Generally, since the stability criterion is a local condition, by ensuring stability, the diffusion term in equation 3.64 will always tend to be too large. Consequently, in the Lax method, and generally in all first-order methods, decay of the modes or diffusion occurs (contrast this with the differential system). This is a difficult and severe problem in dealing with hyperbolic equations.

The occurrence of *diffusion* and of *dispersion* (where modes of differing wavelengths propagate with different velocities) is a general phenomenon associated with difference equations. The von Neumann stability condition has very wide applicability and permits us to obtain a stability criterion in the simplest manner. It tells us little, however, of the more detailed properties associated with a particular difference scheme and, in particular, little of the important properties of diffusion and dispersion. Ideally, providing the problem is not too complicated mathematically, we may obtain the dispersion relation of the difference scheme, relating the frequency of a Fourier mode on the mesh to a particular wavenumber,

$$\omega = \omega(k, \Delta, \Delta t) \tag{3.65}$$

By comparing the dispersion relation of the difference scheme with the dispersion relation of the corresponding differential system, the applicability and accuracy of the difference scheme may be analysed in detail.

To illustrate both the approach and the properties of diffusion and dispersion in the Lax method, we shall obtain the dispersion relations for the differential advective equation and for the difference scheme (equation 3.58). We consider a Fourier mode in time and space,

$$u(x, t) = \hat{u}\,e^{i(\omega t - kx)}$$

and, as before (equation 3.26) for the differential advective equation,

$$\omega = vk \tag{3.66}$$

Consequently, in the differential advective equation, ω is only real so that no damping of any mode occurs and all wavenumbers have the same phase velocity (v) and group velocity (v) so that no dispersion of the modes occurs.

We turn now to the difference scheme (equation 3.58) and apply it to the same Fourier mode:

$$\hat{u}\,e^{i[\omega(t^n + \Delta t) - kx_j]} = \frac{1}{2}\hat{u}\,e^{i(\omega t^n - kx_j)}\left\{(e^{ik\Delta} + e^{-ik\Delta}) - \frac{v\,\Delta t}{\Delta}(e^{ik\Delta} - e^{-ik\Delta})\right\} \tag{3.67}$$

Again, by cancelling the common factors in this equation, we obtain,

$$e^{i\omega\Delta t} = \cos(k\Delta) - \frac{v\,\Delta t}{\Delta}\,i\sin(k\Delta) \tag{3.68}$$

This is the relation we seek, the dispersion relation for the difference scheme, though it is an implicit function of ω. In general, ω is complex, but, by writing $\omega = \Omega + i\gamma$, we may equate the real and imaginary parts of the dispersion relation:

$$\tan(\Omega\,\Delta t) = \frac{v\,\Delta t}{\Delta}\tan(k\Delta) \tag{3.69}$$

$$e^{-2\gamma\Delta t} = \cos^2(k\Delta) + \left(\frac{v\,\Delta t}{\Delta}\right)^2 \sin^2(k\Delta) \tag{3.70}$$

In the special case where $v\,\Delta t/\Delta = 1$, γ is zero and,

$$\Omega = vk \tag{3.71}$$

which is in precise agreement with the dispersion relation of the differential system (equation 3.66). More generally, however, γ exists so that diffusion on the mesh occurs and, furthermore, the phase and group velocities are both functions of the wavenumber so that dispersion occurs. The solutions $\Omega(k)$ and $\gamma(k)$ are drawn in Figure 3.8 for varying values of $v\,\Delta t/\Delta$. Clearly,

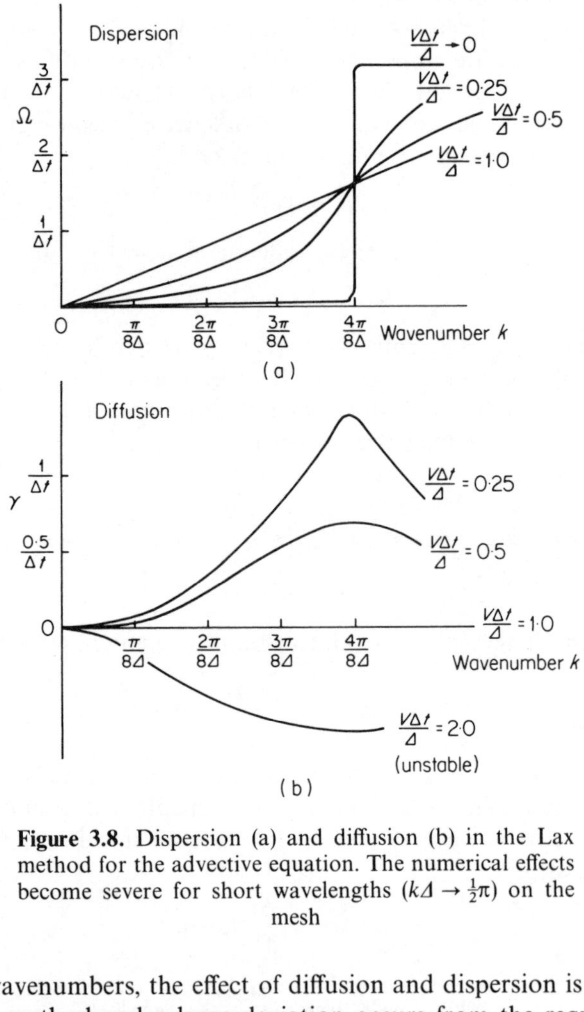

Figure 3.8. Dispersion (a) and diffusion (b) in the Lax method for the advective equation. The numerical effects become severe for short wavelengths ($k\Delta \rightarrow \frac{1}{2}\pi$) on the mesh

for large wavenumbers, the effect of diffusion and dispersion is very severe in the Lax method and a large deviation occurs from the required result. At small wavenumbers, where wavelengths are much larger than the mesh-step length, agreement with the dispersion relation of the differential system becomes reasonable.

Anomalous diffusion and dispersion, particularly for short wavelengths on a difference mesh, are general effects which result from replacing differential equations with difference equations. In first-order methods of integration in the time step, such numerical phenomena are particularly severe and, to mitigate them, methods of higher-order accuracy must generally be employed.

6 Conservation on a Difference Mesh

When a system of partial differential equations is nonlinear, we may define a particular difference method, but there still remains a variety of ways of differencing nonlinear terms. We have seen that many of the partial differential equations of physics arise from principles of conservation (Section 3.1b) and are said to be conservative, and it would be useful to demand that the corresponding difference equations were equally conservative. More specifically, we seek difference equations, which identically conserve, say, the energy, the mass, the momentum or the magnetic flux of the system, irrespective of the errors incurred by the finite-difference lattice.

Although the concept of conservation applies equally to any number of dimensions, for the purpose of this discussion it is sufficient to formulate the problem in a two-dimensional rectangular region R, bounded by the boundary B. The region R is divided into a set of elementary rectangular cells each of 'volume' $\Delta\tau$ (Figure 3.9) and in two dimensions there are IJ rectan-

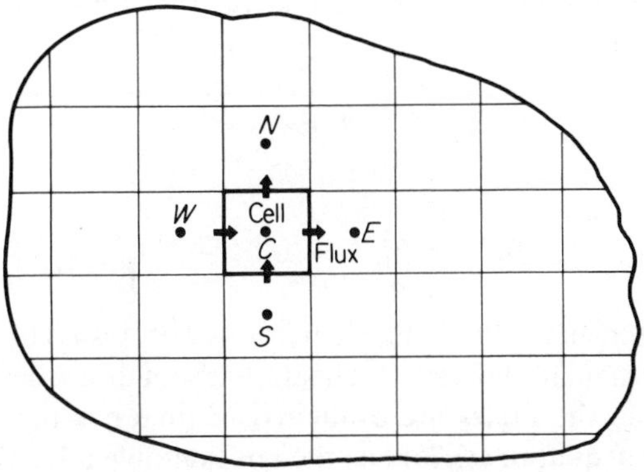

Figure 3.9. Conservation in two space dimensions on a difference mesh. The flux from cell C to cell E is exactly equal and opposite to the flux from cell E to cell C

gular boxes. The system of conservative partial differential equations (equation 3.16),

$$\frac{\partial \mathbf{u}}{\partial t} + \nabla . \mathbf{f} = 0$$

may be integrated over each space-time box of volume $\Delta\tau\ \Delta t$, between the space-like surfaces t^n and t^{n+1}. For example, integrating over the volume of cell C of surface A, we obtain:

$$\int_{t^n}^{t^{n+1}} \mathrm{d}t \iint_{\mathrm{C}} \frac{\partial \mathbf{u}}{\partial t} \mathrm{d}\tau = -\int_{t^n}^{t^{n+1}} \mathrm{d}t \iint_{\mathrm{C}} \nabla . \mathbf{f}\, \mathrm{d}\tau$$

$$\iint_{\mathrm{C}} \mathbf{u}^{n+1}\, \mathrm{d}\tau - \iint_{\mathrm{C}} \mathbf{u}^n\, \mathrm{d}\tau = -\int_{t^n}^{t^{n+1}} \mathrm{d}t \oint_A \mathbf{f} . \mathbf{dS} \tag{3.72}$$

The left-hand side has been integrated over time and, on the right-hand side, the divergence theorem has been applied so that the volume integral has been transformed to a surface integral over the 'area' A of cell C. Now, on the mesh, instead of defining the intensive variables of, say, density, or momentum density, the corresponding integrated quantities of total mass or total momentum, respectively, in each box or cell may be defined:

$$\Delta\tau\, \tilde{\mathbf{u}}_{ij}^n = \iint_{\mathrm{C}} \mathbf{u}^n\, \mathrm{d}\tau \tag{3.73}$$

In addition, the fluxes $\mathbf{f} . \mathbf{dS}$ are only defined on the surfaces of each cell. For the cell ij, there are four fluxes related to the four surrounding cells (α = E, S, N, W) (see Figure 3.9),

$$\sum_\alpha \mathbf{F}_{\alpha ij} = \oint_A \mathbf{f} . \mathbf{dS} \tag{3.74}$$

$$\tilde{\mathbf{u}}_{ij}^{n+1} = \tilde{\mathbf{u}}_{ij}^n - \int_{t^n}^{t^{n+1}} \mathrm{d}t \frac{1}{\Delta\tau} \sum_\alpha \mathbf{F}_{\alpha ij} \tag{3.75}$$

A difference scheme in the form of equations 3.75, is said to be *conservative*. Although a particular difference scheme has not been defined, since the manner in which the fluxes are evaluated in time has not been specified, the formulation (equations 3.75) has the considerable advantage that, if the differential system conserves the variables $\int_R \mathbf{u}\, \mathrm{d}\tau$, the difference scheme, equally, conserves these variables identically. In accordance with the formulation (equations 3.75), any flux $\mathbf{F}_{\alpha ij}$ not defined on the boundary B of the region, is shared equally between two cells, so that for example:

$$\mathbf{F}_{\mathrm{E}ij} = -\mathbf{F}_{\mathrm{W}i+1\,j} \tag{3.76}$$

It follows that, if we sum equations 3.75 over all the boxes ij in R, the fluxes cancel in pairs, leaving only the contributions from the boundary,

$$\sum_{i=1}^{I} \sum_{j=1}^{J} \tilde{\mathbf{u}}_{ij}^{n+1} = \sum_{i=1}^{I} \sum_{j=1}^{J} \tilde{\mathbf{u}}_{ij}^n - \sum_B \int_{t^n}^{t^{n+1}} \mathrm{d}t \frac{1}{\Delta\tau} \mathbf{F}_\alpha \tag{3.77}$$

Consequently, and independently of truncation errors in the difference scheme, the variables $\int_R \mathbf{u}\,d\tau$ are *identically conserved*, except for the contributions crossing the boundary B of R. By now defining the fluxes $\mathbf{F}_{\alpha ij}$ in time, a large set of conservative difference schemes may be obtained.

7 Conservative Methods for Hyperbolic Equations

Using the concepts of conservation, order of accuracy, stability and explicit solution, we are now in a position to define a number of important methods for the difference solution of hyperbolic equations. A number of important methods of integrating in time have been defined in Section 2.6 and they are applicable, equally, to partial differential equations. For simplicity we shall define the algorithms in one dimension, though they may be applied to any number of space dimensions.

Only explicit methods will be outlined here, though we have seen that the advantages of an implicit formulation are considerable, particularly as unconditional stability can be achieved and large time steps may consequently be employed (Section 2.6). For hyperbolic equations, however, the suitability of implicit methods is not clear and certainly the success of an implicit method depends upon the properties of the particular problem and the careful formulation of the descriptive equations (Chapter 10). There is the added difficulty too, in more than one space dimension, of resolving complex matrix equations at each time step (Chapter 4). We shall also restrict ourselves here to the application of the methods to *Eulerian* meshes, namely space meshes which remain fixed in time. Again the use of *Lagrangian* methods, in which a space mesh moves with the local characteristic velocity of the particular problem, is a powerful approach in one dimension (Chapter 9), but a multidimensional Lagrangian mesh rapidly becomes excessively distorted and accuracy is quickly lost.

The general conservative hyperbolic equations in one dimension are analysed,

$$\frac{\partial \mathbf{u}}{\partial t} + \frac{\partial \mathbf{F}}{\partial x} = 0 \tag{3.78}$$

For each explicit method, stability is achieved, if at all, with the general Courant–Friedrichs–Lewy condition (Courant *et al.* 1928) imposed on the time step Δt,

$$\Delta t \leqslant \frac{\Delta}{|v|} \tag{3.79}$$

where v is the fastest propagation velocity anywhere on the space mesh.

(*a*) *The Conservative Lax Method*. The Lax method has been defined previously (Section 3.4). In conservative form, the Lax method is defined by the algorithm:

$$\mathbf{u}_j^{n+1} = \tfrac{1}{2}(\mathbf{u}_{j+1}^n + \mathbf{u}_{j-1}^n) - (\mathbf{F}_{j+1}^n - \mathbf{F}_{j-1}^n)\frac{\Delta t}{2\Delta} \tag{3.80}$$

The example of the advective equation $\mathbf{F} = v\mathbf{u}$, where v is a velocity on the mesh, has been examined in detail (Sections 3.4 and 3.5) and, as before (equation 3.62), the condition for the time step is:

$$\Delta t \leqslant \frac{\Delta}{|v|}$$

The advective equation is a particularly simple example of the equations of interest but, to illustrate the wider application of the method, we implement the Lax method for a pair of coupled equations describing one-dimensional electromagnetic waves. In vacuo, using Maxwell's equations, a plane polarized wave is defined by the differential equations,

$$\begin{aligned} \frac{\partial B}{\partial t} + c\frac{\partial E}{\partial x} &= 0 \\ \frac{\partial E}{\partial t} + c\frac{\partial B}{\partial x} &= 0 \end{aligned} \tag{3.81}$$

where the electric field E lies in the y direction, and the magnetic field lies in the z direction. Implementing the Lax algorithm, with $\mathbf{u} = (E, B)$ and the fluxes $\mathbf{F} = (cB, cE)$, the new field components are obtained at each step:

$$\begin{aligned} E_j^{n+1} &= \frac{1}{2}(E_{j+1}^n + E_{j-1}^n) - \frac{c\,\Delta t}{2\Delta}(B_{j+1}^n - B_{j-1}^n) \\ B_j^{n+1} &= \frac{1}{2}(B_{j+1}^n + B_{j-1}^n) - \frac{c\,\Delta t}{2\Delta}(E_{j+1}^n - E_{j-1}^n) \end{aligned} \tag{3.82}$$

In this vector equation, it is necessary now for the purpose of stability analysis to apply a vector Fourier mode

$$\mathbf{u}^n = (\hat{E}^n\,\mathrm{e}^{ikx}, \hat{B}^n\,\mathrm{e}^{ikx})$$

and, inserting this mode in equations 3.82, the amplification matrix for the scheme is obtained (Section 3.2),

$$\mathbf{G} = \begin{bmatrix} \cos(k\Delta) & \dfrac{-\mathrm{i}c\,\Delta t}{\Delta}\sin(k\Delta) \\ \dfrac{-\mathrm{i}c\,\Delta t}{\Delta}\sin(k\Delta) & \cos(k\Delta) \end{bmatrix} \tag{3.83}$$

In accordance with the von Neumann stability condition (equation 3.47), we are required to determine the eigenvalues g of the matrix:

$$\{\cos(k\Delta) - g\}^2 + \frac{c^2(\Delta t)^2}{\Delta^2}\sin^2(k\Delta) = 0$$

or,

$$g_{1,2} = \cos(k\Delta) \pm \frac{ic\,\Delta t}{\Delta}\sin(k\Delta) \tag{3.84}$$

There are two eigenvalues of the amplification matrix, corresponding to the two coupled equations, and they relate to waves travelling in the positive x-direction, or the negative x-direction. The magnitude of each eigenvalue is, however, the same,

$$|g|^2 = \cos^2(k\Delta) + \frac{c^2(\Delta t)^2}{\Delta}\sin^2(k\Delta)$$

$$= 1 - \sin^2(k\Delta)\left(1 - \frac{c^2(\Delta t)^2}{\Delta^2}\right) \tag{3.85}$$

Thus the von Neumann stability condition is satisfied for both modes if

$$\Delta t \leqslant \frac{\Delta}{c} \tag{3.86}$$

where c is the velocity of the waves. The stability criterion, therefore, is essentially that for the advective equation. The velocity of importance, c, is now a phase velocity and not an advective velocity, but again the Courant–Friedrichs–Lewy condition is obtained since c is the fastest propagation velocity on the mesh.

The examples illustrated have been linear, though the formulation (equation 3.80) is general and applies equally to more interesting examples of nonlinear equations. The scheme is, however, not time-centred and errors of the order of the time step and of the space step occur (Section 3.5).

(b) The Conservative Leapfrog Method. If hyperbolic equations are integrated over a time step Δt to first-order accuracy in Δt only (the time derivative is not time-centred), an effective destabilizing term is introduced into the equations and the only way to ensure stability is to add a larger effective spatial diffusion term (Section 3.5). These methods, of which the Lax method is an example, produce numerical solutions which are greatly smoothed.

We can overcome these difficulties by 'time centring' the equations to define methods which are accurate to second order in the time step Δt. Clearly this requires a more complex difference scheme and more 'work' on the computer since a *three-level* formula must now be used. The simplest of these schemes is the leapfrog method, which is quite analogous to the leapfrog method for ordinary differential equations (Section 2.6b), though it is now cast in conservative form on a space mesh.

In the memory of the computer we store the dependent variables **u** at two time levels at any point in the calculation (the storage requirements are therefore twice those of a first-order method). The calculation is pursued by defining the fluxes **F** at the intermediate time levels t^n,

$$\mathbf{F}_{j+1}^n = \mathbf{F}(\mathbf{u}_{j+1}^n)$$

$$\mathbf{u}_j^{n+1} = \mathbf{u}_j^{n-1} - \frac{\Delta t}{\Delta}(\mathbf{F}_{j+1}^n - \mathbf{F}_{j-1}^n) \tag{3.87}$$

and the dependent variables, so obtained, are used to define the intermediate fluxes to 'leapfrog' to the next time level:

$$\mathbf{u}_{j+1}^{n+2} = \mathbf{u}_{j+1}^n - \frac{\Delta t}{\Delta}(\mathbf{F}_{j+2}^{n+1} - \mathbf{F}_j^{n+1})$$

We represent this scheme in one dimension on the time-space lattice of Figure 3.10. It is to be noted that the values of the dependent variables at the

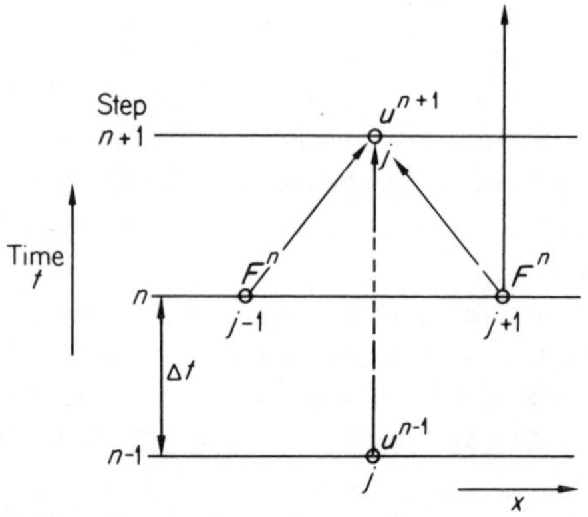

Figure 3.10. The lattice of points in the $(x - t)$ plane used for second-order accuracy in the conservative leapfrog method. Variables at the intermediate points (n, j) are not defined

intermediate time-space points $\mathbf{u}_j^n$ need not be, and are not in general, defined. If all points on the mesh are defined, there are two interlocking meshes which are uncoupled, so that solutions on the two meshes (Figure 3.10) may drift out of phase.

To investigate the stability of the scheme, we again use the advective equation 3.24, where,

$$F = vu$$

For this simple one-dependent variable case, on applying a Fourier mode of wavenumber k in space, we obtain an equation for the amplification factor g for the scheme (equation 3.87):

$$g^2 = 1 - i\frac{2\,\Delta t}{\Delta} v \sin(k\Delta) g \tag{3.88}$$

A quadratic equation for the amplification factor results from the leapfrog scheme, since a three-level formula in time is specified. Two roots for the amplification factor are obtained,

$$g = i\alpha \pm \sqrt{(-\alpha^2 + 1)} \tag{3.89}$$

where $\alpha = (\Delta t/\Delta)v \sin(k\Delta)$. Thus provided the factor under the square root is positive, the magnitude of g for both roots is identically equal to one,

$$|g| = 1 \qquad \text{iff } |\alpha| \leqslant 1 \tag{3.90}$$

Consequently, to ensure stability in the use of this explicit method for all wavenumbers, a condition on the time step is again obtained:

$$\Delta t \leqslant \frac{\Delta}{|v|} \tag{3.91}$$

As before, the condition imposed on the time step is the Courant–Friedrichs–Lewy condition.

It is to be noted that the two roots obtained for the amplification factor correspond to the two uncoupled meshes defined in the leapfrog scheme. Although both roots provide stable solutions under the same conditions, the two modes, related to the two meshes, can drift apart. We avoid such a decoupling of the meshes, either, if possible, by defining the dependent variables only on one mesh, or by employing a small diffusion term to couple the two meshes (see Section 9.6).

(c) *The Two-step Lax–Wendroff Scheme.* We may obtain second-order accuracy in the time step, with the avoidance of large numerical diffusion, by a generalization of the explicit two-step method (Section 2.6c) based on a Taylor expansion in the time step. Just as for ordinary differential equations, the two-step Lax–Wendroff method (Lax and Wendroff 1960, Richtmyer 1962, Richtmyer and Morton 1967) time-centres the integration by defining temporary or intermediate values of the dependent variables at the half time steps $t^{n+\frac{1}{2}}$. The Lax method is used in the first step or auxiliary calculation, at each time step, *auxiliary:*

$$\mathbf{u}_{j+\frac{1}{2}}^{n+\frac{1}{2}} = \tfrac{1}{2}(\mathbf{u}_j^n + \mathbf{u}_{j+1}^n) - \frac{\Delta t}{2\Delta}(\mathbf{F}_{j+1}^n - \mathbf{F}_j^n) \tag{3.92}$$

These variables are now used to define the fluxes at the intermediate time and space points,

$$\mathbf{F}_{j+\frac{1}{2}}^{n+\frac{1}{2}} = \mathbf{F}(\mathbf{u}_{j+\frac{1}{2}}^{n+\frac{1}{2}}) \tag{3.93}$$

Consequently, in the main step of the calculation, a time-centred and space-centred integration formula is obtained (Figure 3.11),

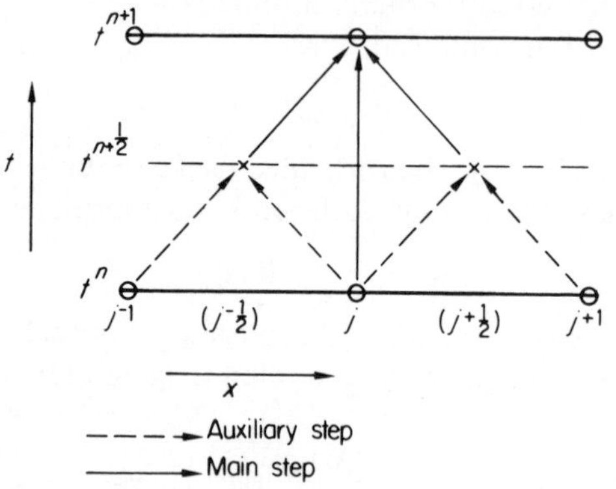

Figure 3.11. The conservative two-step Lax–Wendroff method, represented on the mesh of points in the $(x - t)$ plane. The Lax method is used to obtain temporary variables at the half time steps $(n + \frac{1}{2}, j + \frac{1}{2})$. These variables are used to time-centre the integration for second-order accuracy

main:

$$\mathbf{u}_j^{n+1} = \mathbf{u}_j^n - \frac{\Delta t}{\Delta}(\mathbf{F}_{j+\frac{1}{2}}^{n+\frac{1}{2}} - \mathbf{F}_{j-\frac{1}{2}}^{n+\frac{1}{2}}) \tag{3.94}$$

After each main step the intermediate values of the dependent variables $\mathbf{u}_{j+\frac{1}{2}}^{n+\frac{1}{2}}$ are discarded and form no part of the solution.

The stability of the method may be examined by referring again to the advective equation where $F = vu$. For the Fourier mode $\exp(ikx)$, the amplification factor g defined by equations 3.92 and 3.94 satisfies

$$\begin{aligned} g &= 1 - \frac{\Delta tv}{\Delta}\left[\mathrm{i}\sin(k\Delta) - \frac{\Delta tv}{\Delta}\{\cos(k\Delta) - 1\}\right] \\ &= 1 - \mathrm{i}\alpha\sin(k\Delta) + \alpha^2\{\cos(k\Delta) - 1\} \end{aligned} \tag{3.95}$$

where $\alpha = \Delta t v/\Delta$. The amplification factor is again complex and we consider its magnitude in the complex plane,

$$gg^* = [1 + \alpha^2\{\cos(k\Delta) - 1\}]^2 + \alpha^2 \sin^2(k\Delta)$$
$$= 1 - \alpha^2(1 - \alpha^2)\{1 - \cos(k\Delta)\}^2 \tag{3.96}$$

It follows that the amplification factor for the two-step Lax–Wendroff method will be smaller than, or equal to, unity and, correspondingly, the von Neumann stability condition will be satisfied for all wavenumbers k if,

$$\alpha^2 \leqslant 1$$
$$\Delta t \leqslant \frac{\Delta}{|v|} \tag{3.97}$$

which is again the Courant–Friedrichs–Lewy condition.

In general in the Lax–Wendroff scheme, except in the case where $\alpha = 1$, the magnitude of the amplification factor is smaller than unity, so that damping of the Fourier modes on the mesh occurs. The effect, however, is small for, in the equation for the magnitude of the amplification factor (equation 3.96), if the cosine is expanded for small wavenumbers ($k\Delta$ small), we obtain:

$$gg^* = 1 - \alpha^2(1 - \alpha^2)\frac{k^4\Delta^4}{4} + o(k^4\Delta^4) \tag{3.98}$$

Thus numerical diffusion occurs only to fourth order in the wavenumber, so that long wavelengths are only minimally affected, while the effect of fourth-order diffusion is in fact useful in smoothing discontinuities on the mesh (Chapter 9). In addition it is to be noted that no extraneous and additional numerical modes are introduced by the Lax–Wendroff method. These advantageous properties of the method have lead to its wide application.

(*d*) *Quasi-second-order Method.* Approximate second-order accuracy may be achieved with a particularly simple method to implement, by an extension of the Adams–Bashforth method (Table 2.2). Applying a Taylor expansion in the time step to second order, we shall define dependent variables at the new time step,

$$\mathbf{u}^{n+1} = \mathbf{u}^n + \left(\frac{\partial \mathbf{u}}{\partial t}\right)_n \Delta t + \left(\frac{\partial^2 \mathbf{u}}{\partial t^2}\right)_n \frac{\Delta t^2}{2} \tag{3.99}$$

and, maintaining second-order accuracy, we obtain:

$$\mathbf{u}^{n+1} = \mathbf{u}^n + \left(\frac{\partial \mathbf{u}}{\partial t}\right)_n \Delta t + \left\{\left(\frac{\partial \mathbf{u}}{\partial t}\right)_n - \left(\frac{\partial \mathbf{u}}{\partial t}\right)_{n-1}\right\}\frac{\Delta t}{2}$$

We apply this approach to the conservative hyperbolic equations 3.78,

$$\mathbf{u}_j^{n+1} = \mathbf{u}_j^n - \left(\frac{3}{2} + \epsilon\right)\frac{\Delta t}{2\Delta}(\mathbf{F}_{j+1}^n - \mathbf{F}_{j-1}^n) + \left(\frac{1}{2} + \epsilon\right)\frac{\Delta t}{2\Delta}(\mathbf{F}_{j+1}^{n-1} - \mathbf{F}_{j-1}^{n-1}) \tag{3.100}$$

where ϵ is a small number. Thus, again, a three-level formula is used for second-order accuracy but the fluxes $\mathbf{F}_{j+1}^{n-1}$, rather than the dependent variables, are used from the lowest time level. At each time step in the calculation, the fluxes need only be calculated once and, instead of storing the dependent variables at two time levels, the fluxes are stored directly for one time level. A particular advantage of this scheme is that difficult, alternating, or 'sodium chloride' type meshes need not be defined.

We examine the amplification factor of the scheme in equation 3.100 for the Fourier mode exp (ikx) where, for the advective equation, $F = vu$:

$$g^2 = g - (\tfrac{3}{2} + \epsilon)i\alpha g + (\tfrac{1}{2} + \epsilon)i\alpha \tag{3.101}$$

with

$$\alpha = \frac{\Delta t v}{\Delta}\sin(k\Delta) \tag{3.102}$$

The amplification factor may be determined from equation 3.101 by an expansion in small α:

$$\text{for } |\alpha| \leqslant \tfrac{1}{2}, \qquad |g| < 1, \qquad \text{if } \epsilon > \tfrac{1}{4}\alpha^2 + \tfrac{1}{2}\alpha^4 + o(\alpha^4) \tag{3.103}$$

Thus for small α, the method will be stable for all modes if the number ϵ is sufficiently large. For example, if we choose a time step

$$\Delta t = \frac{1}{2}\frac{\Delta}{|v|} \tag{3.104}$$

then for stability we must use a value ϵ sufficiently large,

$$\epsilon \geqslant \tfrac{3}{32}$$

while certainly for smaller time steps, ϵ rapidly becomes vanishingly small. The method is in fact only second-order accurate when the parameter ϵ is identically equal to zero. However, since ϵ may be taken arbitrarily small (at the expense of small Δt), the method is quasi-second-order accurate in the time step.

The conservative explicit methods for hyperbolic equations outlined in this section are particularly useful. A variety of other methods exist and are summarized in Table 3.1. More complex schemes than those of second-order accuracy in the time step are certainly desirable to minimize the effects of numerical dispersion and diffusion, though increasing complexity limits the application and the number of arithmetic operations required for a solution increases rapidly. Roberts and Weiss (1966) have developed a fourth-order accurate method in the time step.

Table 3.1 Explicit conservative methods for hyperbolic equations

$$\frac{\partial \mathbf{u}}{\partial t} + \frac{\partial \mathbf{F}}{\partial x} = 0$$

where $\mathbf{F} = \mathbf{F}(\mathbf{u})$

Example:

$$F = vu, \qquad \alpha = \frac{\Delta t}{\Delta} v$$

Method	Algorithm
1. First-order explicit	$\mathbf{u}_j^{n+1} = \mathbf{u}_j^n - (\mathbf{F}_{j+1}^n - \mathbf{F}_{j-1}^n)\frac{\Delta t}{2\Delta}$
	$g = 1 + i\alpha \sin(k\Delta)$
	Always unstable
2. Lax method	$\mathbf{u}_j^{n+1} = \frac{1}{2}(\mathbf{u}_{j+1}^n + \mathbf{u}_{j-1}^n) - (\mathbf{F}_{j+1}^n - \mathbf{F}_{j-1}^n)\frac{\Delta t}{2\Delta}$
	$g = \cos(k\Delta) + i\alpha \sin(k\Delta)$
	Stable for $\Delta t \leqslant \frac{\Delta}{\lvert v \rvert}$
3. Lelevier method	$\mathbf{u}_j^{n+1} = \mathbf{u}_j^n - \{(v\mathbf{u})_{j+1}^n - (v\mathbf{u})_j^n\}\frac{\Delta t}{\Delta}$ if $v_j^n \leqslant 0$
	$= \mathbf{u}_j^n - \{(v\mathbf{u})_j^n - (v\mathbf{u})_{j-1}^n\}\frac{\Delta t}{\Delta}$ if $v_j^n \geqslant 0$
	$g = 1 - \lvert\alpha\rvert + \lvert\alpha\rvert \cos(k\Delta) + i\alpha \sin(k\Delta)$
	Stable for $\Delta t \leqslant \frac{\Delta}{\lvert v \rvert}$, only applicable to advection
4. Two-step Lax–Wendroff	$\mathbf{u}_{j+\frac{1}{2}}^{n+\frac{1}{2}} = \frac{1}{2}(\mathbf{u}_{j+1}^n + \mathbf{u}_j^n) - (\mathbf{F}_{j+1}^n - \mathbf{F}_j^n)\frac{\Delta t}{2\Delta}$
	$\mathbf{u}_j^{n+1} = \mathbf{u}_j^n - (\mathbf{F}_{j+\frac{1}{2}}^{n+\frac{1}{2}} - \mathbf{F}_{j-\frac{1}{2}}^{n+\frac{1}{2}})\frac{\Delta t}{\Delta}$
	$g = 1 - i\alpha \sin(k\Delta) + \alpha^2\{\cos(k\Delta) - 1\}$
	Stable for $\Delta t \leqslant \frac{\Delta}{\lvert v \rvert}$

Table 3.1 (*continued*)

Method	Algorithm
5. Lax–Wendroff (one step)	$\mathbf{u}_j^{n+1} = \mathbf{u}_j^n - \frac{\Delta t}{2\Delta}(\mathbf{F}_{j+1}^n - \mathbf{F}_{j-1}^n) + \frac{\Delta t^2}{2\Delta^2}$ $\times \{\mathbf{C}_{j+\frac{1}{2}}^n(\mathbf{F}_{j+1}^n - \mathbf{F}_j^n) - \mathbf{C}_{j-\frac{1}{2}}^n(\mathbf{F}_j^n - \mathbf{F}_{j-1}^n)\}$ where the matrix $\mathbf{C}$ is the Jacobian, $\mathbf{C}_{\mu\nu} = \frac{\partial F\mu}{\partial u_\nu}$ and $\mathbf{C}_{j+\frac{1}{2}}^n = \mathbf{C}(\frac{1}{2}(\mathbf{u}_{j+1}^n + \mathbf{u}_j^n))$ $g = 1 - i\alpha \sin(k\Delta) + \alpha^2\{\cos(k\Delta) - 1\}$ Stable for $\Delta t \leqslant \Delta/\lvert v\rvert$
6. Leapfrog	$\mathbf{u}_j^{n+1} = \mathbf{u}_j^{n-1} - \frac{\Delta t}{\Delta}(\mathbf{F}_{j+1}^n - \mathbf{F}_{j-1}^n)$ $g = i\alpha \sin k\Delta \pm \sqrt{(-\alpha^2 \sin^2 k\Delta + 1)}$ Stable for $\Delta t \leqslant \Delta/\lvert v\rvert$
7. Quasi Second-Order	$\mathbf{u}_j^{n+1} = \mathbf{u}_j^n - \left(\frac{3}{2} + \epsilon\right)\frac{\Delta t}{2\Delta}(\mathbf{F}_{j+1}^n - \mathbf{F}_{j-1}^n)$ $+ \left(\frac{1}{2} + \epsilon\right)\frac{\Delta t}{2\Delta}(\mathbf{F}_{j+1}^{n-1} - \mathbf{F}_{j-1}^{n-1})$ Stable for $\alpha \leqslant \frac{1}{2}$, if $\epsilon > \frac{1}{4}\alpha^2 + \frac{1}{2}\alpha^4$ namely, $\Delta t \leqslant \frac{1}{2}\frac{\Delta}{\lvert v\rvert}$, if $\epsilon > \frac{1}{4}\frac{\Delta t^2 v^2}{\Delta^2} + \frac{1}{2}\frac{\Delta t^4 v^4}{\Delta^4}$.

8 Multidimensional Explicit Methods

The explicit conservative methods defined in one space dimension in Section 3.7 extend naturally to equations in two or more space dimensions, and it is particularly in multidimensional problems that the methods have their greatest application. The space meshes now defined in two or more dimensions do have greater complexity, and it is necessary to ensure that interlocking meshes do not become uncoupled.

The Courant–Friedrichs–Lewy condition imposed on the time step to ensure stability (equation 3.79) is broadly applicable to a variety of methods and in N space dimensions is modified as,

$$\Delta t \leqslant \frac{\Delta}{\lvert\mathbf{v}\rvert\sqrt{N}} \tag{3.105}$$

where the vector velocity $\mathbf{v}$ is the fastest propagation velocity on the mesh and equal space steps Δ are assumed along each dimension.

We illustrate the approach particularly for the Lax method, where in N dimensions the algorithm takes the form:

$$\mathbf{u}^{n+1} = \frac{1}{2N} \sum_{\alpha=-N}^{N} \mathbf{u}_{\alpha}^{n} - \sum_{\alpha=1}^{N} (\mathbf{F}_{\alpha}^{n} - \mathbf{F}_{-\alpha}^{n}) \frac{\Delta t}{2\Delta} \tag{3.106}$$

The subscripts defining the particular mesh point have been dropped, and the equation is taken to apply at each mesh point where the subscript α refers to the $2N$ adjacent points in N space dimensions. The mesh employed in the Lax method in two space dimensions is illustrated in Figure 3.12, where the effect of spatially averaging the old dependent variable at each time step is to couple the space meshes by numerical diffusion.

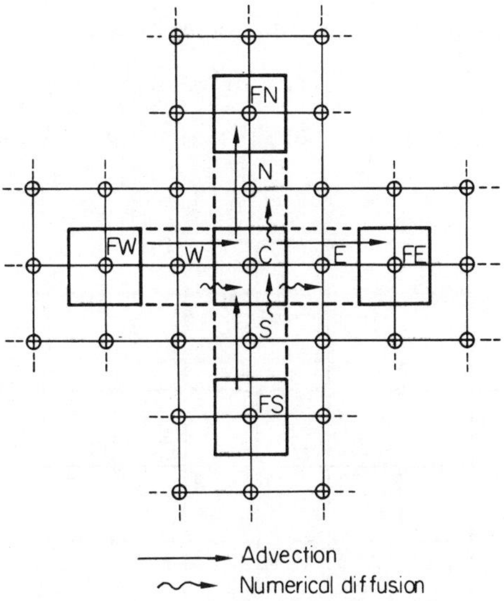

Figure 3.12. The Lax scheme in two space-dimensions. The cell C is coupled to the cells at FN, FE, FS, FW by the fluxes, while numerical diffusion is used to couple the cells at N, E, S, W to the cell at C

For the purpose of analysing the stability of a multidimensional scheme, or indeed of analysing the dispersion relation of the scheme, we must now apply the technique of Fourier analysis along each space dimension separately. In a two-dimensional formulation, Fourier modes in both the x and y dimensions are taken, $\hat{\mathbf{u}} \exp(ik_x x + ik_y y)$, and we use as an example the Lax method (equation 3.106) for the advective equation where $F_x = v_x u$ and $F_y = v_y u$,

$$u_{ij}^{n+1} = \tfrac{1}{4}(u_{i+1,j}^n + u_{i-1,j}^n + u_{i,j+1}^n + u_{i,j-1}^n) - \frac{\Delta t}{2\Delta} v_x (u_{i+1,j}^n - u_{i-1,j}^n)$$

$$- \frac{\Delta t}{2\Delta} v_y (u_{i,j+1}^n - u_{i,j-1}^n) \tag{3.107}$$

Using the notation $\theta_x = \Delta t v_x/\Delta$, $\theta_y = \Delta t v_y/\Delta$, $\alpha = k_x\Delta$ and $\beta = k_y\Delta$, the amplification factor g of the double Fourier mode is:

$$g = \tfrac{1}{2}\cos\alpha + \tfrac{1}{2}\cos\beta - i\theta_x \sin\alpha - i\theta_y \sin\beta \tag{3.108}$$

This is complex so that we examine the magnitude of the amplification factor in the complex plane,

$$gg^* = (\tfrac{1}{2}\cos\alpha + \tfrac{1}{2}\cos\beta)^2 + (\theta_x \sin\alpha + \theta_y \sin\beta)^2$$

which after rearranging the terms, may be written as

$$gg^* = 1 - (\sin^2\alpha + \sin^2\beta)\{\tfrac{1}{2} - (\theta_x^2 + \theta_y^2)\}$$

$$- \tfrac{1}{4}(\cos\alpha - \cos\beta)^2 - (\theta_y \sin\alpha - \theta_x \sin\beta)^2 \tag{3.109}$$

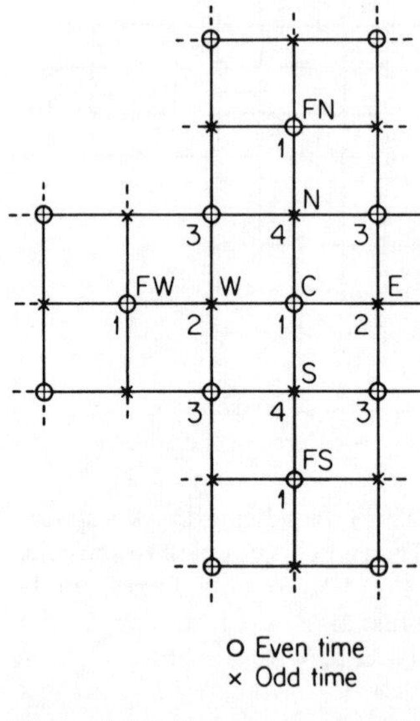

Figure 3.13. The staggered mesh used in both the Lax–Wendroff and leapfrog methods. The mesh is analogous to a sodium-chloride ion lattice. The cross points are defined at intermediate times compared to the dotted points

The last two terms here are always negative, and the von Neumann stability condition (equation 3.47) is certainly satisfied if,

$$\tfrac{1}{2} - (\theta_x^2 + \theta_y^2) \geqslant 0$$

or,

$$\Delta t \leqslant \frac{\Delta}{(v_x^2 + v_y^2)^{\frac{1}{2}}\sqrt{2}} \tag{3.110}$$

which is the appropriate two-dimensional Courant–Friedrichs–Lewy condition (3.105). Clearly the approach applies equally to a greater number of space dimensions.

For the leapfrog, Lax–Wendroff and quasi-second-order schemes, equivalent considerations prevail and the space meshes employed in the first two of these schemes are included in Figures 3.13 and 3.14. In some problems a weak decoupling can occur in a multidimensional mesh (Figure 3.14), which, nevertheless, can be overcome by a small angled diffusion (see Chapter 9).

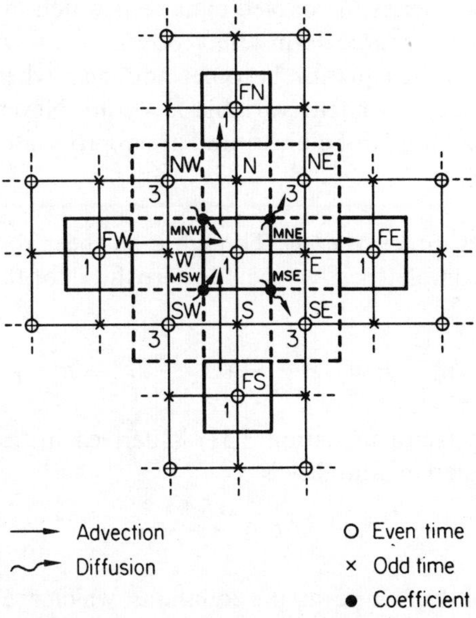

Figure 3.14. Coupling of the meshes in the Lax–Wendroff method. In two space dimensions, there are two distinguishable meshes and a weak instability can occur between mesh 1 and mesh 3. If the differential equations include diffusion terms, meshes 1 and 3 may be coupled by the angled five-point diffusion operator; otherwise, an angled artificial diffusion term must be added

9 Summary of Methods for Parabolic Equations

We have examined hyperbolic equations and their difference formulations in some detail and we turn now to parabolic equations, where, although inherently the same principles of the difference method apply, the particular algorithms which prove satisfactory are different. The principles of conservation and the application to coupled equations and to multidimensional problems are equally relevant for parabolic equations. As an example, however, we shall consider the simple diffusion equation in one dimension (equation 3.28),

$$\frac{\partial u}{\partial t} - \kappa \frac{\partial^2 u}{\partial x^2} = 0 \tag{3.111}$$

where, for the purpose of illustration, a constant coefficient is assumed. This is not to deny the extensive application of the methods which we shall define to more complex problems. Indeed in problems of interest, parabolic equations frequently include wave-like properties or 'hyperbolic terms', and they can occur as a series of coupled equations which must be solved self-consistently. Equally source terms may occur on the right-hand side of equation 3.111, and the equation becomes nonlinear when the conductivity is a function of the dependent variable $\kappa = \kappa(u)$. Nevertheless, the same techniques and methods to be illustrated are appropriate to more complex systems.

(*a*) *Explicit First-order Method.* Though it has been discussed previously (Section 3.3), for completeness we include here the algorithm for the simplest method:

$$u_j^{n+1} = u_j^n + \frac{\kappa \, \Delta t}{\Delta^2}(u_{j+1}^n - 2u_j^n + u_{j-1}^n) \tag{3.112}$$

The amplification factor (equation 3.51) is derived in Section 3.3 and the limit on the time step for stability is

$$\Delta t \leqslant \frac{0{\cdot}5\Delta^2}{\kappa}$$

It is to be noted that for parabolic equations, which are solved by explicit methods subject to a stability criterion, the time step is limited by the square of the space step, rather than by the first power of the space step as in hyperbolic equations. Doubling the accuracy in space, therefore, quarters the allowable time step and consequently, in parabolic equations, the dependence of the time step on a stability condition can be a particularly severe limitation. In addition the explicit method described here is only first-order accurate in the time step. These limitations may be overcome by implementing an implicit approach.

(*b*) *The Crank–Nicholson Implicit Method.* The Crank–Nicholson method (Crank and Nicholson 1947, Richtmyer and Morton 1967) is the analogue, for parabolic equations, of the implicit second-order method defined for ordinary differential equations (equation 2.95, Section 2.6d). We average the spatial diffusion term in time,

$$u_j^{n+1} = u_j^n + \frac{\kappa\,\Delta t}{2\Delta^2}(u_{j+1}^{n+1} - 2u_j^{n+1} + u_{j-1}^{n+1}) + \frac{\kappa\,\Delta t}{2\Delta^2}(u_{j+1}^n - 2u_j^n + u_{j-1}^n) \tag{3.113}$$

We analyse the method for a Fourier mode $\hat{u}\exp(ikx)$,

$$\hat{u}^{n+1} = \hat{u}^n - \frac{\kappa\,\Delta t}{\Delta^2}\{1 - \cos(k\Delta)\}\hat{u}^{n+1} - \frac{\kappa\,\Delta t}{\Delta^2}\{1 - \cos(k\Delta)\}\hat{u}^n$$

and it follows that the amplification factor satisfies the equation,

$$g = 1 - \frac{\kappa\,\Delta t}{\Delta^2}\{1 - \cos(k\Delta)\}g - \frac{\kappa\,\Delta t}{\Delta^2}\{1 - \cos(k\Delta)\}$$

$$g = \frac{1 - \dfrac{2\kappa\,\Delta t}{\Delta^2}\sin^2\left(\dfrac{k\Delta}{2}\right)}{1 + \dfrac{2\kappa\,\Delta t}{\Delta^2}\sin^2\left(\dfrac{k\Delta}{2}\right)} \tag{3.114}$$

g is a real number and, for all wavenumbers and all time steps, the magnitude of g is always smaller than unity. It follows that the von Neumann condition is always satisfied, so that the Crank–Nicholson method is *unconditionally stable*. In addition, it is accurate to second order in both the time step and the space step and, with these advantageous properties, the method may be applied very extensively.

The accuracy and stability of the scheme have been obtained, however, at the expense of a more complex set of equations for the determination of the variables u_j^{n+1} for all j. In the algorithm (equation 3.113), the new dependent variables u_j^{n+1} are not defined explicitly and we are still left with the problem of resolving a matrix equation at each time step (see Chapter 4).

(*c*) *The Unstable Leapfrog Method.* A seemingly useful method—as it is for ordinary differential equations (Section 2.6b) and hyperbolic equations (Section 3.7b)—is the three-level leapfrog method, which, for the diffusion equation, takes the form:

$$u_j^{n+1} = u_j^{n-1} + 2\frac{\kappa\,\Delta t}{\Delta^2}(u_{j+1}^n - 2u_j^n + u_{j-1}^n) \tag{3.115}$$

Analysis of the stability of the method yields the equation for the amplification factor,

$$g = -\alpha \pm \sqrt{(\alpha^2 + 1)} \tag{3.116}$$

where

$$\alpha = \frac{4\kappa\,\Delta t}{\Delta^2}\sin^2\left(\frac{k\Delta}{2}\right)$$

Since α is real, one of the roots of the amplification factor is always smaller than minus one and therefore the method, although consistent, is *unconditionally unstable.*

(*d*) *The Dufort–Frankel Method.* By employing a three-level formula, and slightly altering the leapfrog method, a method for parabolic equations, which is exceptional for being both explicit and unconditionally stable, is obtained (Dufort and Frankel 1953),

$$u_j^{n+1} = u_j^{n-1} + \frac{2\kappa\,\Delta t}{\Delta^2}\{u_{j+1}^n - (u_j^{n+1} + u_j^{n-1}) + u_{j-1}^n\} \tag{3.117}$$

The dependent variable under the diffusion operator at the central time and space point has been replaced by a time average, so that the intermediate mesh points need not be defined. We may solve explicitly for the new dependent variable u_j^{n+1} at each mesh point,

$$u_j^{n+1} = \left(\frac{1-\alpha}{1+\alpha}\right)u_j^{n-1} + \frac{\alpha}{1+\alpha}(u_{j+1}^n + u_{j-1}^n) \tag{3.118}$$

where

$$\alpha = 2\frac{\kappa\,\Delta t}{\Delta^2}$$

For the three-level formula, a quadratic equation is obtained for the amplification factor:

$$g^2 = \frac{1-\alpha}{1+\alpha} + \frac{2\alpha}{1+\alpha}\cos(k\Delta)g$$

$$g = \frac{1}{1+\alpha}[\alpha\cos(k\Delta) \pm \sqrt{\{1-\alpha^2\sin^2(k\Delta)\}}] \tag{3.119}$$

To interpret this equation we note that there are two cases. If $\alpha^2\sin^2(k\Delta) \leqslant 1$ (small time steps), the amplification factor is real and always smaller than unity for both roots. In the case where $\alpha^2\sin^2(k\Delta) > 1$ (large time steps), the amplification factor becomes complex and its magnitude is given by

$$|g| = \frac{1-\alpha}{1+\alpha} \quad \text{for } \alpha^2\sin^2(k\Delta) > 1$$

Thus, for any time step and for all wavenumbers on the mesh, the magnitude of g is always smaller than unity so that the scheme is both explicit and stable. Clearly the method is powerful, but it is to be noted that, for large time steps, the amplification factor becomes complex and it follows that, unlike the differential equation, in the limit of large time steps the Dufort–Frankel method leads to an oscillation, though not a growing one. This is because the errors in the scheme include errors of the form $\Delta/\Delta t$ and, though stable, the method is not accurate for large time steps.

The methods described here apply to multidimensional problems, and are summarized in Table 3.2.

Table 3.2 Methods for parabolic equations

$$\frac{\partial u}{\partial t} = \kappa \frac{\partial^2 u}{\partial x^2} = 0$$

Amplification factor g. let $\beta = \dfrac{\kappa \, \Delta t}{\Delta^2}$

Method	Algorithm
1. Explicit first-order	$u_j^{n+1} = u_j^n + \dfrac{\kappa\Delta t}{\Delta^2}(u_{j+1}^n - 2u_j^n + u_{j-1}^n)$ $g = 1 - 4\beta \sin^2\left(\dfrac{k\Delta}{2}\right)$ Stable for $\Delta t \leqslant \dfrac{1}{2}\dfrac{\Delta^2}{\kappa}$
2. Crank–Nicholson	$u_j^{n+1} = u_j^n + \dfrac{\kappa \, \Delta t}{2\Delta^2}(u_{j+1}^{n+1} - 2u_j^{n+1} + u_{j-1}^{n+1})$ $+ \dfrac{\kappa \, \Delta t}{2\Delta^2}(u_{j+1}^n - 2u_j^n + u_{j-1}^n)$ $g = \dfrac{1 - 2\beta \sin^2(\frac{1}{2}k\Delta)}{1 + 2\beta \sin^2(\frac{1}{2}k\Delta)}$ Always stable
3. Leapfrog	$u_j^{n+1} = u_j^{n-1} + \dfrac{2\kappa \, \Delta t}{\Delta^2}(u_{j+1}^n - 2u_j^n + u_{j-1}^n)$ $g = -4\beta \sin^2(\frac{1}{2}k\Delta) \pm \sqrt{\{16\beta^2 \sin^4(\frac{1}{2}k\Delta) + 1\}}$ Always unstable
4. Dufort–Frankel	$u_j^{n+1} = u_j^{n-1} + \dfrac{2\kappa \, \Delta t}{\Delta^2}\{u_{j+1}^n - (u_j^{n+1} + u_j^{n-1}) + u_{j-1}^n\}$ explicity, $u_j^{n+1} = \left(\dfrac{1 - 2\beta}{1 + 2\beta}\right)u_j^{n-1} + \dfrac{2\beta}{1 + 2\beta}(u_{j+1}^n + u_{j-1}^n)$ $g = \dfrac{1}{1 + 2\beta}\left[2\beta \cos(k\Delta) \pm \sqrt{\{1 - 4\beta^2 \sin^2(k\Delta)\}}\right]$ Always stable

CHAPTER IV

Numerical Matrix Algebra

1 Introduction

The application of the finite difference calculus transforms the differential equations of physics into a finite algebraic form which allows the manipulation of numerical information. Through the use of the difference calculus, a function is transformed to a vector of finite dimension, a differential operator is transformed to a matrix operator and differential equations are transformed to finite matrix equations. We are therefore left with the problem of having to resolve such matrix equations and, particularly, to resolve them as effectively and elegantly as possible. Certainly in some situations, such as the explicit formulation of the initial-value problem, solutions are explicitly or directly obtainable, and the algebraic manipulation required is trivial. More generally, however, and particularly for problems of the boundary-value type, we may only obtain the results of interest as solutions to implicit matrix equations and clearly there is a necessity to solve such matrix equations as rapidly as possible. This is true, too, in many initial-value problems, where we must solve matrix equations at each time step, either because the time-dependent problem has been formulated implicitly in the finite difference calculus (Section 2.4), or because the time-dependent equations are solved at each time step, subject to a solution from an elliptic equation. Numerical matrix algebra is concerned with developing the most effective algorithms for solving matrix equations and with quantitatively determining the properties of matrices. By effective algorithms, we mean algorithms which both minimize the number of individual arithmetic or logical operations and which minimize the storage requirements in obtaining solutions.

There are two problems in matrix algebra which concern us particularly in physics and in computational physics: the solution of the matrix equation and the determination of the eigenvalues and eigenvectors pertaining to particular matrices. In the former, the problem arises in requiring solutions to a set of simultaneous equations, which relate the unknown variables u, to the known variables w,

$$
\begin{aligned}
a_{11}u_1 + a_{12}u_2 + a_{13}u_3 + \cdots + a_{1n}u_n &= w_1 \\
a_{21}u_1 + a_{22}u_2 + \quad \cdots \quad + a_{2n}u_n &= w_2 \\
\vdots \qquad\qquad\qquad\qquad & \vdots \\
a_{m1}u_1 + a_{m2}u_2 + \quad \cdots \quad + a_{mn}u_n &= w_m
\end{aligned} \tag{4.1}
$$

This is a set of m equations in the n unknown variables $u_1, \ldots, u_n$. It is convenient to write this system of equations as a 'matrix' **A** of elements a_{ij} operating on a 'column vector' **u** of unknown elements u_j*

$$\mathbf{Au} = \mathbf{w} \tag{4.2}$$

or,

$$
\begin{bmatrix}
a_{11} & a_{12} & \cdots & a_{1n} \\
a_{21} & a_{22} & \cdots & a_{2n} \\
\vdots & & & \\
a_{m1} & & \cdots & a_{mn}
\end{bmatrix}
\begin{bmatrix} u_1 \\ u_2 \\ \vdots \\ u_n \end{bmatrix}
=
\begin{bmatrix} w_1 \\ w_2 \\ \vdots \\ w_n \end{bmatrix} \tag{4.3}
$$

In general, the matrix **A** has m rows and n columns, but where the number of rows is equal to the number of columns, $m = n$, we would like to obtain the solutions:

$$\mathbf{u} = \mathbf{A}^{-1}\mathbf{w} \tag{4.4}$$

where $\mathbf{A}^{-1}$ is the inverse of the matrix **A**. Conventional methods in matrix algebra define such an inverse, for example by Cramer's rule†

$$(\mathbf{A}^{-1})_{ij} = (-1)^{i+j}\frac{|A|^{ji}}{|A|} \tag{4.5}$$

where $|A|$ is the determinant of **A** and $|A|^{ji}$ is the determinant of the $(n-1) \times (n-1)$ matrix formed by striking out the ith row and jth column of the matrix **A**. We may define the determinant recursively,

$$|A| = \sum_{j=1}^{n} (-1)^{i+j}|A|^{ij}a_{ij} \tag{4.6}$$

A simple examination of these conventional algorithms demonstrates that a very large number of arithmetic operations are involved. For example, the calculation of the determinant alone for an $n \times n$ matrix involves $n!$ multiplications (equation 4.6), but we are concerned with extremely large matrices where the dimension (n) of the problem may be at least 100 and as large as 100,000. Clearly the direct application of Cramer's rule for these problems would be out of the question. It is essential for the problems of our

* It is assumed that the reader has a basic knowledge of the elementary principles of matrix algebra (see, for example, Birkhoff and Mac Lane 1965).

† ibid, p. 306.

interest to develop algorithms which are very rapid and certainly more effective than, for example, Cramer's rule.

It is evident that matrix equations arise in computational physics independently of the finite difference calculus : an example which occurs is in the description of the motions of the ions in a classical lattice. We are concerned with the problem of solving the general matrix equation in which the elements of the matrix are arbitrary and for which there exists a number of direct algorithms (Section 4.5). But the matrices which result from the finite difference calculus have special properties and the importance of the finite difference calculus demands careful attention to such properties. The fundamental property here is that although they are frequently of very large dimension, matrices from the difference method are 'sparse' in the sense that they have very few elements which are nonzero. This property arises from physical problems because local points tend to be intimately coupled, whereas distant points in space are weakly coupled. The property of being sparse is common to matrices which result from problems formulated in any number of space dimensions. In one dimension, however, the resultant matrices frequently have the additional property of being *tridiagonal*, where only elements along the three leading diagonals are nonzero. These special matrix properties of being sparse, or tridiagonal, or of having constant elements along diagonals allow particular algorithms, whereby solutions can be obtained with great efficiency (Sections 4.3 and 4.4).

It is the sparse nature of many matrices in particular, which allows the effective use of iterative methods, frequently described as inexact methods, when a solution is guessed and used to obtain successively improved solutions (Section 4.6). A similar approach can be used in determining the eigenvectors and eigenvalues of sparse matrices (Section 4.7).

The subject of numerical matrix algebra and, more generally, numerical linear algebra is of application to many fields and is in the process of rapid development. Justice cannot be done to the subject in one chapter alone and for a more general discussion the interested reader is referred to Fox (1964), Faddeev and Faddeeva (1963), Fox and Mayers (1968), Varga (1962) and Wilkinson (1964). In this chapter we shall restrict ourselves to those methods which are essential to problems arising from computational mechanics.

2 Matrix Equations from the Finite Difference Calculus

Many of the effective methods used in numerical matrix algebra are effective because they make use of the particular properties of special and commonly recurring matrices. To demonstrate these properties, we shall consider the form of the matrices which arise from the finite difference calculus and, in particular, we shall find that the form of such matrices depends on the number of space dimensions in which the problem is formulated.

(*a*) *Boundary-value Problems.* As a simple but important example of a matrix equation resulting from the difference calculus, we examine the solution of a one-dimensional Poisson's equation,

$$\frac{d^2\Phi}{dx^2} = -\rho \tag{4.7}$$

where $\rho(x)$ is the known source function, while $\Phi(x)$ is the unknown potential function. On an equally spaced difference mesh (space step Δ) $1 \leqslant j \leqslant J$, the function $\rho(x)$ is replaced by a vector of elements ρ_j, defined at the mesh points x_j, and, as before (Section 2.3), we approximate the differential operator d^2/dx^2 by the difference operator Δ_x'' (equation 2.27). Then the differential equation 4.7 is approximated by the set of equations,

$$\Phi_{j+1} - 2\Phi_j + \Phi_{j-1} = -\Delta^2 \rho_j \tag{4.8}$$

which is to apply at every internal point j on the mesh. Assuming that the boundary conditions specify the value of the potential at the end points, $\Phi_1 = w_1$ and $\Phi_J = w_J$, then equations 4.8 form a set of simultaneous linear equations,

$$\begin{aligned}
&\Phi_1 &&= w_1 \\
&\Phi_1 - 2\Phi_2 + \Phi_3 &&= -\Delta^2 \rho_2 \\
&\quad \Phi_2 - 2\Phi_3 + \Phi_4 &&= -\Delta^2 \rho_3 \\
&\qquad \cdots && \\
&\qquad\quad \Phi_{J-2} - 2\Phi_{J-1} + \Phi_J &&= -\Delta^2 \rho_{J-1} \\
&\qquad\qquad\qquad \Phi_J &&= w_J
\end{aligned} \tag{4.9}$$

The unknown variables Φ_j for all j may be represented as a vector $\mathbf{\Phi}$, while the right-hand side of the set of equations 4.9 consists of a known vector $\mathbf{w}$. Then we wish to solve the equations,

$$\mathbf{A\Phi} = \mathbf{w} \tag{4.10}$$

where the matrix $\mathbf{A}$ has the form,

$$\mathbf{A} = \begin{bmatrix} 1 & 0 & 0 & 0 & \cdots & 0 \\ 1 & -2 & 1 & 0 & \cdots & \\ 0 & 1 & -2 & 1 & & \\ & & \cdots & & & \\ & & & 1 & -2 & 1 \\ 0 & \cdots & & 0 & 0 & 1 \end{bmatrix} \tag{4.11}$$

$\mathbf{A}$ has nonzero elements along only the central three diagonals of the matrix and is termed a *tridiagonal* matrix. If we choose a difference mesh of great resolution by taking J very large, typically for example $J = 10{,}000$, the matrix **A** consists of many elements, but is said to be *sparse*, in that very few of its elements are nonzero. In addition it is to be noted that, in the particular case of Poisson's equation, the elements along each diagonal have constant values.

Tridiagonal matrices arise in general from considering one-dimensional problems in space. We formulate the general boundary-value problem in one dimension,

$$f\frac{\mathrm{d}^2u}{\mathrm{d}x^2} + g\frac{\mathrm{d}u}{\mathrm{d}x} + hu = w \tag{4.12}$$

where $f(x)$, $g(x)$, $h(x)$ and $w(x)$ are known functions of x, and we are required to find solutions for the dependent variable u. Again, we difference the equation (Section 2.3) on an equally spaced mesh $1 \leqslant j \leqslant J$,

$$\frac{f_j}{\Delta^2}(u_{j+1} - 2u_j + u_{j-1}) + \frac{g_j}{2\Delta}(u_{j+1} - u_{j-1}) + h_ju_j = w_j$$

which may be rewritten as,

$$\alpha_j u_{j+1} + \beta_j u_j + \gamma_j u_{j-1} = w_j \tag{4.13}$$

where,

$$\alpha_j = \frac{f_j}{\Delta^2} + \frac{g_j}{2\Delta}$$

$$\beta_j = h_j - \frac{2f_j}{\Delta^2}$$

$$\gamma_j = \frac{f_j}{\Delta^2} - \frac{g_j}{2\Delta} \tag{4.14}$$

Equations 4.13 pertain to the general point in the domain, $1 < j < J$, and boundary conditions are applied at the end points x_1 and x_J. In differential form, the general boundary conditions may be written as,

$$\left(a\frac{\mathrm{d}u}{\mathrm{d}x} + bu\right)_{x_1} = c$$

$$\left(a'\frac{\mathrm{d}u}{\mathrm{d}x} + b'u\right)_{x_J} = c' \tag{4.15}$$

where a, b, c, a', b' and c' are prescribed constants. In the case where $a = 0$, the dependent variable u is specified at the end points, while if $b = 0$, the derivatives at the end points are defined. We may write these general boundary conditions in difference form as,

$$au_2 + (b\Delta - a)u_1 = c\Delta$$

$$-a'u_{J-1} + (b'\Delta + a')u_J = c'\Delta \tag{4.16}$$

or,

$$\alpha_1 u_2 + \beta_1 u_1 = w_1$$

$$\gamma_J u_{J-1} + \beta_J u_J = w_J \tag{4.17}$$

where $\alpha_1, \beta_1, w_1, \gamma_J, \beta_J$ and w_J are defined as in equation 4.16 by the particular boundary conditions.

Thus again we have transformed the general differential equation 4.12 to a set of simultaneous linear equations (4.13, 4.17). The matrix **A** now has the form:

$$\mathbf{A} = \begin{bmatrix} \beta_1 & \alpha_1 & 0 & 0 & \cdots & 0 \\ \gamma_2 & \beta_2 & \alpha_2 & 0 & \cdots & 0 \\ 0 & \gamma_3 & \beta_3 & \alpha_3 & \cdots & 0 \\ 0 & 0 & & \cdots & & 0 \\ & & & & & \cdots \\ & & \cdots & & \gamma_J & \beta_J \end{bmatrix} \tag{4.18}$$

which is again tridiagonal, but with variable nonzero elements.

If we consider the boundary-value problem in two dimensions in space, the matrix **A** is similarly *sparse*, but is now not simply tridiagonal. As an example, we shall formulate Poisson's equation in two dimensions and in Cartesian coordinates:

$$\frac{\partial^2 \Phi}{\partial x^2} + \frac{\partial^2 \Phi}{\partial y^2} = -\rho \tag{4.19}$$

If it is required to solve this equation in, say, the rectangular region R, $0 \leqslant x \leqslant X, 0 \leqslant y \leqslant Y$, we may divide the region R by a two-dimensional mesh (Figure 4.1). The operator $\partial^2/\partial x^2$ is approximated by Δ_x'' and $\partial^2/\partial y^2$ is

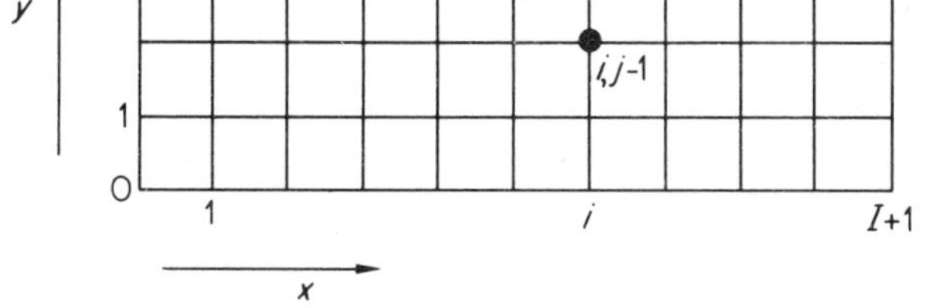

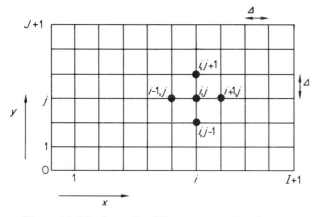

Figure 4.1. The five-point difference approximation to Poisson's equation in two dimensions

approximated by Δ_y'' (Section 2.3). If $\rho_{i,j}$ is defined at every lattice point (i, j), we wish to determine the potential $\Phi_{i,j}$ at every lattice point (i, j) in R, where, from the differential equation 4.19, the potentials $\Phi_{i,j}$ satisfy,

$$(\Phi_{i+1,j} - 2\Phi_{i,j} + \Phi_{i-1,j}) + (\Phi_{i,j+1} - 2\Phi_{i,j} + \Phi_{i,j-1}) = -\Delta^2 \rho_{i,j}$$

or,

$$(\Phi_{i+1,j} + \Phi_{i-1,j} + \Phi_{i,j+1} + \Phi_{i,j-1}) - 4\Phi_{i,j} = -\Delta^2 \rho_{i,j} \tag{4.20}$$

at every point (i, j) on the mesh. It is useful now to replace all the variables $\Phi_{i,j}$ by singly subscripted elements of the vector **u** which has dimension IJ,

$$u_k = u_{iJ+j} = \Phi_i \tag{4.21}$$

and similarly we define the vector **w** from $\rho_{i,j}$

$$w_k = w_{iJ+j} = -\Delta^2 \rho_{i,j} \tag{4.22}$$

The elements of the vector **u** are obtained as solutions to the simultaneous equations 4.20 and, therefore, they satisfy the matrix equation

$$\mathbf{A}\mathbf{u} = \mathbf{w}$$

where the matrix **A** has the form illustrated in Figure 4.2.

Clearly from this two-dimensional problem, the tridiagonal property of the matrix has been lost, but on the other hand the matrix remains sparse with very few nonzero elements. Matrices arising from problems in two dimensions in space have the quindiagonal form of Figure 4.2, though the five diagonals with finite elements are not leading diagonals. It is to be noted, therefore, that the coupling in two dimensions is considerably more complex and consequently solutions are more difficult to obtain. We notice that the matrix may usefully be partitioned, where each partitioned matrix relates to different columns of the two-dimensional space mesh. In the special case of Poisson's equation there is considerable symmetry, in that the elements of each row are constant, a property which is used to good effect in solving Poisson's equation (Section 4.5). Conversely in the general elliptic equation in two dimensions, as in one dimension, those matrix elements which are finite vary from mesh point to mesh point.

It is not difficult to extend the formulation of boundary-value problems to three or more 'space' dimensions where the resulting forms of the matrices are self-evident.

(*b*) *Implicit Formulation of Initial-value Problems.* Large sets of algebraic simultaneous equations arise when the initial-value problem for a continuous medium is integrated *implicitly* in time (equation 2.46). Again, the resulting matrices are sparse and have completely analogous forms to those arising from boundary-value problems.

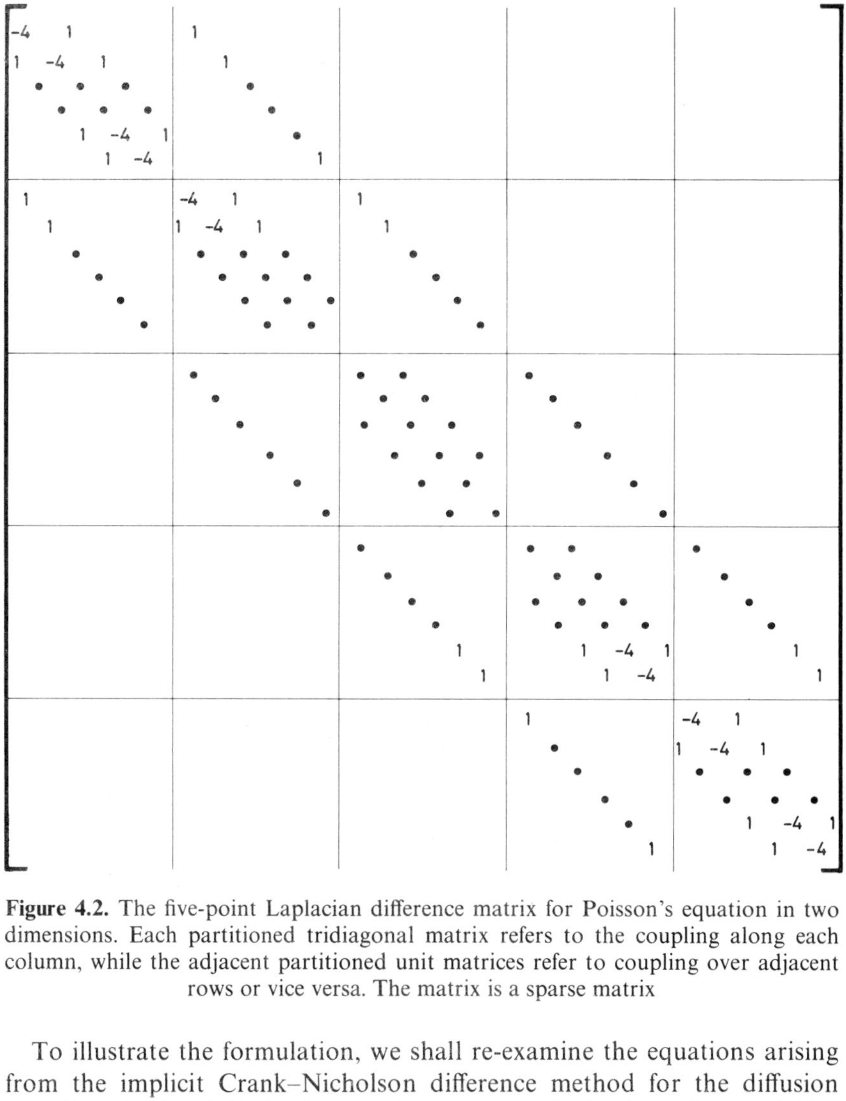

Figure 4.2. The five-point Laplacian difference matrix for Poisson's equation in two dimensions. Each partitioned tridiagonal matrix refers to the coupling along each column, while the adjacent partitioned unit matrices refer to coupling over adjacent rows or vice versa. The matrix is a sparse matrix

To illustrate the formulation, we shall re-examine the equations arising from the implicit Crank–Nicholson difference method for the diffusion equation (Section 3.9). At the new time step $n + 1$, the dependent variables u_j^{n+1} at local points on the space mesh j are coupled (equation 3.113),

$$u_j^{n+1} = u_j^n + \frac{\kappa\,\Delta t}{2\Delta^2}(u_{j+1}^{n+1} - 2u_j^{n+1} + u_{j-1}^{n+1}) + \frac{\kappa\,\Delta t}{2\Delta^2}(u_{j+1}^n - 2u_j^n + u_{j-1}^n)$$

and defining $\alpha = \kappa\,\Delta t/2\Delta^2$,

$$\begin{aligned} -\alpha u_{j+1}^{n+1} + (1 + 2\alpha)u_j^{n+1} - \alpha u^{n+1} &= u_j^n + \alpha(u_{j+1}^n - 2u_j^n + u_{j-1}^n) \\ &= w_j^n \end{aligned} \tag{4.23}$$

The right-hand side of these simultaneous equations forms a known vector $\mathbf{w}^n$ and, at each time step, we are required to solve a matrix equation for the dependent variables u_j^{n+1} which form a vector $\mathbf{u}^{n+1}$. Again, since the problem occurs in one space dimension, the coupling matrix is a tridiagonal matrix of the form 4.18.

Similarly, in the implicit formulation of multidimensional problems, sparse matrices are obtained and the same situation prevails for the implicit solution of hyperbolic equations. Since these matrix equations must be solved at every time step and since the dimension of the vector on the space mesh is usually large, the equations must be solved efficiently, if an unacceptably long calculation is to be avoided.

3 Special Matrices: Recurrence Solution of the Tridiagonal Matrix Equation

In view of the importance of problems in one space dimension and the consequent frequent occurrence of tridiagonal matrix equations, we consider the method of solution of such problems in some detail. It turns out that simple and extremely rapid algorithms exist for the solution of tridiagonal matrix equations and we examine the method of solution from both an algebraic and a matrix point of view.

(*a*) *Algebraic Approach.* The general one-dimensional problem has been formulated in Section 4.2 (equation 4.13),

$$\alpha_j u_{j+1} + \beta_j u_j + \gamma_j u_{j-1} = w_j \quad \text{in } 1 < j < J \tag{4.24}$$

with the boundary conditions (equation 4.17) and we wish to solve for the unknown variables u_j. We look for a recursive solution such that, given the value of the variable u_j at the point j, we can obtain the variable u_{j+1} at the point $j + 1$. Namely, we seek intermediate variables x_j and y_j such that,

$$u_{j+1} = x_j u_j + y_j \tag{4.25}$$

If such an algorithm is consistent, how are the variables x_j and y_j to be determined? Clearly, if formula 4.25 is applicable at the point j, then for consistency it must be applicable at every other point j'. In particular we may show that the same formula applies at the point $j - 1$, by substituting for u_{j+1} (equation 4.25) in the original equation (4.24),

$$\alpha_j(x_j u_j + y_j) + \beta_j u_j + \gamma_j u_{j-1} = w_j$$

or,

$$u_j = \frac{-\gamma_j}{(\alpha_j x_j + \beta_j)} u_{j-1} + \frac{(w_j - \alpha_j y_j)}{(\alpha_j x_j + \beta_j)} \tag{4.26}$$

It is clear that this equation has precisely the form as the recursive algorithm from points j to $j + 1$ (equation 4.25), namely:

$$u_j = x_{j-1}u_{j-1} + y_{j-1} \tag{4.27}$$

Since the boundary conditions at the end points (equation 4.17) are defined by precisely the same algorithm, we have shown by induction that the recursive relation is valid at every point j. These equations must apply to any arbitrary vector **u** and, therefore, we must equate each term in equations 4.26 and 4.27 separately:

$$x_{j-1} = \frac{-\gamma_j}{(\alpha_j x_j + \beta_j)}, \qquad y_{j-1} = \frac{w_j - \alpha_j y_j}{(\alpha_j x_j + \beta_j)} \tag{4.28}$$

These relations provide the required values of all x_j and y_j and, therefore, equations 4.25 and 4.28 define a double recursive procedure for obtaining solutions to tridiagonal sets of equations. Starting at the point J, the mesh is scanned *downwards* in j to $j = 1$, using equations 4.28 to obtain successive values of x_j and y_j on the 'mesh'. Having obtained all values of x_j and y_j, we scan the mesh *upwards*, from $j = 1$ to $j = J$, to obtain successive values of u_j from equation 4.25. The boundary conditions at $j = J$ define starting values x_{J-1} and y_{J-1}, while the boundary conditions at $j = 1$ define the first value u_1. Specifically, we compare the boundary conditions (4.16, 4.17) at the point J, with the general recursion formula (equation 4.25),

$$x_{J-1} = \frac{-\gamma_J}{\beta_J} = \frac{a'}{b'\Delta + a'}$$

$$y_{J-1} = \frac{w_J}{\beta_J} = \frac{c'\Delta}{b'\Delta + a'} \tag{4.29}$$

There are two particular boundary conditions of frequent occurrence. First when the dependent variable u is explicity defined at a boundary ($a' = 0$) so that $x_{J-1} = 0$ and $y_{J-1} = u_J$ and second when the derivative at the boundary is zero ($b' = 0$), so that $x_{J-1} = 1$ and $y_{J-1} = 0$.

Having obtained the values x_j and y_j for all values j down to x_1 and y_1, we use the consistency of the recursion relation (equation 4.25) with the boundary condition to define the first value of the dependent variable u_1,

$$u_1 = \frac{w_1 - \alpha_1 y_1}{\beta_1 + \alpha_1 x_1}$$

$$= \frac{-c\Delta + ay_1}{a - b\Delta - ax_1} \tag{4.30}$$

The remaining values of the dependent variables u_j may now be determined.

To illustrate the method, we solve the simple example of Poisson's equation on a small eight-point mesh. The differenced form of Poisson's equation is written as

$$\Phi_{j+1} - 2\Phi_j + \Phi_{j-1} = m_j \quad \text{for } 1 < j < J \tag{4.31}$$

where we take the source function as

$$\mathbf{m} = (0, 0, 2, 0, 1, 0, 0, 0)$$

with the boundary conditions:

$$\Phi = 0 \text{ at } j = 1, \qquad \Phi = 0 \text{ at } j = J$$

Thus the first values, x_{J-1} and y_{J-1}, are defined,

$$x_7 = 0, \qquad y_7 = 0$$

The numerical procedure and results are included in Table 4.1.

Table 4.1 Example of the recurrence solution of a tridiagonal matrix equation

solutions to $\Phi_{j+1} - 2\Phi_j + \Phi_{j-1} = m_j$

$J = 8$, boundary conditions $\Phi_1 = 0, \Phi_J = 0$

(1) Source function m_j	(2) First scan x_j	(3) y_j	(4) Solution j
$m_8 = 0$			$\Phi_1 = 0{\cdot}000$
$m_7 = 0$	$x_7 = 0$	$y_7 = 0{\cdot}000$	$\Phi_2 = 1{\cdot}860$
$m_6 = 0$	$x_6 = 0{\cdot}5$	$y_6 = 0{\cdot}000$	$\Phi_3 = 3{\cdot}716$
$m_5 = 1$	$x_5 = 0{\cdot}667$	$y_5 = 0{\cdot}000$	$\Phi_4 = 3{\cdot}569$
$m_4 = 0$	$x_4 = 0{\cdot}750$	$y_4 = 0{\cdot}750$	$\Phi_5 = 3{\cdot}429$
$m_3 = 2$	$x_3 = 0{\cdot}800$	$y_3 = 0{\cdot}600$	$\Phi_6 = 2{\cdot}283$
$m_2 = 0$	$x_2 = 0{\cdot}833$	$y_2 = 2{\cdot}167$	$\Phi_7 = 1{\cdot}142$
$m_1 = 0$	$x_1 = 0{\cdot}857$	$y_1 = 1{\cdot}860$	$\Phi_8 = 0{\cdot}0$

In column 1 the given source function is listed. In columns 2 and 3, the results of the first recurrence step $x_{j-1} = \gamma_j/(\alpha_j x_j + \beta_j)$ and $y_{j-1} = (w_j - \alpha_j y_j)/(\alpha_j x_j + \beta_j)$ respectively are listed; the values x_7 and y_7 are defined by the 'right' boundary conditions. In column 4 the solutions $\Phi_{j+1} = x_j\Phi_j + y_j$ are listed; Φ_1 is defined by the 'left' boundary condition.

(*b*) *Matrix Approach.* To stress and to illustrate the method of inverting a tridiagonal matrix equation, we examine the method from a matrix point of view. Given a known vector **w**, we are required to determine the unknown vector **u**, which satisfies the matrix equation

$$\mathbf{Au} = \mathbf{w}$$

where the matrix **A** is tridiagonal. It is useful to define matrix ladder operators $\mathbf{L}_+$ and $\mathbf{L}_-$, such that, for every component of a vector **u**,

$$\mathbf{L}_+\{u_j\} = \{u_{j+1}\} \tag{4.32}$$

$$\mathbf{L}_-\{u_j\} = \{u_{j-1}\} \tag{4.33}$$

Since the matrix **A** is tridiagonal, the matrix equation may be written as,

$$(\mathbf{PL}_+ + \mathbf{Q} + \mathbf{RL}_-)\mathbf{u} = \mathbf{w} \tag{4.34}$$

where **P**, **Q** and **R** are diagonal matrices.

Equation 4.34 defines a double recursion which it is convenient to split into two stages by defining a diagonal matrix **X** and a vector **y**,

$$\mathbf{L}_+\mathbf{u} = \mathbf{Xu} + \mathbf{y} \tag{4.35}$$

Multiplying this equation by the diagonal matrix **P**, equation 4.35 is rewritten,

$$(\mathbf{PL}_+ - \mathbf{PX} + \mathbf{0})\mathbf{u} = \mathbf{Py} \tag{4.36}$$

where **0** is the null matrix. Subtracting equation 4.36 from 4.34,

$$(\mathbf{Q} + \mathbf{PX})\mathbf{u} = -\mathbf{RL}_-\mathbf{u} + \mathbf{w} - \mathbf{Py} \tag{4.37}$$

$$\mathbf{u} = -(\mathbf{Q} + \mathbf{PX})^{-1}\mathbf{RL}_-\mathbf{u} + (\mathbf{Q} + \mathbf{PX})^{-1}(\mathbf{w} - \mathbf{Py}) \tag{4.38}$$

This equation is precisely of the form (4.35) and, indeed, to be consistent with the problem at hand (equation 4.34), it must be identical so that we may equate terms separately:

$$\mathbf{L}_-\mathbf{X} = -(\mathbf{Q} + \mathbf{PX})^{-1}\mathbf{R}$$

$$\mathbf{L}_-\mathbf{y} = (\mathbf{Q} + \mathbf{PX})^{-1}(\mathbf{w} - \mathbf{Py}) \tag{4.39}$$

Equations 4.35 and 4.39 form the required two-stage recursion procedure. We may compare these results with those obtained using a scalar approach, since, by definition, **P**, **Q**, **R** and **X** are all diagonal matrices whose diagonal elements at each column j are α_j, β_j, γ_j and x_j respectively. It is clear that the results are identical with the algebraic equations 4.25 and 4.28.

It is to be noted that the tridiagonal inversion procedure is extremely simple and efficient, since, for an $n \times n$ tridiagonal matrix equation operating on a vector of dimension n, only $9n$ arithmetic operations are required to solve the system.

4 Special Matrices: The 'Exact' Solution of Poisson's Equation

Poisson's equation is of such frequent occurrence in physics that particular attention has been paid to its solution. Generally, elliptic equations in two or more space dimensions, and the consistent matrix equations which result, are difficult and time-consuming to solve, but the symmetry implicit in Poisson's equation admits of a number of rapid methods which cannot be used for the general problem. The symmetry of Poisson's equation arises from the properties of the Laplacian operator which can be written as the sum of commuting operators. For example, in two space dimensions, the Laplacian operator is:

$$\nabla^2 = \frac{\partial^2}{\partial x^2} + \frac{\partial^2}{\partial y^2}$$

where,

$$\left(\frac{\partial^2}{\partial x^2}, \frac{\partial^2}{\partial y^2}\right) = 0 \tag{4.40}$$

It follows that the eigenfunctions of the multidimensional Laplacian operator may be written as the product of the separate eigenfunctions of the Laplacian operator along each space dimension. Consequently, the solution of a multidimensional Poisson's equation may always be reduced to the separate solution of a series of one-dimensional problems. This is equally true of the matrix equation which results from the finite difference form of Poisson's equation. The eigenfunctions along each dimension are simply the harmonic functions or Fourier modes, the particular choice of which is defined by the boundary conditions for the problem in hand.

It is sufficient, here, to consider the two-dimensional Poisson's equation, which in finite difference form has been expressed previously (Section 4.2, equation 4.20, Figure 4.2). Three approaches for the rapid solution of the matrix equation have been developed. In the simplest method, the complete matrix equation is diagonalized by applying the methods of Fourier analysis along each space dimension (Boris and Roberts 1969). In its simplest form, the method of Hockney (1970) employs Fourier analysis along only one space dimension; from this, uncoupled tridiagonal matrix equations, which may readily be solved, are obtained along the other space dimension. Finally a method of cyclic reduction, or tridiagonal inversion, may be applied along each space dimension (Buneman 1969).

These solutions to Poisson's equation are exact, not in the sense that errors are not incurred by approximating the differential equations by finite difference equations, but in the sense that an exact solution is found by directly solving the finite matrix equation.

(*a*) *Multiple Fourier Analysis.* We consider the five-point difference formulation of Poisson's equation in two Cartesian space dimensions (equation 4.20),

$$(\Phi_{i+1,j} - 2\Phi_{i,j} + \Phi_{i-1,j}) + (\Phi_{i,j+1} - 2\Phi_{i,j} + \Phi_{i,j-1}) = -\Delta^2 \rho_{i,j}$$

at all points

$$(i, j) \quad (0 < i < I, 0 < j < J) \tag{4.41}$$

where $\rho_{i,j}$ and $\Phi_{i,j}$ are the known source and unknown potential functions respectively. We represent the five-point operator on the left-hand side by P,

$$\mathrm{P}\Phi_{i,j} = (\Phi_{i+1,j} - 2\Phi_{i,j} + \Phi_{i-1,j}) + (\Phi_{i,j+1} - 2\Phi_{i,j} + \Phi_{i,j-1}) \tag{4.42}$$

Without considering boundary conditions and, as an example, we show first that the double Fourier harmonic,

$$s_{ij}(k, l) = \hat{\Phi}(k, l) \sin\frac{\pi kj}{J} \sin\frac{\pi li}{I} \tag{4.43}$$

is an eigenfunction of the operator P, when k and l are integers,

$$\begin{aligned}
\mathrm{P}s_{ij}(k, l) &= \hat{\Phi}(k, l) \sin\frac{\pi kj}{J}\left\{\sin\frac{\pi l(i+1)}{I} - 2\sin\frac{\pi li}{I} + \sin\frac{\pi l(i-1)}{I}\right\} \\
&\quad + \hat{\Phi}(k, l) \sin\frac{\pi li}{I}\left\{\sin\frac{\pi k(j+1)}{J} - 2\sin\frac{\pi kj}{J} + \sin\frac{\pi k(j-1)}{J}\right\} \\
&= \hat{\Phi}(k, l) \sin\frac{\pi kj}{J}\left\{2\cos\frac{\pi l}{I}\sin\frac{\pi li}{I} - 2\sin\frac{\pi li}{I}\right\} \\
&\quad + \hat{\Phi}(k, l) \sin\frac{\pi li}{I}\left\{2\cos\frac{\pi k}{J}\sin\frac{\pi kj}{J} - 2\sin\frac{\pi kj}{J}\right\}
\end{aligned} \tag{4.44}$$

where, within each pair of curly brackets we have taken the sum of two sine functions. Rearranging the right-hand side,

$$\begin{aligned}
\mathrm{P}s_{ij}(k, l) &= \hat{\Phi}(k, l) \sin\frac{\pi kj}{J}\sin\frac{\pi li}{I}\left\{2\cos\frac{\pi l}{I} + 2\cos\frac{\pi k}{J} - 4\right\} \\
&= \alpha_{kl}s_{ij}(k, l)
\end{aligned} \tag{4.45}$$

This is the eigenvalue equation and $s_{ij}(k, l)$ (equation 4.43) is an eigenfunction with eigenvalue,

$$\alpha_{kl} = 2\cos\frac{\pi l}{I} + 2\cos\frac{\pi k}{J} - 4 \tag{4.46}$$

Thus, we may solve Poisson's equation (4.4) for each eigenfunction of P separately.

The method of double Fourier analysis for the solution of Poisson's equation therefore involves five steps. The known source functions ($\rho_{i,j}$) are analysed, first along one direction, on each 'row' i.

For all $k, 0 < k < J$:

$$\hat{\rho}_i(k) = \frac{2}{J}\sum_{j=1}^{J} \rho_{i,j} \sin\frac{\pi kj}{J} \tag{4.47}$$

for each Fourier mode of wavelength $2J\Delta/k$. Similarly the other dimension may then be analysed.

For all $l, 0 < l < I$:

$$\hat{\rho}(k, l) = \frac{2}{I} \sum_{i=1}^{I} \hat{\rho}_i \sin \frac{\pi l i}{I} \tag{4.48}$$

The amplitude of the potential for each double Fourier mode may now be determined separately according to the eigenvalues (4.46),

$$\hat{\Phi}(k, l) = \frac{\hat{\rho}(k, l)\Delta^2}{4 - 2\cos(\pi k/J) - 2\cos(\pi l/J)} \tag{4.49}$$

Finally Fourier synthesis is performed along each dimension.

For all $i, 0 < i < I$:

$$\hat{\Phi}_i(k) = \sum_{l=1}^{I} \hat{\Phi}(k, l) \sin \frac{\pi l i}{I} \tag{4.50}$$

For all $j, 0 < j < J$:

$$\Phi_{ij} = \sum_{k=1}^{J} \hat{\Phi}_i(k) \sin \frac{\pi k j}{J} \tag{4.51}$$

This is the general procedure, but we must choose particular harmonics according to the boundary conditions and the Fourier half-sine analysis (equation 4.47) and synthesis (equation 4.51) is used when the potential is zero or constant at the boundaries. The half-cosine series is used when the derivative of the potential is zero at the boundary:

synthesis:
$$\Phi_j = \tfrac{1}{2}\hat{\Phi}(0) + \sum_{k=1}^{J-1} \hat{\Phi}(k) \cos \frac{\pi k j}{J} \tag{4.52}$$

analysis:
$$\hat{\Phi}(k) = \frac{2}{J} \sum_{j=1}^{J-1} \Phi_j \cos \frac{\pi k j}{J} \tag{4.53}$$

And for periodic boundary conditions the full Fourier series, using both sine and cosines, must be used (Table 4.2),

synthesis:
$$\Phi_j = \tfrac{1}{2}\hat{\Phi}^c(0) + \sum_{k=1}^{\frac{1}{2}J-1} \hat{\Phi}^c(k) \cos \frac{2\pi k j}{J} + \hat{\Phi}^s(k) \sin \frac{2\pi k j}{J} \tag{4.54}$$

analysis:
$$\hat{\Phi}^c(k) = \frac{2}{J} \sum_{j=1}^{J-1} \Phi_j \cos \frac{2\pi k j}{J}$$

$$\hat{\Phi}^s(k) = \frac{2}{J} \sum_{j=1}^{J-1} \Phi_j \sin \frac{2\pi k j}{J} \tag{4.55}$$

In the latter case the eigenvalues for Poisson's equation in two dimensions are:

$$\alpha_{kl} = 2\cos \frac{2\pi k}{J} + 2\cos \frac{2\pi l}{I} - 4 \tag{4.56}$$

Table 4.2 The eigenfunctions (Fourier harmonics) of the tridiagonal matrix with elements (1, − 4, 1)

j = space point $0 \leqslant j \leqslant J$
k = integer wavenumber $0 \leqslant k \leqslant J$

	Boundary conditions		
	$\Phi = 0$ at $j = 0, j = J$	$\dfrac{\partial \Phi}{\partial x} = 0$ at $j = 0, j = J$	Periodic boundary conditions $\Phi_0 = \Phi_J$
1. Synthesis	$\Phi_j = \sum_{k=1}^{J-1} \hat{\Phi}_k \sin \frac{\pi k j}{J}$	$\Phi_j = \frac{1}{2}\hat{\Phi}_0 \sum_{k=1}^{J-1} \hat{\Phi}_k \cos \frac{\pi k j}{J}$	$\Phi_j = \frac{1}{2}\hat{\Phi}_0 \sum_{k=1}^{\frac{1}{2}J-1} \hat{\Phi}_k^c \cos \frac{2\pi k j}{J} + \hat{\Phi}_k^s \sin \frac{2\pi k j}{J}$
2. Analysis	$\hat{\Phi}_k = \frac{2}{J} \sum_{j=1}^{J} \Phi_j \sin \frac{\pi k j}{J}$	$\hat{\Phi}_k = \frac{2}{J} \sum_{j=1}^{J-1} \Phi_j \cos \frac{\pi k j}{J}$	$\hat{\Phi}_k^c = \frac{2}{J} \sum_{j=1}^{J-1} \Phi_j \cos \frac{2\pi k j}{J}$ $\hat{\Phi}_k^s = \frac{2}{J} \sum_{j=1}^{J-1} \Phi_j \sin \frac{2\pi k j}{J}$
3. Eigenvalue for wavenumber k	$-4 + 2 \cos \frac{\pi k}{J}$	$-4 + 2 \cos \frac{\pi k}{J}$	$-4 + 2 \cos \frac{2\pi k}{J}$

In conclusion it must be stressed that, in the application of the method, the numerical evaluation of sine and cosine functions is avoided by the use of fast Fourier transform techniques, where the harmonic vectors are generated by simple recurrence relations (Hockney 1970, Cooley and Tukey 1965). The total number of arithmetic operations required to solve Poisson's equation, say on a two-dimensional mesh ($I \times J$ points), is of the order $4IJ(I + J)$.

(*b*) *Fourier Analysis and Cyclic Reduction.* An extremely rapid method and a method of wide application in the solution of a two-dimensional Poisson's equation is to apply Fourier analysis along only one space direction, say along the 'columns' $0 < i < I$. Consequently the problem is reduced to a set of $J - 2$ tridiagonal matrix equations, which couple the variables across columns $0 \leqslant i \leqslant I$ and which may each be solved separately (Hockney 1970).

We consider the equations in two Cartesian coordinates (equations 4.41), though the method may equally be applied in cylindrical coordinates. As an example, taking the case where the boundary conditions specify the potential as constant at the end of each column (say $\Phi_{i0} = \Phi_{iJ} = 0$, for all i), we may expand the potential and source functions in terms of the Fourier half-sine series (Table 4.2),

$$\Phi_{ij} = \sum_{k=1}^{J} \hat{\Phi}_i(k) \sin\frac{\pi kj}{J}, \qquad \rho_{ij} = \sum_{k=1}^{J} \hat{\rho}_i(k) \sin\frac{\pi kj}{J} \tag{4.57}$$

Inserting these expressions for the potential and the source function into the five-point finite formulation of Poisson's equation (4.41), we obtain:

$$\sum_{k=1}^{J} \{\hat{\Phi}_{i+1}(k) - 2\hat{\Phi}_i(k) + \hat{\Phi}_{i-1}(k)\} \sin\frac{\pi kj}{J} + \left\{2\cos\frac{\pi k}{J} - 2\right\}\hat{\Phi}_i(k)\sin\frac{\pi kj}{J}$$
$$+ \Delta^2 \hat{\rho}_i(k) \sin\frac{\pi kj}{J} = 0 \tag{4.58}$$

which must be valid at every point j. Hence in these equations, we may relate the amplitude of each Fourier harmonic separately to zero: for each 'wave-number' k,

$$\hat{\Phi}_{i+1}(k) - \left(4 - 2\cos\frac{\pi k}{J}\right)\hat{\Phi}_i(k) + \hat{\Phi}_{i-1}(k) = -\Delta^2 \hat{\rho}_i(k) \tag{4.59}$$

Thus for each Fourier mode k, a tridiagonal matrix equation is obtained which acts on the unknown vector $\{\hat{\Phi}_i(k)\}$ and each tridiagonal matrix equation may in turn be solved by one of a number of methods (for example tri-diagonal inversion, Section 4.3).

The procedure is therefore to Fourier analyse the source function ρ_{ij} along one dimension only, to obtain the Fourier coefficients for each i,

$$\hat{\rho}_i(k) = \frac{2}{J} \sum_{j=1}^{J} \rho_{ij} \sin \frac{\pi k j}{J} \tag{4.60}$$

The tridiagonal equations (4.59) with the known right-hand side are obtained for each wavenumber. Hockney (1970) has solved the tridiagonal equations by a method of cyclic reduction, rather than the inversion method described in Section 4.3. At three adjacent points, the tridiagonal equations (4.59) are of the form:

$$\hat{\Phi}_{i+2}(k) - \left(4 - 2\cos\frac{\pi k}{J}\right)\hat{\Phi}_{i+1}(k) + \hat{\Phi}_i(k) = -\Delta^2 \hat{\rho}_{i+1}(k)$$

$$\hat{\Phi}_{i+1}(k) - \left(4 - 2\cos\frac{\pi k}{J}\right)\hat{\Phi}_{i}(k) + \hat{\Phi}_{i-1}(k) = -\Delta^2 \hat{\rho}_{i}(k)$$

$$\hat{\Phi}_{i}(k) - \left(4 - 2\cos\frac{\pi k}{J}\right)\hat{\Phi}_{i-1}(k) + \hat{\Phi}_{i-2}(k) = -\Delta^2 \hat{\rho}_{i-1}(k) \tag{4.61}$$

By multiplying the second of these equations by the coefficient $(4 - 2\cos \pi k/J)$ and adding, we eliminate intermediate points,

$$\hat{\Phi}_{i+2}(k) - \left\{\left(4 - 2\cos\frac{\pi k}{J}\right)^2 - 2\right\}\hat{\Phi}_i(k) + \hat{\Phi}_{i-2}(k) = -\Delta^2 \hat{\rho}_{i+1}(k)$$
$$- \left(4 - 2\cos\frac{\pi k}{J}\right)\Delta^2 \hat{\rho}_i(k) - \Delta^2 \hat{\rho}_{i-1}(k) \tag{4.62}$$

Thus the system of I tridiagonal equations has been reduced to $\frac{1}{2}I$ tridiagonal equations, which are essentially of the same form as before (equation 4.59). Similarly, the same procedure may be applied recursively until the central point $\frac{1}{2}I$ is related only to the boundary points $i = 0$ and $i = I$, where the boundary conditions enable the variable at the central point to be determined. The reverse procedure is now carried out by determining the variables at the points $\frac{1}{4}I$ and $\frac{3}{4}I$ etc., until the whole mesh is defined.

Consequently, after solving each of the tridiagonal equations, the variables $\hat{\Phi}_i(k)$ are determined for all i and for all wavenumbers k. The potential at every mesh point (i, j) is obtained by Fourier synthesis is applied to each column i:

$$\Phi_{ij} = \sum_{k=1}^{J} \hat{\Phi}_i(k) \sin \frac{\pi k j}{J} \tag{4.63}$$

This method, of Fourier analysis along one direction followed by cyclic reduction (or tridiagonal inversion) along the other direction, is rapid and requires of the order $IJ \log_2 J$ operations. It has the advantage too, of being applicable to cylindrical or more complex geometries, since Fourier analysis need only be applied along one dimension. Some improvement in the efficiency of the method is obtained by implementing one level of cyclic reduction along the rows and eliminating even lines before Fourier analysing the source functions along the rows.

(*c*) *Double Cyclic Reduction*. Fourier analysis may be avoided entirely, by again using the commutative properties of the two parts of the Laplacian operator in two dimensions. Cyclic reduction may be applied to the *vector* of variables $\{\Phi_i\}_j$ for each row j (Buneman 1969). We write the five-point difference formulation of Poisson's equation (equation 4.41) in the form,

$$\{\Phi_i\}_{j+1} + \mathbf{A}^{(0)}\{\Phi_i\}_j + \{\Phi_i\}_{j-1} = \{\tilde{\rho}_i^{(0)}\}_j \tag{4.64}$$

where $\mathbf{A}^{(0)}$ is a tridiagonal matrix with tridiagonal elements $(1, -4, 1)$, which operates on the elements Φ_{ij} of each vector $\{\Phi_i\}_j$. The right-hand side is proportional to the known source function. Similar equations are valid for every row j, and we may eliminate intermediate rows $j+1$, $j-1$, etc. by cyclic reduction to obtain:

$$\{\Phi_i\}_{j+2} + (\mathbf{A}^{(0)2} - 2\mathbf{I})\{\Phi_i\}_j + \{\Phi_i\}_{j-2} = \mathbf{A}^{(0)}\{\tilde{\rho}_i^{(0)}\}_j - \{\tilde{\rho}_i^{(0)}\}_{j+1} - \{\tilde{\rho}_i^{(0)}\}_{j-1} \tag{4.65}$$

These reduced equations $(J/2)$ have precisely the tridiagonal form as before,

$$\{\Phi_i\}_{j+2} + \mathbf{A}^{(1)}\{\Phi_i\}_j + \{\Phi_i\}_{j-2} = \{\tilde{\rho}_i^{(1)}\}_j \tag{4.66}$$

and it follows that cyclic reduction may be continued until, after p steps, the central row is defined in terms of the source functions $\{\tilde{\rho}_i^{(p)}\}_j$ and the boundary values,

$$\mathbf{A}^{(p)}\{\Phi_i\}_{J/2} = \{\tilde{\rho}_i^{(p)}\}_{J/2} \tag{4.67}$$

We are left, therefore, with the reduced matrix equation which couples only I, rather than IJ, unknown variables through the matrix $\mathbf{A}^{(p)}$. $\mathbf{A}^{(p)}$ itself is no longer a tridiagonal matrix but a banded or full matrix, and the reduced equations may be solved by Gauss elimination (Section 5). In simple cases, the matrix $\mathbf{A}^{(p)}$ can be factorized as a series of tridiagonal matrices which may be solved in succession (Buneman 1969). Finally the inverse process to cyclic reduction defines all the variables on the intermediate rows.

(*d*) *Summary*. Three methods of solving Poisson's equation for multi-dimensional problems have been defined. The methods employ Fourier analysis followed by synthesis, cyclic reduction (or tridiagonal inversion), or a combination of Fourier analysis and cyclic reduction. But the success of each of these direct methods depends on the property that the difference Laplacian operator with simple boundaries may be written as the sum of commuting operators (equation 4.40), each of which operates along only one space dimension. The methods are extremely fast, involving of the order $n \log_2 n$ operations, where n is the total number of mesh points, but they may not be applied to more complex problems and, in particular, to the multidimensional finite difference form of the general elliptic equation (Section 3.2f).

5 The Exact Solution of the General Matrix Equation

In the algorithms for the solution of the matrix equation that have been described previously, rapid solutions have been obtained by employing to good effect the particular properties of the operative matrix. In the absence of special properties, the solution of the general matrix equation is a large and time-consuming problem involving many arithmetic operations. We shall illustrate the approach with one particular method, the Gauss elimination method, though a number of other variations based on the Gauss method exist (Fox 1964). Certainly for large matrices, no one method is markedly more efficient than another.

Generally it is our purpose to solve by a direct or 'exact' method the set of n unknown variables $u_1, \ldots, u_n$, which satisfy the set of n simultaneous inhomogeneous equations,

$$\begin{aligned}
a_{11}u_1 + a_{12}u_2 + \cdots + a_{1n}u_n &= w_1 \\
a_{21}u_1 + a_{22}u_2 + \cdots + a_{2n}u_n &= w_2 \\
\cdots\cdots\cdots\cdots\cdots\cdots\cdots\cdots\cdots \\
a_{i1}u_1 + \quad \cdots \quad + a_{in}u_n &= w_i \\
\cdots\cdots\cdots\cdots\cdots\cdots\cdots\cdots\cdots \\
a_{n1}u_1 + \quad \cdots \quad + a_{nn}u_n &= w_n
\end{aligned} \tag{4.68}$$

In the Gauss elimination method, the dependent variables $u_1, \ldots, u_n$ are eliminated successively, until a single variable remains. One of the equations (say the first) is chosen to define one variable, say u_1, in terms of the remaining $n - 1$ variables. This equation, the pivotal equation is used to eliminate u_1 successively from the remaining $n - 1$ equations. If the last $n - 1$ equations are multiplied by a_{11}/a_{i1} and subtracted from the pivotal equation, the variable u_1 is eliminated from each equation. After the first step the equations have been reduced to the form:

$$\begin{aligned}
a_{11}u_1 + a_{12}u_2 + a_{13}u_3 + \cdots + a_{1n}u_n &= w_1 \\
a'_{22}u_2 + a'_{23}u_3 + \cdots + a'_{2n}u_n &= w'_2 \\
a'_{32}u_2 + a'_{33}u_3 + \cdots + a'_{3n}u_n &= w'_3 \\
\vdots \qquad\qquad\qquad\qquad \vdots \quad &\quad \vdots \\
a'_{n2}u_2 + \quad \cdots \quad + a'_{nn}u_n &= w'_n
\end{aligned} \tag{4.69}$$

By eliminating the variable u_1 in the last $n - 1$ equations, the system has been reduced to $(n - 1)$ equations in the $(n - 1)$ unknowns, $u_2, \ldots, u_n$. The new coefficients a', and the known right-hand vector are given by:
for $i, j \geqslant 2$,

$$a'_{ij} = \frac{a_{11}}{a_{i1}} a_{ij} - a_{1j}, \qquad w'_i = \frac{a_{11}}{a_{i1}} w_i - w_1 \tag{4.70}$$

Precisely the same elimination technique can now be applied to this reduced set of equations and indeed the elimination procedure may be continued until a single equation in the single variable u_n is obtained. The system of equations has now been transformed into 'triangular' form,

$$\begin{aligned} a_{11}u_1 + a_{12}u_2 + a_{13}u_3 + \cdots + a_{1n}u_n &= w_1 \\ a'_{22}u_2 + a'_{23}u_3 + \cdots + a'_{2n}u_n &= w'_2 \\ a''_{33}u_3 + \cdots + a''_{3n}u_n &= w''_3 \\ \cdots \\ a_{nn}^{(n-1)}u_n &= w_n^{(n-1)} \end{aligned} \tag{4.71}$$

It follows that u_n can be evaluated and that each of the variables from u_n to u_1 can be successively evaluated from the triangular system of equations (4.71).

In matrix terms, it is clear that the process of solution involves the factorization of the matrix **A** into two triangular matrices **L** (lower) and **U** (upper):

$$\mathbf{A} = \mathbf{LU} \tag{4.72}$$

where **L** has the form,

$$\mathbf{L} = \begin{bmatrix} \mathrm{x} & & & & \\ \mathrm{x} & \mathrm{x} & & 0 & \\ \mathrm{x} & \mathrm{x} & \mathrm{x} & & \\ & & & \ddots & \\ \mathrm{x} & \mathrm{x} & \cdots & \mathrm{x} & \mathrm{x} \end{bmatrix}$$

and **U** is,

$$\mathbf{U} = \begin{bmatrix} \mathrm{x} & \mathrm{x} & \mathrm{x} & \cdots & \mathrm{x} \\ & \mathrm{x} & \mathrm{x} & & \mathrm{x} \\ & & \mathrm{x} & & \vdots \\ & 0 & & \ddots & \vdots \\ & & & & \mathrm{x} \end{bmatrix}$$

When such matrices are inverted, they retain their form and, hence, the solution is obtained by inverting each of **L** and **U** in turn,

$$\mathbf{LUu} = \mathbf{w}$$

$$\mathbf{Uu} = \mathbf{L}^{-1}\mathbf{w} \tag{4.73}$$

The matrix equation (4.73) is the set of equations (4.71): finally the variables **u** are each evaluated,

$$\mathbf{u} = \mathbf{U}^{-1}\mathbf{L}^{-1}\mathbf{w} \tag{4.74}$$

If the unknown vector **u** is of length n, simple inspection shows that, in general, the number of arithmetic operations required to obtain solutions is of the order $\frac{2}{3}n^3$.

A number of variations on the Gauss elimination method exist, which marginally reduce the number of arithmetic operations required for the solution of the equations, but all such methods scale with the cube of the length of the unknown vector.* For large matrices, and particularly for those which arise from the difference formulation of multidimensional problems, a very large number of operations is required to obtain a solution and, indeed, on large space-meshes an unacceptable length of computer time is demanded.

6 'Inexact' or Iterative Methods for the Solution of the Matrix Equation

It has been shown that the exact solution of the general matrix equation,

$$\mathbf{Au} = \mathbf{w} \tag{4.75}$$

where **w** is a known vector and **u** is to be determined, involves of the order of n^3 arithmetic operations, where n is the dimension of the vector **u**. For the large sparse matrices which frequently result from physical problems, the computer time associated with the performance of such a large number of arithmetic operations proves unacceptable. In the particular cases where the matrix **A** is tridiagonal, or where **A** arises from the finite difference formulation of a multidimensional Poisson's equation, direct and rapid algorithms exist (Sections 4.3 and 4.4). But more generally, where **A** arises from a multidimensional general boundary-value problem, or from an implicit formulation of the initial-value problem, direct methods require many operations.

As an alternative approach to solving the matrix equation 4.75 directly and exactly, we may guess some initial value of **u**, say $\mathbf{u}^{(0)}$, and proceed in successive steps p to attempt to improve our *inexact* solution $\mathbf{u}^{(p)}$. In the simplest form, this procedure may be represented by the matrix equation,

$$\mathbf{u}^{(p+1)} = \mathbf{P}\mathbf{u}^{(p)} + \mathbf{c} \tag{4.76}$$

where **P** is the iteration matrix, and an improved solution $\mathbf{u}^{(p+1)}$ is derived explicitly from the previous solution $\mathbf{u}^{(p)}$. The matrix **P** is related to the given

* The methods of Jordan, Aitken (Fox 1964).

matrix **A**, and the constant vector **c** is related to the known vector **w**. If, in the limit of many steps (p is very large), convergence is achieved so that,

$$\lim_{p\to\infty} \mathbf{u}^{(p)} = \mathbf{P}\mathbf{u}^{(p)} + \mathbf{c} \tag{4.77}$$

then the vector $\mathbf{u}^{(\infty)}$ satisfies the matrix equation,

$$(\mathbf{I} - \mathbf{P})\mathbf{u}^{(\infty)} = \mathbf{c} \tag{4.78}$$

Thus $\mathbf{u}^{(\infty)}$ will be the solution we require if this matrix equation (4.78) is consistent with the equation of interest (4.75). It follows that if the vector **c** is related to the known vector **w** by the nonsingular matrix **T**,

$$\mathbf{c} = \mathbf{T}\mathbf{w}$$

then the iteration matrix must satisfy,

$$\mathbf{T}^{-1}(\mathbf{I} - \mathbf{P}) = \mathbf{A}$$

or,

$$\mathbf{P} = \mathbf{I} - \mathbf{T}\mathbf{A} \tag{4.79}$$

The benefit of this procedure is that each improved value $\mathbf{u}^{(p+1)}$ is obtained *explicitly*. It is informative to associate a physical significance with the step-by-step improvement of the 'solution'. For example, where the matrix equation 4.75 results from the difference formulation of a boundary-value problem, it has been illustrated previously (Section 3.1f) that the equation represents an infinite speed of propagation of information across the mesh. With an iteration procedure, we insert a 'pseudo-time' in the equation in order that an explicit formulation of the equation may be obtained by ensuring that information propagates with a finite speed on the mesh. We suppose, for example, that the matrix equation 4.75 is the finite difference formulation of Poisson's equation (Section 2) and the matrix **A** is the finite difference Laplacian operator Δ'' (Figure 4.1). Clearly, one method of arriving at the solution of Poisson's equation is to employ a diffusion equation to the limit of a steady state,

$$\frac{\partial u}{\partial t} - \nabla^2 u = \rho$$

where ρ is the known source function. When a steady state is reached, the variable u satisfies Poisson's equation. The equivalent explicit difference form is:

$$\mathbf{u}^{(p+1)} = \mathbf{u}^{(p)} + \Delta t\, \Delta''\mathbf{u}^{(p)} + \rho\, \Delta t \tag{4.80}$$

An iterative procedure may be regarded as an explicit solution to convergence in imaginary iteration time. Clearly the technique is simple to apply. In particular, for sparse matrices, it involves only a small number of arithmetic operations at each iteration step. On the other hand, many iteration steps (p) may be required before a reasonable solution is obtained.

(*a*) *The Properties and Formulation of an Iterative Matrix.* In the application of the iterative method, we must question first, whether convergence will be achieved for any solution guess $\mathbf{u}^{(0)}$ and second, how rapidly the solution vectors $\mathbf{u}^{(p)}$ will converge. Rather than considering the absolute solution vector $\mathbf{u}^{(p)}$, it is useful to examine the error vector $\boldsymbol{\epsilon}^{(p)}$ which remains at any step: thus if $\mathbf{u}$ is the exact solution of the matrix equation,

$$\boldsymbol{\epsilon}^{(p)} = \mathbf{u}^{(p)} - \mathbf{u} \tag{4.81}$$

For consistency of the iterative scheme (equation 4.76) with the matrix equation 4.75, we require that the exact solution $\mathbf{u}$ satisfies,

$$\mathbf{u} = \mathbf{P}\mathbf{u} + \mathbf{c} \tag{4.82}$$

It follows from equations 4.76 and 4.82 that the error vectors $\boldsymbol{\epsilon}$ are related at successive steps,

$$\boldsymbol{\epsilon}^{(p+1)} = \mathbf{P}\boldsymbol{\epsilon}^{(p)} \tag{4.83}$$

For convergence, we demand that the magnitude of the error vector decays,

$$|\boldsymbol{\epsilon}^{(p+1)}| < |\boldsymbol{\epsilon}^{(p)}| \tag{4.84}$$

Therefore we define a quantity, the spectral *norm* $\|\mathbf{P}\|$,

$$\|\mathbf{P}\| = \frac{|\mathbf{P}\boldsymbol{\epsilon}^{(p)}|}{|\boldsymbol{\epsilon}^{(p)}|} \tag{4.85}$$

and require that the norm of $\mathbf{P}$ be smaller than unity, in order that the solutions $\mathbf{u}^{(p)}$ will converge.

We shall examine this requirement more specifically, by analysing the error vectors in terms of the eigenvectors of the iteration matrix $\mathbf{P}$. If $\mathbf{P}$ has a set (n) of distinct eigenvalues ρ_i, then the starting error $\boldsymbol{\epsilon}^{(0)}$ in the solution-vector $\mathbf{u}^{(0)}$ can be expanded in terms of the eigenvectors $\mathbf{s}^{(i)}$ of $\mathbf{P}$,

$$\boldsymbol{\epsilon}^{(0)} = \sum_{i=1}^{n} \alpha_i \mathbf{s}^{(i)} \tag{4.86}$$

α_i is the amplitude of the initial error vector along each eigenvector $\mathbf{s}^{(i)}$. By using equation 4.83, which relates successive error vectors, we may relate the error at step p to the initial error,

$$\begin{aligned}\boldsymbol{\epsilon}^{(p)} &= \mathbf{P}\boldsymbol{\epsilon}^{(p-1)} \\ &= \mathbf{P}^p\boldsymbol{\epsilon}^{(0)}\end{aligned} \tag{4.87}$$

We employ the expansion (equation 4.86),

$$\begin{aligned}\boldsymbol{\epsilon}^{(p)} &= \mathbf{P}^p \sum_{i=1}^{n} \alpha_i \mathbf{s}^{(i)} \\ &= \sum_{i=1}^{n} \alpha_i \rho_i^p \mathbf{s}^{(i)}\end{aligned} \tag{4.88}$$

Consequently, after p iterative steps, the amplitude of the error vector along each eigenvector is proportional to the pth power of each eigenvalue. For large p, the error is clearly related to the eigenvector with eigenvalue of largest modulus and it follows that, for convergence, the modulus of the largest eigenvalue $\rho = |\rho_m|$ of $\mathbf{P}$ must be smaller than unity. ρ is the spectral radius of the matrix $\mathbf{P}$.

$$\lim_{p\to\infty} \boldsymbol{\epsilon}^{(p)} = \alpha_m \rho_m^p \mathbf{s}^{(m)} \tag{4.89}$$

where,

$$|\rho_m| = \max\,(|\rho_1|, |\rho_2|, \ldots, |\rho_n|). \tag{4.90}$$

In addition, the asymptotic (large p) rate of convergence depends on the spectral radius ρ of the iteration matrix; the smaller ρ, the more rapid will be the convergence. Therefore, in choosing an iterative method and correspondingly an iterative matrix, we shall be concerned with minimizing the spectral radius of the iterative matrix.

It is useful in the definition of various iterative methods and in the analysis of their properties, to divide the given matrix $\mathbf{A}$ into a diagonal matrix $\mathbf{D}$, an upper triangular matrix $\mathbf{U}$ and a lower triangular matrix $\mathbf{L}$,

$$\underset{\mathbf{A}}{\begin{bmatrix} x & x & x & x & \cdots & x \\ x & x & x & & & \\ x & x & x & & & \\ & & & \ddots & & \\ x & & & & x & x \\ x & & & & x & x \end{bmatrix}} \tag{4.91}$$

$$= \begin{bmatrix} x & & & & 0 \\ & x & & & \\ & & x & & \\ & & & \ddots & \\ 0 & & & & x \end{bmatrix} + \begin{bmatrix} 0 & x & x & x & \cdots & x \\ & 0 & x & x & \cdots & x \\ 0 & & 0 & x & & \\ & & & \ddots & & \\ 0 & & & & 0 & x \\ & & & & & 0 \end{bmatrix} + \begin{bmatrix} 0 & & & & & \\ x & 0 & & & 0 & \\ x & x & 0 & & & \\ x & x & x & 0 & & \\ \vdots & & & & \ddots & \\ x & x & & & x & 0 \end{bmatrix}$$

$$= \qquad \mathbf{D} \qquad + \qquad \mathbf{L} \qquad + \qquad \mathbf{U}$$

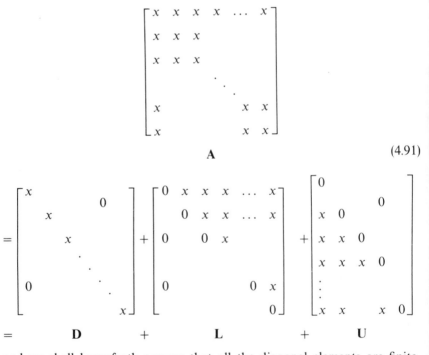

and we shall henceforth assume that all the diagonal elements are finite, so that we may always write the matrix equation 4.75 as,*

$$\mathbf{A}'\mathbf{u} = \mathbf{w}' \tag{4.92}$$

where,

$$\mathbf{A}' = \mathbf{D}^{-1}\mathbf{A}$$

$$\mathbf{A}' = \mathbf{I} + \mathbf{U}' + \mathbf{L}'$$

The matrix $\mathbf{B} = -(\mathbf{U}' + \mathbf{L}')$ has a specific significance and is known as the block Jacobi matrix. In particular it is clear that each of the eigenvalues μ_i of the block Jacobi matrix $\mathbf{B}$ are simply related to the eigenvalues λ_i of $\mathbf{A}'$ by, †

$$\mu_i = 1 - \lambda_i \tag{4.93}$$

To devise some simple iterative schemes, we shall imagine the evolution of the solution in an imaginary time,

$$\frac{\partial \mathbf{u}}{\partial t} = -\mathbf{A}'\mathbf{u} + \mathbf{w}' \tag{4.94}$$

so that the solutions at the new step $(p + 1)$ are contained explicitly over an iteration time step $\Delta t = \omega$,

$$\mathbf{u}^{(p+1)} = \mathbf{u}^{(p)} - \omega\mathbf{A}'\mathbf{u}^{(p)} + \omega\mathbf{w}'$$

$$\mathbf{u}^{(p+1)} = (\mathbf{I} - \omega\mathbf{A}')\mathbf{u}^{(p)} + \omega\mathbf{w}' \tag{4.95}$$

Such an iterative procedure will clearly be effective if the matrix $\mathbf{A}'$ is diagonally dominant, and if we choose an appropriate or best value of ω.

To illustrate the approach and as a model problem, we shall use the example of Poisson's equation in Cartesian coordinates on an equally spaced mesh. The resulting operative matrix has been illustrated in the Figure 4.2. For the five-point Laplacian operator in two dimensions, it is useful to adopt the notation of compass points (Figure 4.3) so that Poisson's equation may be written as:

$$u_C - \tfrac{1}{4}(u_N + u_S + u_E + u_W) = \tfrac{1}{4}\rho_C \Delta^2 = w_C \tag{4.96}$$

where the point C, the central point, is used to represent an ordered scanning over all the mesh-points. On a rectangular mesh of J rows and I columns, we know all the eigenvalues for this problem (Table 4.1),

$$1 - \frac{1}{2}\cos\frac{\pi}{J} - \frac{1}{2}\cos\frac{\pi}{I} \leqslant \lambda_i \leqslant 1 \tag{4.97}$$

*We shall be particularly concerned with matrices $\mathbf{A}$ which are diagonally dominant so that $\mathbf{D}$ is large (Varga, 1962).

† The relation $\mu_i = 1 - \lambda_i$ may easily be seen, by using a similarity transform.

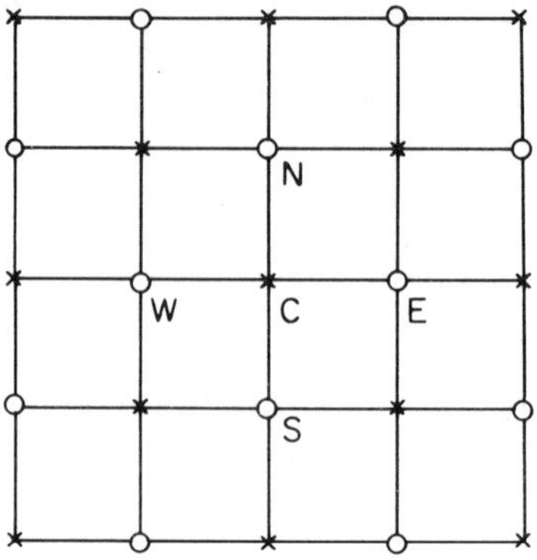

Figure 4.3. The notation of compass points used in two space dimensions to label adjacent points. In the Chebyshev method for the solution of Poisson's equation, solutions on the crossed points are obtained alternately with solutions on the dotted points

The model problem of Poisson's equation may be used to contrast methods, though in practice it can be solved directly.

For the general iterative method (equation 4.95) Poisson's equation (4.96) takes the form:

$$u_C^{(p+1)} = (1 - \omega)u_C^{(p)} + \tfrac{1}{4}\omega(u_N^{(p)} + u_S^{(p)} + u_E^{(p)} + u_W^{(p)}) + \omega w_C \qquad (4.98)$$

In effect, this is the explicit form of the diffusion equation, with a 'time step' $\frac{1}{4}\omega$, a unit space step and unit conductivity (Section 3.3), and in two dimensions the requirement for stability (and thus convergence here) is:

$$\tfrac{1}{4}\omega \leqslant \tfrac{1}{4}$$

Thus certainly, for the convergence of such an iterative scheme, the parameter ω must satisfy the condition:

$$0 \leqslant \omega \leqslant 1 \qquad (4.99)$$

(*b*) *The Jacobi Method.* In the special case where $\omega = 1$, the iterative algorithm (4.95) is known as Jacobi iteration,

$$\mathbf{u}^{(p+1)} = (\mathbf{I} - \mathbf{A}')\mathbf{u}^{(p)} + \mathbf{w}' \qquad (4.100)$$

Since the matrix $\mathbf{A}'$ may be written as the sum of the unit matrix and upper and lower triangular matrices (equation 4.92), Jacobi iteration is of the form,

$$\mathbf{u}^{(p+1)} = -(\mathbf{L}' + \mathbf{U}')\mathbf{u}^{(p)} + \mathbf{w}' \qquad (4.101)$$

We may view the Jacobi method, therefore, as the equating of the new diagonal terms to the old off-diagonal terms, and the iterative matrix **P** is simply the block Jacobi matrix $\mathbf{B} = -(\mathbf{L}' + \mathbf{U}')$. According to the asymptotic limit of the decay of the error (equation 4.89), the rate of convergence of the method is

$$\lim_{p \to \infty} \delta\boldsymbol{\epsilon}^{(p)} = \boldsymbol{\epsilon}^{(p+1)} - \boldsymbol{\epsilon}^{(p)}$$

$$= (\rho_m - 1)\boldsymbol{\epsilon}^{(p)}$$

$$\frac{|\delta\boldsymbol{\epsilon}^{(p)}|}{|\boldsymbol{\epsilon}^{(p)}|} = |\rho_m - 1| \tag{4.102}$$

where ρ_m is the eigenvalue of largest modulus of the iteration matrix **P**, or in this case the maximum eigenvalue of the block Jacobi matrix **B**, namely μ_m (equation 4.93).

For the model problem of Poisson's equation solved by the Jacobi method, the maximum eigenvalue of the block Jacobi matrix is (equation 4.97):

$$\mu_m = \frac{1}{2}\cos\frac{\pi}{J} + \frac{1}{2}\cos\frac{\pi}{I} \tag{4.103}$$

or, for large J and I,

$$\mu_m \sim 1 - \frac{\pi^2}{4J^2} - \frac{\pi^2}{4I^2} \tag{4.104}$$

so that the rate of convergence (equation 4.102) is very small,

$$\frac{|\delta\boldsymbol{\epsilon}^{(p)}|}{|\boldsymbol{\epsilon}^{(p)}|} = \frac{\pi^2}{4J^2} + \frac{\pi^2}{4I^2} \tag{4.105}$$

(*c*) *Gauss–Seidel Iteration.* In each iteration step, each component of the vector $\mathbf{u}^{(p+1)}$ is obtained in sequence. For example, in Poisson's equation (4.98), when the scan reaches the point C, the new iteration values $u_S^{(p+1)}$ (at the 'south' point) and $u_W^{(p+1)}$ (at the 'west' point) have already been determined. Since they are an improvement of the old values $u_S^{(p)}$ and $u_W^{(p)}$, to enhance convergence it is reasonable to use the new values in the iteration procedure. If, again, the parameter ω is taken equal to unity,

$$u_C^{(p+1)} = \tfrac{1}{4}(u_N^{(p)} + u_S^{(p+1)} + u_E^{(p)} + u_W^{(p+1)}) + w_C' \tag{4.106}$$

the method is known as the Gauss–Seidel method. In the general terms of the matrix equation (4.92), the Gauss–Seidel method may be written as

$$(\mathbf{I} + \mathbf{L}')\mathbf{u}^{(p+1)} = -\mathbf{U}'\mathbf{u}^{(p)} + \mathbf{w}' \tag{4.107}$$

or,

$$\mathbf{u}^{(p+1)} = -(\mathbf{I} + \mathbf{L}')^{-1}\mathbf{U}'\mathbf{u}^{(p)} + (\mathbf{I} + \mathbf{L}')^{-1}\mathbf{w}' \tag{4.108}$$

since, at any point in the iteration step $p + 1$, the new variables $\mathbf{L}'\mathbf{u}^{(p+1)}$ are known. It can be shown that the rate of convergence of the Gauss–Seidel method is greater than that of the Jacobi method, though only marginally so (Varga 1962).

(*d*) *Successive Over-relaxation.* For convergence in the iterative solution of Poisson's equation (4.98), it has been shown that the parameter ω must lie between zero and one for the Jacobi method. In the Gauss–Seidel method, ω was taken as unity though in practice the fastest rate of convergence is not necessarily achieved with ω equal to unity and is frequently increased by taking ω greater than unity, when the method is said to be one of over-relaxation. The contrary case of $\omega < 1$ is said to be a method of successive under-relaxation.

The method of successive over-relaxation uses the Gauss–Seidel formulation, but with an optimized constant value of the relaxation parameter ω,

$$(\mathbf{I} + \omega\mathbf{L}')\mathbf{u}^{(p+1)} = (1 - \omega)\mathbf{I}\mathbf{u}^{(p)} - \omega\mathbf{U}'\mathbf{u}^{(p)} + \omega\mathbf{w}' \tag{4.109}$$

or,

$$\mathbf{u}^{(p+1)} = (\mathbf{I} + \omega\mathbf{L}')^{-1}\{(1 - \omega)\mathbf{I} - \omega\mathbf{U}'\}\mathbf{u}^{(p)} + (\mathbf{I} + \omega\mathbf{L}')^{-1}\omega\mathbf{w}' \tag{4.110}$$

For the most rapid convergence, the optimized value of the relaxation factor, ω_b is (Young 1962, Varga 1962):

$$\omega_b = \frac{2}{1 + \sqrt{(1 - \mu_m^2)}} \tag{4.111}$$

where μ_m as before is the spectral radius, or eigenvalue of largest modulus, of the block Jacobi matrix $\mathbf{B} = -(\mathbf{L}' + \mathbf{U}')$. In addition, the asymptotic rate of convergence of the successive over-relaxation method (equation 4.110) is:

$$\rho = \frac{|\boldsymbol{\epsilon}^{(p+1)}|}{|\boldsymbol{\epsilon}^{(p)}|} = \omega_b^{-1}$$

or,

$$\frac{|\delta\boldsymbol{\epsilon}^{(p)}|}{|\boldsymbol{\epsilon}^{(p)}|} = \sqrt{(1 - \mu_m^2)} \tag{4.112}$$

For example, we consider the model problem of Poisson's equation where, with the notation as before, the iteration formula is the simple algorithm:

$$u_C^{(p+1)} = (1 - \omega)u_C^{(p)} + \tfrac{1}{4}\omega(u_N^{(p)} + u_S^{(p+1)} + u_E^{(p)} + u_W^{(p+1)}) + \omega w_C \tag{4.113}$$

The eigenvalues of the block Jacobi matrix are (equations 4.93, 4.97)

$$0 \leqslant \mu_i \leqslant \frac{1}{2}\cos\frac{\pi}{J} + \frac{1}{2}\cos\frac{\pi}{I}$$

where J and I are the numbers of rows and columns on the mesh. Thus,

$$\omega_b = \frac{2}{1 + \sqrt{\{1 - (\frac{1}{2}\cos \pi/J + \frac{1}{2}\cos \pi/I)^2\}}} \tag{4.114}$$

and, for large I and J,

$$\omega_b \simeq \frac{2}{1 + \pi\sqrt{\left(\dfrac{1}{2J^2} + \dfrac{1}{2I^2}\right)}}$$

Thus, the larger the mesh the more closely ω_b approaches the limiting value, 2. Similarly the asymptotic rate of convergence in the model problem (equation 4.112) is:

$$\frac{|\delta\epsilon^{(p)}|}{|\epsilon^{(p)}|} \simeq \pi\left(\frac{1}{2J^2} + \frac{1}{2I^2}\right)^{\frac{1}{2}} \tag{4.115}$$

This is a considerably enhanced rate of convergence compared to the Gauss–Seidel or Jacobi methods (equation 4.105), varying inversely as the number of mesh points rather than the square of the mesh points.

We may introduce a variation on the point successive over-relaxation method, by noticing that in the matrix equations of interest, and particularly in those derived from difference formulations, there is a symmetry over alternate points. For example, in Poisson's equation, each point C is coupled to its four neighbouring points (N, S, E, W; Figure 4.3); that is, the variables at the even points are directly coupled only to the variables at the odd points and vice versa. It is useful, therefore, to define a double or cyclic iteration where all the variables on the even points are improved first and are then used to improve all the variables on the odd mesh. We redefine the vector of unknown variables $\mathbf{u}$, as two vectors $\mathbf{u}_0$ and $\mathbf{u}_1$ where the first vector is defined on the even points and the second on the odd points. Similarly, the known vector $\mathbf{w}$ is split as $\mathbf{w}_0$ and $\mathbf{w}_1$. Then the cyclic successive over-relaxation method may be defined as:

over successive steps p, $\nu = 0, 1$:

$$\mathbf{u}_\nu^{(p+1)} = -\omega(\mathbf{L}' + \mathbf{U}')\mathbf{u}_{1-\nu}^{(p)} + (1 - \omega)\mathbf{u}_\nu^{(p-1)} + \omega\mathbf{w}'_\nu \tag{4.116}$$

The optimum value of the relaxation factor $\omega = \omega_b$ is as before (equation 4.111).

(*e*) *Cyclic Chebyshev Method.* In the method of successive over-relaxation, while the asymptotic rate of convergence is extremely rapid (equation 4.115), during the early stages of the iteration the convergence may be slow and, indeed, it is possible for errors to grow temporarily. In the cyclic Chebyshev method, convergence during the early stages of the iteration is considerably improved by varying the relaxation parameter from step to step. The idea has arisen from the concept of using Chebyshev polynomials to accelerate convergence (Varga 1962, Golub and Varga 1961). The method is the same as that used in cyclic successive over-relaxation (equation 4.116), but now a variable relaxation parameter ω is used:

over successive steps,

$$\mathbf{u}_v^{(p+1)} = -\omega_p(\mathbf{L}' + \mathbf{U}')\mathbf{u}_{1-v}^{(p)} + (1 - \omega_p)\mathbf{u}_v^{(p-1)} + \omega_p \mathbf{w}'_v \tag{4.117}$$

where,

$$v = \begin{cases} 0, & \text{if } p \text{ is even} \\ 1, & \text{if } p \text{ is odd} \end{cases}$$

Over successive steps, the parameter ω_p is varied to optimize the average rate of convergence,

$$\begin{aligned} \omega_0 &= 1 \\ \omega_1 &= 1/(1 - \tfrac{1}{2}\mu_m^2) \\ \text{for } p \geqslant 1: \quad \omega_{p+1} &= 1/(1 - \tfrac{1}{4}\mu_m^2\omega_p) \end{aligned} \tag{4.118}$$

where, as before, μ_m is the eigenvalue of largest modulus of the block Jacobi matrix, $-(\mathbf{L}' + \mathbf{U}')$, defined over the whole vector. It is to be noted that the first step corresponds to the Gauss–Seidel method ($\omega = 1$) and, thereafter, the relaxation factor ω gradually increases. In the asymptotic limit, the relaxation factor becomes identical with the optimum value for successive over-relaxation ($\omega_\infty = \omega_b$; equation 4.111), so that the asymptotic properties of the Chebyshev method are identical with those of successive over-relaxation. However, in the early stages of the iteration, convergence is considerably

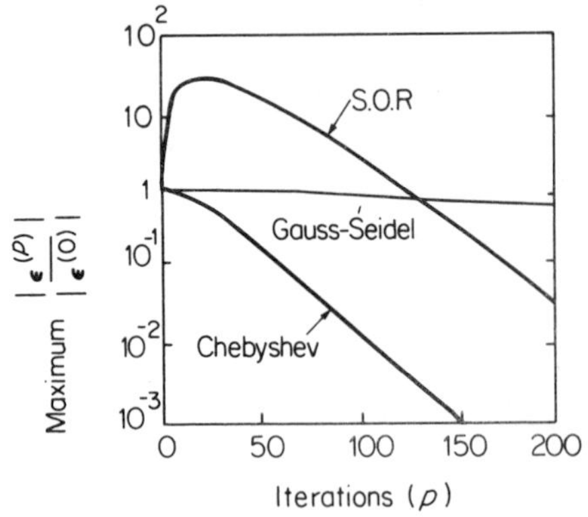

Figure 4.4. Convergence rates for the iterative solution of Poisson's equation in two dimensions (from Hockney, R. W., 'The Potential Calculation and Some Applications' in *Methods in Computational Physics*, **9**, Alder, Fernbach and Rotenberg (eds.), Academic Press, by permission)

improved. The rates of convergence for the Gauss–Seidel, Chebyshev and successive over-relaxation methods are illustrated in Figure 4.4 for the model problem of Poisson's equation on a 128×128-point mesh (Hockney 1970).

(*f*) *Alternating-direction Implicit Method.* We have seen that, where matrix equations arise from the difference formulation of one-dimensional boundary-value problems or implicitly formulated parabolic equations, the operator is tridiagonal (Section 4.2a). While tridiagonal matrix equations are particularly simple to resolve as illustrated in Section 4.3, in the analogous multidimensional problems the matrix coupling is considerably more complex (Section 4.2b) and, apart from the simplest example of Poisson's equation, the matrix equation is difficult to resolve. Nevertheless we may usefully employ the properties of tridiagonal matrices in such multidimensional problems by noticing that the coupling matrix $\mathbf{A}'$ (equation 4.92) may be written as the sum of tridiagonal matrices.

We shall take the example of a problem in two space dimensions on the difference mesh of J rows and I columns, though the method applies equally to any number of space dimensions. As discussed in Section 2b, a vector $\mathbf{u}$ may be defined, with components from every point of the space mesh, and the components may be ordered by counting along each column in succession or along each row in succession:

$$\begin{aligned} \mathbf{u} &= \{u_{iJ+j}\} = \{u_{ij}\} \\ \mathbf{v} &= \{u_{jI+i}\} = \{u_{ij}\} \end{aligned} \tag{4.119}$$

The vectors $\mathbf{u}$ and $\mathbf{v}$ are essentially equivalent, except in so far as their elements are ordered according to rows or according columns, which we may represent by the reordering matrix $\mathbf{N}$,

$$\mathbf{v} = \mathbf{N}\mathbf{u}$$

Then, we may rewrite the matrix $\mathbf{A}'$ in the coupling equation,

$$\mathbf{A}'\mathbf{u} = \mathbf{w}'$$

as the sum of two tridiagonal matrices,

$$(\mathbf{C} + \mathbf{R}\mathbf{N})\mathbf{u} = \mathbf{w}'$$

or,

$$\mathbf{C}\mathbf{u} + \mathbf{R}\mathbf{v} = \mathbf{w}' \tag{4.120}$$

where $\mathbf{C}$ is a tridiagonal matrix acting along the columns of the mesh, and $\mathbf{R}$ is a tridiagonal matrix acting along the rows of the mesh. The difference form of Poisson's equation in two dimensions is such an example (Figure 4.1). In practice, equation (4.120) is simpler since both $\mathbf{C}$ and $\mathbf{R}$ may be partitioned into independent tridiagonal matrices which act separately on each column and each row.

Since we may solve tridiagonal matrix equations directly and simply, it is evident that an effective iterative method may be devised by solving the tridiagonal equations separately along the columns and then along the rows, namely,

$$(\mathbf{C} + \omega\mathbf{I})\mathbf{u}^{(p+\frac{1}{2})} = (\omega\mathbf{N}^{-1} - \mathbf{R})\mathbf{v}^{(p)} + \mathbf{w}'$$

$$(\mathbf{R} + \omega\mathbf{I})\mathbf{v}^{(p+1)} = (\omega\mathbf{N} - \mathbf{C})\mathbf{u}^{(p+\frac{1}{2})} + \mathbf{w}' \qquad (4.121)$$

where ω is to be chosen to optimize the iteration. Hence, we solve implicitly for the tridiagonal equations, along alternating directions of the mesh (Peaceman and Rachford 1955, Varga 1962).

To examine the rate of convergence of the method, and the choice of the relaxation parameter ω, it is clear (from equation 4.121) that the errors over the double step are related by,

$$\boldsymbol{\epsilon}^{(p+1)} = (\mathbf{R} + \omega\,\mathbf{I})^{-1}(\omega\mathbf{N} - \mathbf{C})(\mathbf{C} + \omega\mathbf{I})^{-1}(\omega\mathbf{N}^{-1} - \mathbf{R})\boldsymbol{\epsilon}^{(p)}$$

$$\boldsymbol{\epsilon}^{(p+1)} = \mathbf{P}\boldsymbol{\epsilon}^{(p)} \qquad (4.122)$$

As before, to optimize the convergence we wish to minimize the largest eigenvalue of the iterative matrix **P**. The matrix **N** is simply a reordering operator (equation 4.119) and has unit eigenvalues and, by considering a similarity transform, it is clear that the eigenvalues ρ of **P** are related to the eigenvalues r of **R** and c of **C**,

$$\rho = \frac{\omega - c}{\omega + c} \cdot \frac{\omega - r}{\omega + r} \qquad (4.123)$$

If λ_1 and λ_2 are the minimum and maximum eigenvalues of **A**, or **C** + **R**, then the optimum choice of the relaxation factor ω to minimize $|\rho_m|$, the spectral radius of the iteration operator **P**, is (Varga 1962),

$$\omega_b = \sqrt{(\lambda_1 \lambda_2)} \qquad (4.124)$$

For example, in the model problem of Poisson's equation, if $I = J$, then,

$$\omega_b \sim \frac{\pi}{J} \qquad (4.125)$$

for J large. With this optimum choice of relaxation factor, it may be verified that the asymptotic rate of convergence is:

$$\frac{|\delta\boldsymbol{\epsilon}^{(p)}|}{|\boldsymbol{\epsilon}^{(p)}|} \sim \frac{\pi}{J} \qquad (4.126)$$

and indeed the asymptotic rate of convergence is precisely the same as that for the method of successive over-relaxation or the cyclic Chebyshev method.

The methods discussed in this section are summarized in Table 4.3.

Table 4.3 Iterative methods for the solution of the matrix equation $\mathbf{Au} = \mathbf{w}$

$\mathbf{A}$ is taken to be of the form $\mathbf{A} = \mathbf{I} + \mathbf{L} + \mathbf{U}$
$\mathbf{B}$ is the block Jacobi matrix $\mathbf{B} = -(\mathbf{L} + \mathbf{U})$

p = iteration step
μ = eigenvalue of maximum modulus of $\mathbf{B}$
ω = relaxation factor

Method	Iteration formula	Optimum relaxation factor	Asymptotic rate of convergence	Ex: Poisson's Equation on $J \times J$ points: asymptotic error
1. Jacobi	$\mathbf{u}^{(p+1)} = -(\mathbf{L} + \mathbf{U})\mathbf{u}^{(p)} + \mathbf{w}$	none	$1 - \mu$	$\frac{\lvert\epsilon^{(p)}\rvert}{\lvert\epsilon^{(0)}\rvert} \sim \exp\left(-\frac{\pi^2}{2J^2}p\right)$
2. Gauss–Seidel	$\mathbf{u}^{(p+1)} = -(\mathbf{I} + \mathbf{L})^{-1}\mathbf{U}\mathbf{u}^{(p)} + (\mathbf{I} + \mathbf{L})^{-1}\mathbf{w}$	none	$1 - \mu^2$	$\frac{\lvert\epsilon^{(p)}\rvert}{\lvert\epsilon^{(0)}\rvert} \sim \exp\left(-\frac{\pi^2}{J^2}p\right)$
3. Successive over-relaxation	$\mathbf{u}^{(p+1)} = (\mathbf{I} + \omega\mathbf{L})^{-1}\{(1 - \omega)\mathbf{I} - \omega\mathbf{U}\}\mathbf{u}^{(p)} + (\mathbf{I} + \omega\mathbf{L})^{-1}\mathbf{w}$	$\omega_b = \frac{2}{1 + 1 - \mu^2}$	$\omega_b - 1 = \sqrt{(1 - \mu^2)}$	$\frac{\lvert\epsilon^{(p)}\rvert}{\lvert\epsilon^{(0)}\rvert} \sim \exp\left(-\frac{\pi}{J}p\right)$
4. Chebyshev	$\mathbf{u}^{(p+1)} = (\mathbf{I} + \omega_{2p}\mathbf{L})^{-1}\{(1 - \omega_{2p})\mathbf{I} - \omega_{2p}\mathbf{U}\}\mathbf{u}^{(p)} + (\mathbf{I} + \omega_{2p}\mathbf{L})^{-1}\mathbf{w}$	$\omega_0 = 1$ $\omega_1 = 1/(1 - \frac{1}{2}\mu^2)$ $\omega_p = 1/(1 - \frac{1}{4}\mu^2\omega_{p-1})$ $\omega_\infty = \frac{2}{1 + \sqrt{(1 - \mu^2)}}$	$\omega_\infty - 1 = \sqrt{(1 - \mu^2)}$	$\frac{\lvert\epsilon^{(p)}\rvert}{\lvert\epsilon^{(0)}\rvert} \sim \exp\left(-\frac{\pi}{J}p\right)$ Early iterations converge more rapidly than in S.O.R.
5. Alternating-direction implicit	$\mathbf{A} = \mathbf{C} + \mathbf{RN}$, where $\mathbf{v} = \mathbf{Nu}$ $(\mathbf{C} + \omega\mathbf{I})\mathbf{u}^{(p+\frac{1}{2})} = (\omega\mathbf{N}^{-1} - \mathbf{R})\mathbf{v}^{(p)} + \mathbf{w}$ $(\mathbf{R} + \omega\mathbf{I})\mathbf{v}^{(p+1)} = (\omega\mathbf{N} - \mathbf{C})\mathbf{u}^{(p+\frac{1}{2})} + \mathbf{w}$	$\omega = \sqrt{(\lambda_1\lambda_2)}$ λ_1, λ_2 = minimum, maximum eigenvalues of $\mathbf{A}$	$1 - \lambda_1^2\lambda_2^2$	$\frac{\lvert\epsilon^{(p)}\rvert}{\lvert\epsilon^{(0)}\rvert} \sim \exp\left(-\frac{\pi}{J}p\right)$

7 Two Inexact Methods for the Determination of Eigenvectors and Eigenvalues

The $n \times n$ matrix $\mathbf{A}$ has n eigenvalues λ_i, each associated with an eigenvector $\mathbf{s}^{(i)}$,

$$\mathbf{A}\mathbf{s}^{(i)} = \lambda_i \mathbf{s}^{(i)} \tag{4.127}$$

It is assumed that the matrix is nondegenerate: the eigenvectors are independent and there are n distinct eigenvalues. The eigenvalues λ_i satisfy the determinant equation,

$$|\mathbf{A} - \lambda \mathbf{I}| = 0 \tag{4.128}$$

and, to determine the eigenvalues, this determinant could be expanded to provide an nth-order polynomial in λ. Such an approach will, however, involve a very large number of arithmetic operations and, even when the expansion is completed, we are left with the non-trivial problem of solving a very high-order polynomial equation. Direct transformation procedures can be adopted (Wilkinson 1964). These are complex, however, and we restrict ourselves here to two very simple iterative procedures.

(*a*) *The Maximum Eigenvalue.* We consider the arbitrary vector $\mathbf{v}^{(0)}$, which contains at least a small amplitude α_i of each eigenvector $\mathbf{s}^{(i)}$ of $\mathbf{A}$. We may expand $\mathbf{v}^{(0)}$,

$$\mathbf{v}^{(0)} = \sum_{i=1}^{n} \alpha_i \mathbf{s}^{(i)} \tag{4.129}$$

We shall operate the matrix $\mathbf{A}$ directly on $\mathbf{v}$ a number of times p.

$$\begin{aligned} \mathbf{v}^{(p)} &= \mathbf{A}\mathbf{v}^{(p-1)} \\ &= \mathbf{A}^p \mathbf{v}^{(0)} \end{aligned} \tag{4.130}$$

To consider the effect of this operation, we use the expansion (equation 4.129)

$$\begin{aligned} \mathbf{v}^{(p)} &= \mathbf{A}^p \sum_{i=1}^{n} \alpha_i \mathbf{s}^{(i)} \\ &= \sum_{i=1}^{n} \alpha_i \mathbf{A}^p \mathbf{s}^{(i)} \\ &= \sum_{i=1}^{n} \alpha_i \lambda_i^p \mathbf{s}^{(i)} \end{aligned} \tag{4.131}$$

Thus after p operations, the vector $\mathbf{v}^{(p)}$ has an amplitude along each eigenvector which depends on the pth power of the corresponding eigenvalue λ_i. It follows that, if λ_m is the eigenvalue of greatest modulus, after a large number of operations p, $\mathbf{v}^{(p)}$ consists almost entirely of the eigenvector $\mathbf{s}^{(m)}$.

Hence the eigenvector corresponding to the maximum eigenvalue is very easily obtained. In the actual practice of this method, of course, to avoid $\mathbf{v}^{(p)}$ becoming very large or very small, we renormalize $\mathbf{v}^{(p)}$ at each step p. For p large, the growth of any component will then just be the eigenvalue λ_m.

(*b*) *Determination of All Eigenvalues and Eigenvectors.* The foregoing method is clearly of the utmost simplicity. It does not, however, provide a general method of determining any but the eigenvalue (and the associated eigenvector) of maximum modulus.

Any or all of the eigenvalues and eigenvectors may be determined by inverse iteration. Again choosing some arbitrary vector $\mathbf{v}^{(0)}$, we now *solve* for $\mathbf{v}^{(p)}$ at the pth step:

$$(\mathbf{A} - \mathbf{I}k)\mathbf{v}^{(p)} = \mathbf{v}^{(p-1)} \tag{4.132}$$

where the number k is chosen to provide the particular eigenvalue and eigenvector in which we are interested. From formula 4.132,

$$\begin{aligned}\mathbf{v}^{(p)} &= (\mathbf{A} - \mathbf{I}k)^{-1}\mathbf{v}^{(p)} \\ &= (\mathbf{A} - \mathbf{I}k)^{-p}\mathbf{v}^{(0)}\end{aligned} \tag{4.133}$$

To observe the effect of this operation, we shall again expand the initial vector $\mathbf{v}^{(0)}$ in terms of the eigenvectors of $\mathbf{A}$ (equation 4.129),

$$\begin{aligned}\mathbf{v}^{(p)} &= (\mathbf{A} - \mathbf{I}k)^{-p}\sum_{i=1}^{n}\alpha_i\mathbf{s}^{(i)} \\ &= \sum_{i=1}^{n}\alpha_i(\mathbf{A} - \mathbf{I}k)^{-p}\mathbf{s}^{(i)} \\ &= \sum_{i=1}^{n}\alpha_i(\lambda_i - k)^{-p}\mathbf{s}^{(i)}\end{aligned} \tag{4.134}$$

Clearly, when p is large, $\mathbf{v}^{(p)}$ consists almost entirely of the eigenvector $\mathbf{s}^{(m)}$, whose eigenvalue gives the smallest modulus to $\lambda_m - k$,

$$|\lambda_m - k| < |\lambda_i - k| \quad (\text{all } i \neq m) \tag{4.135}$$

By choosing different values of k, the condition (equation 4.135) can be applied to each eigenvalue in turn. The method is optimized when k is a good guess for a particular eigenvalue λ_m.

To solve the successive matrix equations (equation 4.132), an exact method such as Gauss elimination may be applied.

CHAPTER V

Particles: *N*-body Action at a Distance

1 Particles and Assemblies of Particles

The material world is constructed of an aggregation of particles, and the behaviour of any macroscopic matter is dependent on the microscopic interactions between the constitutive particles. To determine the evolution or structure of a macroscopic system, therefore, the fundamental approach is to construct or synthesize the system as an assembly of particles. We may certainly formulate the description of such an assembly quantitatively. In the first place, the laws governing the motion of individual particles, whether they are classical, quantum, or relativistic, are well defined and equally the force laws operative between many types of particles are well established. But, although the problem may be formulated for assemblies of particles, explicit solutions of the motion prove impossible.

We may classify assemblies of particles into few-body and many-body problems, where the approach in each case is quite different. To illustrate the class of problems involved, one might include as examples of the few-body problem, the atom, the nucleus, the molecule, the solar system, star clusters. Each of these systems is an assembly of a small number of particles and, in computational mechanics, we might expect to resolve the structure and internal motion of each system exactly.

On the other hand, in what we may describe as many-body problems, the system of interest is an assembly of so many particles that it is inconceivable that the complete internal motion may be resolved exactly. Examples of such many-body problems are numerous and diverse and include liquids, gases, stars, galaxies, plasmas, classical fluids, the electrons in an ion lattice, ion lattices and super-fluids. Typically for a laboratory liquid, for example, 10^{23} particles are involved. Nevertheless, in assemblies of many particles, the macroscopic properties of the whole system may be rather simple and dependent on average properties of the particles. While statistical physics is of great importance in such problems, it is frequently the case that problems become nonlinear and intractable. These systems are, however, amenable to the application of computational mechanics, since we may infer macroscopic properties by studying the *self-consistent* motion of 'typical' particles.

In this chapter, we shall first examine the motion of single particles in external fields. The direct and exact solution of few-particle assemblies are then considered where, by the exact solution, we mean that the system is described by the direct application and resolution of the fundamental laws of motion to the separate particles interacting through two-particle force laws.

2 Single-Particle Motion in Conservative Fields

Two forces of fundamental importance are the electric and gravitational, both of which are long-range and conservative. We may describe the motion of a single particle under the action of a given externally applied field **E** and, since the field is conservative, it is derivable from a scalar potential Φ,

$$\mathbf{E} = -\nabla\Phi \tag{5.1}$$

The 'state' of a point particle of mass m is defined by six coordinates: the vector positions $\mathbf{x} = (x, y, z)$ and the vector velocity $\mathbf{v} = (v_x, v_y, v_z)$. The coordinates of the particles satisfy the equations of motion:

$$\frac{\mathrm{d}\mathbf{x}}{\mathrm{d}t} = \mathbf{v}$$

$$\frac{\mathrm{d}\mathbf{v}}{\mathrm{d}t} = -\frac{e}{m}\nabla\Phi \tag{5.2}$$

where e is the charge of the particle in the case of the electric field, and $e = m$ is the mass of the particle in the case of a gravitational field. It follows that the state of the system is described by the six dependent variables $\mathbf{u} = (\mathbf{x}, \mathbf{v})$. Since the potential Φ is a function of the position coordinates $\mathbf{x}$ and time t only, the right-hand sides of each of the equations 5.2 are independent of their own dependent variable or, more specifically, each of the equations for the coordinates $\mathbf{x}$ depend only on the velocity, while each of the equations for the velocity coordinates are dependent only on the position coordinates. Hence, for this system of equations, the leapfrog difference scheme (Section 2.6b) is particularly appropriate, since we need only define the positions and velocities at alternate times. If the velocities are defined at the time step $n-1$, we determine the new position coordinates only at time step n,

$$\mathbf{x}^n = \mathbf{x}^{n-2} + 2\,\Delta t\mathbf{v}^{n-1} \tag{5.3}$$

and the velocities are only again determined at step $n+1$,

$$\mathbf{v}^{n+1} = \mathbf{v}^{n-1} - 2\,\Delta t\frac{e}{m}\nabla\Phi(\mathbf{x}^n, t^n) \tag{5.4}$$

The leapfrog scheme is particularly simple and elegant, but in the general case, when each dependent variable must be defined at every time step, an arbitrary computational mode is introduced, leading to the decoupling of alternate time points (Section 2.6b). Clearly, no such difficulty arises in this instance, since we do not have to define each variable at every time step.

The motion of a particle in a conservative field of force exhibits the two particularly important properties of time reversibility and conservation of energy. It is important that we examine any particle integration scheme to ensure time reversibility and, at least to a high order of accuracy, the con-conservation of energy. It is clear here, that the difference scheme is time reversible, for if $\Delta t' = -\Delta t$, then

$$\mathbf{v}^{n-1} = \mathbf{v}^{n+1} - 2\,\Delta t' \frac{e}{m} \nabla\Phi(\mathbf{x}^n, t^n)$$

$$\mathbf{x}^{n-2} = \mathbf{x}^n + 2\,\Delta t'\,\mathbf{v}^{n-1} \tag{5.5}$$

and these equations are exactly analogous to the forward time equations (5.3, 5.4). It is evident that we require this property of a particle integration scheme since, if it were violated, an assembly of such particles would generate entropy numerically.

The second consideration, of the conservation of energy, may be examined by rewriting the equation for the velocities (equation 5.4),

$$\mathbf{v}^{n+1} - \Delta t \frac{e}{m} \mathbf{E}^n = \mathbf{v}^{n-1} + \Delta t \frac{e}{m} \mathbf{E}^n \tag{5.6}$$

We take the square of this equation,

$$(v^{n+1})^2 - 2\,\Delta t \frac{e}{m} \mathbf{v}^{n+1} . \mathbf{E}^n + \Delta t^2 \frac{e^2}{m^2} (E^n)^2$$

$$= (v^{n-1})^2 + 2\,\Delta t \frac{e}{m} \mathbf{v}^{n-1} . \mathbf{E}^n + \Delta t^2 \frac{e^2}{m^2} (E^n)^2 \tag{5.7}$$

Thus the difference in the kinetic energy over two time levels is obtained,

$$\tfrac{1}{2}m(v^{n+1})^2 - \tfrac{1}{2}m(v^{n-1})^2 = e(\mathbf{v}^{n+1} + \mathbf{v}^{n-1}) . \mathbf{E}^n\,\Delta t \tag{5.8}$$

The leapfrog difference equation (5.3) for the position coordinates over two time levels is used to replace the velocities on the right-hand side,

$$\tfrac{1}{2}m(v^{n+1})^2 - \tfrac{1}{2}m(v^{n-1})^2 = \tfrac{1}{2}e(\mathbf{x}^{n+2} - \mathbf{x}^{n-2}) . \mathbf{E}^n \tag{5.9}$$

The right-hand side of this equation represents the negative of the change in potential energy of particle between the two time-levels t^{n+1} and t^{n-1} but, since we cannot express the change in the potential energy as the difference between two potentials, the energy of the particle is not identically conserved.

Nevertheless the right-hand side (equation 5.9) is a difference approximation to the integral $\int_{n-1}^{n+1} \mathbf{E}.\,\mathbf{dx}$ to second-order accuracy. Thus, while the energy is not identically conserved, errors in energy conservation are very small and there exists no bias since the scheme is reversible.

To represent the state of the one-particle system, we define the six-dimensional space whose coordinates are the position and velocity coordinates of the particle. At any time t, the state of the system is represented as a point in this six-dimensional phase space. Equally, we represent the evolution of the system in time as a line in phase space, which connects all intermediate states (Figure 5.1).

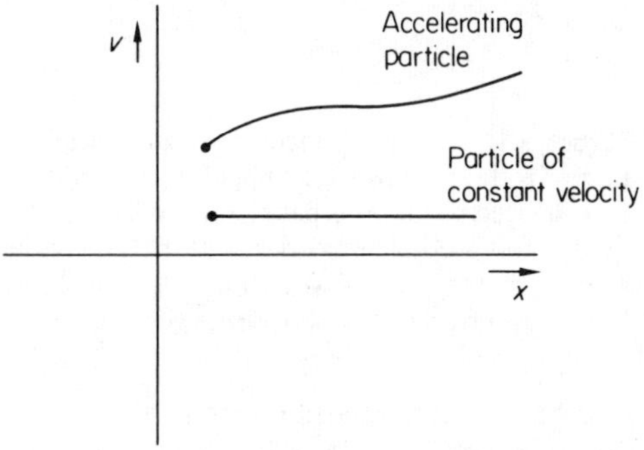

Figure 5.1. The representation of the motion of a particle as a trajectory in phase space

3 Single-Particle Motion in a Perpendicular Magnetic Field

A particle of mass m and charge e in a magnetic field $\mathbf{B}$ moves according to the Lorentz force, so that the equations of motion of the particle are

$$\frac{\mathrm{d}\mathbf{x}}{\mathrm{d}t} = \mathbf{v}$$

$$\frac{\mathrm{d}\mathbf{v}}{\mathrm{d}t} = \frac{e}{mc}\mathbf{v} \wedge \mathbf{B} \tag{5.10}$$

where c is the velocity of light. The simplicity of the leapfrog method, when applied to the difference description of a particle in a conservative field, suggests that the same technique may be applied to a particle in a magnetic field. However, the right-hand sides of the momentum equations of the particle are functions of the velocity of the particle and, in this situation, it is no longer sufficient to define the position and velocity coordinates of the particle alternately in time. Necessarily, therefore, the additional computational mode leads to the creation of different orbiting centres of the particle on even and odd time steps, with the consequence that the particle zig-zags between the circles (Figure 5.2).

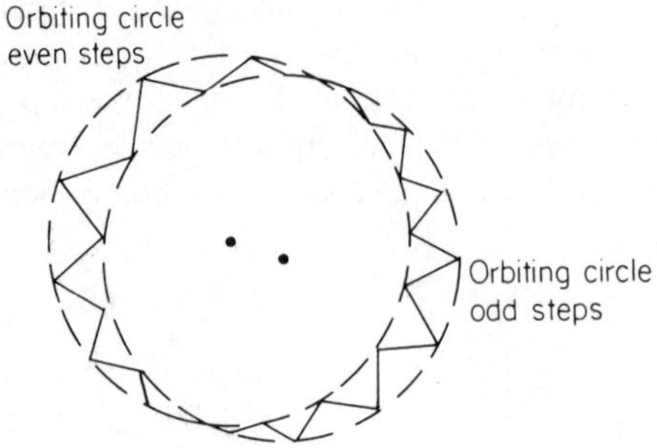

Figure 5.2. The occurrence of a computational mode in the leapfrog scheme when the motion of a charged particle perpendicular to a magnetic field is followed. The positions at the even time steps and the positions at the odd time steps occur on different orbiting centres

To obtain satisfactory difference equations in this case, the second-order accurate implicit method (Section 2.6d) may be applied and the resulting algebraic equations can be solved directly. It is sufficient to consider the two-dimensional problem where the plane of motion, $\mathbf{v} = (v_x, v_y, 0)$, is perpendicular to the magnetic field, $\mathbf{B} = (0, 0, B_z)$. We define the local cyclotron frequency,

$$\Omega = \frac{eB_z}{mc} \tag{5.11}$$

where, in general, $\Omega = \Omega(\mathbf{x}, t)$. By using complex variables for the velocity and position coordinates,

$$\hat{v} = v_x + \mathrm{i}v_y, \qquad \hat{x} = x + \mathrm{i}y \tag{5.12}$$

the notation for the equations of motion (equations 5.10) is simplified,

$$\frac{\mathrm{d}\hat{x}}{\mathrm{d}t} = \hat{v}, \qquad \frac{\mathrm{d}\hat{v}}{\mathrm{d}t} = -\mathrm{i}\hat{v}\Omega \tag{5.13}$$

We integrate the velocity equations between the time levels t^n and $t^{n+1} = t^n + \Delta t$ implicitly, according to the algorithm (equation 2.95):

$$\hat{v}^{n+1} - \hat{v}^n = -\mathrm{i}\phi(\hat{v}^{n+1} + \hat{v}^n), \tag{5.14}$$

where $\phi = \frac{1}{2}\Delta t\Omega$ is half the cyclotron angle through which the particle rotates in the time step Δt. Equation 5.14 may be solved directly for the updated velocity components,

$$\hat{v}^{n+1} = \frac{(1 - \phi^2 - 2i\phi)}{(1 + \phi^2)}\hat{v}^n \tag{5.15}$$

to obtain the new velocity components effectively as explicit solutions. It is informative to write out the real and imaginary components of this equation:

$$\begin{aligned} v_x^{n+1} &= \left(\frac{1 - \phi^2}{1 + \phi^2}\right)v_x^n + \frac{2\phi}{1 + \phi^2}v_y^n \\ v_y^{n+1} &= \left(\frac{1 - \phi^2}{1 + \phi^2}\right)v_y^n - \frac{2\phi}{1 + \phi^2}v_x^n \end{aligned} \tag{5.16}$$

In general, the cyclotron angle through which the particle rotates is a function of the space coordinates and of time and, to centre the equations correctly in time, the angle ϕ, or correspondingly the field B_z, must be evaluated at the intermediate or half points in time,

$$\phi = \phi^{n+\frac{1}{2}} = \frac{\Delta t}{2}\frac{e}{mc}B_z\left(\mathbf{x}^n + \frac{\Delta t}{2}\mathbf{v}^n, t^{n+\frac{1}{2}}\right) \tag{5.17}$$

With the new velocity variables defined (equation 5.16), the position co-ordinates are readily determined,

$$\mathbf{x}^{n+1} = \mathbf{x}^n + (\mathbf{v}^n + \mathbf{v}^{n+1})\frac{\Delta t}{2} \tag{5.18}$$

Such a difference formulation (equations 5.16, 5.18) for a particle in a magnetic field is equally applicable to the three-dimensional problem, with a magnetic field of varying direction. The method is implicit and a numerically stable solution is obtained, even in the event of an arbitrarily large time step where $\phi \gg 1$. In addition, it is to be noted that the difference equations satisfy the essential requirement in this case, that the kinetic energy of the particle is *identically* conserved, for, if we multiply equation 5.15 for the complex velocity by its complex conjugate $\hat{v}^{n+1*}$, we obtain:

$$\begin{aligned} \hat{v}^{n+1}\hat{v}^{n+1*} &= \hat{v}^n\hat{v}^{n*}\left(\frac{1 - \phi^2 - 2i\phi}{1 + \phi^2}\right)\left(\frac{1 - \phi^2 + 2i\phi}{1 + \phi^2}\right) \\ (v_x^{n+1})^2 + (v_y^{n+1})^2 &= (v_x^n)^2 + (v_y^n)^2 \end{aligned} \tag{5.19}$$

Certainly, this is a necessary condition of the difference formulation since, if it were not satisfied, the particle would spiral inwards or outwards from the orbiting centre, contrary to the differential solution.

Finally, if an electric field $\hat{E} = E_x + \mathrm{i}E_y$ exists, it may readily be incorporated in such a difference formulation when equation 5.15 is modified as:

$$\hat{v}^{n+1} = \frac{e}{m}\Delta t\left(\frac{1 - \mathrm{i}\phi}{1 + \phi^2}\right)\hat{E} + \left(\frac{1 - \phi^2 - 2\mathrm{i}\phi}{1 + \phi^2}\right)\hat{v}^n \tag{5.20}$$

It is to be noted that this difference algorithm has the interesting and useful property that, in the presence of a strong magnetic field and with large time steps $\phi \gg 1$, the difference equation 5.20 approximates to:

$$\hat{v}^{n+1} \simeq -\frac{e}{m}\Delta t\frac{\mathrm{i}\hat{E}}{\phi} - \hat{v}^n$$

or

$$\frac{\hat{x}^{n+1} - \hat{x}^n}{\Delta t} = \frac{1}{2}(\hat{v}^{n+1} + \hat{v}^n) = -\mathrm{i}\frac{\hat{E}}{B_z}c \tag{5.21}$$

Namely, and in agreement with the differential solution (Spitzer 1962), the particle experiences a slow drift perpendicular to the electric field (the 'E/B drift') while still retaining a large 'thermal' velocity perpendicular to the magnetic field.

4 *N*-body Direct 'Action-at-a-Distance' Simulation

We are now in a position to formulate the problem for an assembly of interacting particles. To illustrate the point, and as an important class of such problems, we consider an assembly of N particles which interact with an inverse-square law of force (gravitational or electrostatic forces). The state of the system $\mathbf{u}$ is defined as the $6N$-dimensional vector formed by the position and velocity coordinates $\mathbf{x}_\mu, \mathbf{v}_\mu (1 \leqslant \mu \leqslant N)$ of all the particles. Correspondingly and according to Newton's laws of motion, there are $6N$ first-order ordinary differential equations which define the evolution of the system in time:
for all μ, $1 \leqslant \mu \leqslant N$:

$$\frac{\mathrm{d}\mathbf{x}_\mu}{\mathrm{d}t} = \mathbf{v}_\mu$$

$$\frac{\mathrm{d}\mathbf{v}_\mu}{\mathrm{d}t} = \sum_{\substack{\nu=1\\ \nu\neq\mu}}^{N} \frac{e_\mu e_\nu}{m_\mu} \cdot \frac{(\mathbf{x}_\mu - \mathbf{x}_\nu)}{|\mathbf{x}_\mu - \mathbf{x}_\nu|^3} \tag{5.22}$$

where m_μ is the mass of the μth particle and e_μ is the charge of the particle μ in the electrostatic case or proportional to the mass ($e_\mu = m_\mu\sqrt{-G}$) in the gravitational case.

When $N = 2$, we may without much difficulty obtain analytic solutions to this system. For example, the system may frequently be simplified by rewriting the equations in the centre-of-mass frame, whereby invariants of the motion are immediately obtained: namely, the centre-of-mass momenta. Additional invariants include the angular momenta and the energy. With three particles, the problem becomes complex.

One of the difficulties of a system with long-range forces is that every particle may effectively interact with every other particle. In general, therefore, for a system of N particles there are $\frac{1}{2}N(N-1)$ interactions. Hence even computationally there is an upper limit on N of the order of 10^3, since a million interactions must then necessarily be followed. Nevertheless 10^3 particles are sufficient to exhibit statistical behaviour and this gives us the opportunity of studying statistical mechanics 'experimentally'.

We shall first consider the difference formulation of these equations. As for the motion of a single particle in a conservative field, the equations for the position coordinates depend only on the velocity coordinates, while the equations for the velocity coordinates depend only on the position vectors. It follows that the simplest satisfactory approach is to difference the equations by the leapfrog method (Section 5.2):
for all μ, $1 \leqslant \mu \leqslant N$,

$$\mathbf{x}_\mu^n = \mathbf{x}_\mu^{n-2} + \mathbf{v}_\mu^{n-1} 2\,\Delta t,$$
$$\mathbf{v}_\mu^{n+1} = \mathbf{v}_\mu^{n-1} + \sum_{\substack{\nu=1 \\ \nu\neq\mu}}^{N} \frac{2e_\mu e_\nu\,\Delta t}{m_\mu} \cdot \frac{(\mathbf{x}_\mu^n - \mathbf{x}_\nu^n)}{|\mathbf{x}_\mu^n - \mathbf{x}_\nu^n|^3} \tag{5.23}$$

We ensure the identical conservation of energy, and at the same time simplify the computation, by using Newton's third law of motion so that the increase of momentum of the μth particle due to an interaction with the νth is equated with the decrease of momentum of the νth due to the μth.

If the total volume occupied by the system is V, then we may associate a free volume with each particle (n is the number density),

$$\frac{1}{n} = \frac{4\pi}{3}a^3 = V/N \tag{5.24}$$

where a is the typical interaction distance. A sufficiently small time step Δt is to be chosen to satisfy the condition:

$$\Delta t \ll \frac{a}{|\mathbf{v}_\mu|} \quad (\text{all } \mu) \tag{5.25}$$

It is most convenient to represent the state of the system at any time t by plotting each particle as a point in the six-dimensional $(\mathbf{v}, \mathbf{x})$ phase-space. For example, for a system in one space and one velocity dimension, a phase-space plot has the form of Figure 5.3.

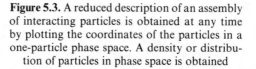

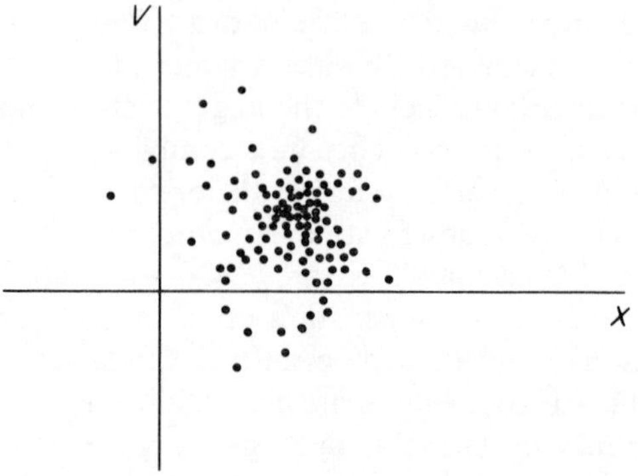

Figure 5.3. A reduced description of an assembly of interacting particles is obtained at any time by plotting the coordinates of the particles in a one-particle phase space. A density or distribution of particles in phase space is obtained

5 Equilibrium Statistical Properties from Particle-Particle Models

The 'exact' simulation of assemblies of particles, which interact directly through the particle force laws, is limited to assemblies of a few particles ($N \sim 1{,}000$) since there are of the order of N^2 interactions which must be calculated directly. There is little scope, therefore, in applying such a model directly to time-dependent many-body problems. The most useful application of such a model on the computer is in the determination of thermodynamic or equilibrium macroscopic properties of the assembly, obtained by averaging over the individual particle states which constitute the assembly. Clearly, if we are to obtain meaningful statistical averages, a sufficient number of particles in the assembly must be used to minimize the fluctuations due to individual particles and, in assemblies where particles interact with short-range forces, of the order of 1,000 particles (a million interactions) are sufficient to exhibit statistical behaviour. It is, therefore, particularly in the study of classical fluids and solids, where molecular forces are short-range (for example of the van der Waals type), that exact particle-particle simulations have had their widest application.

If the individual force laws or potentials operating between the molecules which constitute a classical fluid are known, the computer assembly of interacting molecules would describe thermodynamic properties such as the equation of state of the 'fluid', or phase transitions, or magnetic and dielectric coefficients. Frequently, however, we do not know the forces which operate between molecules, though, by postulating the form of the forces, hypothetical thermodynamic properties can be obtained by particle-particle simulations.

On comparison with experiment, improved molecular potentials may in turn be postulated. The N-body action-at-a-distance model on the computer may be applied, therefore, to provide, first, information on the macroscopic thermodynamic properties of fluids, and second, information on microscopic molecular properties.

Two approaches in particular have proved successful in linking macroscopic thermodynamic properties with microscopic molecular properties. In the method of molecular dynamics (Alder and Wainwright 1959, 1960), the exact evolution of a system of N molecules is followed in time by integrating the deterministic equations of motion of each of the molecules in a stepwise fashion. After a sufficient number of time steps, a thermodynamic equilibrium is assumed and thermodynamic properties are obtained by *time-averaging* the microscopic properties of interest. In the *Monte Carlo* method, the principles of Gibbsian statistical mechanics are applied and thermodynamic properties are obtained by *ensemble averaging* (Metropolis *et al.* 1953). We shall consider both approaches, but it is stressed here that the second method employing the Monte Carlo technique is of general interest to a wide variety of problems and, as the name suggests, solutions are obtained in a finite time by a random sampling technique.

In both methods we shall construct a system of N molecules in some volume V, where usually periodic boundary conditions to approximate an infinite fluid are applied. A two-particle potential will be assumed to operate between all $\frac{1}{2}N(N-1)$ pairs of molecules and, as an example of such a potential, the Lennard–Jones potential may be considered:

$$\Phi(r_{12}) = 4\epsilon\left\{\left(\frac{\sigma}{r_{12}}\right)^{12} - \left(\frac{\sigma}{r_{12}}\right)^{6}\right\} \tag{5.26}$$

r_{12} is the distance of separation between the centres of the molecules 1 and 2, and ϵ and σ are constant. Such a potential is drawn in Figure 5.4 and describes

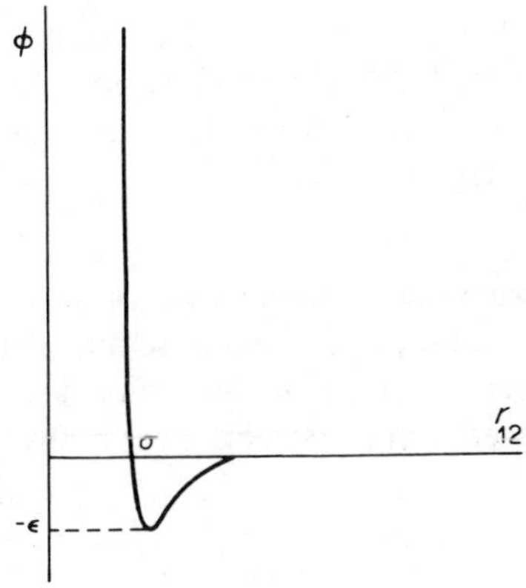

Figure 5.4. The Lennard–Jones potential used to describe the interaction between spherically symmetric atoms or molecules (equation 5.26). The potential describes a repulsion at short distances and a weak attraction at long ranges

an attraction when the molecules are distant and a repulsion when they are close. In this particular case, the potential is central and conservative but, in general, potentials to describe more complex phenomena, such as polar and finite-sized molecules, may readily be employed.

(*a*) *Thermodynamic Properties of Classical Fluids by Time Averaging.* If the potential law between molecules is known or assumed, we evolve the time-development of the system of N molecules according to the deterministic equations of motion for each molecule (Section 5.4). If each of the particles is labelled, say by μ, then the differential equations of motion for the μth particle are:

$$\frac{d\mathbf{x}_\mu}{dt} = \mathbf{v}_\mu$$

$$\frac{d\mathbf{v}_\mu}{dt} = -\frac{1}{m}\frac{\partial}{\partial \mathbf{x}_\mu}\sum_{\substack{\nu=1\\ \nu\neq\mu}}^{N} \Phi(|\mathbf{x}_\mu - \mathbf{x}_\nu|) \tag{5.27}$$

where the space-derivative on the right-hand side of equation 5.27 represents the gradient operator at the coordinates of the μth particle. Computationally, the scalar force $F(r) = -\partial\Phi/\partial r$, rather than the potential, is usually defined since difference derivatives in space may then be avoided.

In this formulation (equations 5.27), the force has the properties of being contral and conservative and consequently, as before (Section 5.4), the leap-frog method is most appropriate in integrating the equations in time. The particle μ has coordinates $\mathbf{x}_\mu^{n-2}, \mathbf{v}_\mu^{n-1}$ at a given time step $n - 1$ and the equations of motion are integrated over the double time step:

$$\mathbf{x}_\mu^n = \mathbf{x}_\mu^{n-2} + \mathbf{v}_\mu^{n-1} 2\,\Delta t$$

$$\mathbf{v}_\mu^{n+1} = \mathbf{v}_\mu^{n-1} + \frac{2\,\Delta t}{m}\sum_{\substack{\nu=1\\ \nu\neq\mu}}^{N} F(|\mathbf{x}_\mu^n - \mathbf{x}_\nu^n|)\frac{\mathbf{x}_\mu^n - \mathbf{x}_\nu^n}{|\mathbf{x}_\mu^n - \mathbf{x}_\nu^n|} \quad (\text{for all } \mu,\ 1 \leqslant \mu \leqslant N) \tag{5.28}$$

Microscopic, or molecular, properties are readily averaged over all the particles and over many time steps to yield the macroscopic thermodynamic properties of interest. As an example, the thermodynamic quantity of internal energy U is defined as:

$$U = \tfrac{3}{2}kT + \bar{\Phi} \tag{5.29}$$

where $\frac{3}{2}kT$ is the thermal or mean kinetic energy per particle (k is Boltzmann's constant and T is the temperature) and $\bar{\Phi}$ is the averaged potential energy per particle. To determine the temperature, we average the total kinetic energy of the system over many time steps p,

$$\tfrac{3}{2}kT = \frac{1}{N}\cdot\frac{1}{p}\sum_{n=1}^{p}\sum_{\mu=1}^{N}\tfrac{1}{2}m(v_\mu^n)^2 \tag{5.30}$$

and equally, we average the total potential energy over many time steps,

$$\bar{\Phi} = \frac{1}{2N}\cdot\frac{1}{p}\sum_{n=1}^{p}\sum_{\mu=1}^{N}\sum_{\substack{\nu=1\\ \nu\neq\mu}}^{N}\Phi(|\mathbf{x}_{\mu}^{n} - \mathbf{x}_{\nu}^{n}|) \tag{5.31}$$

The method has been applied to a wide variety of phenomena in classical fluids, particularly in obtaining the equation of state of non-ideal gases and in studying phase transitions (see Figures 5.5 and 5.6). The method also has application in studying nonequilibrium thermodynamics, where macroscopic rate equations of a system evolving towards equilibrium can be investigated.

The approach has here been demonstrated in terms of a central conservative potential, but the method has equal validity in studying the thermodynamic properties of polar molecules, both for solids and liquids, where rotational degrees of freedom are introduced. This is particularly important in the electric and magnetic properties of materials.

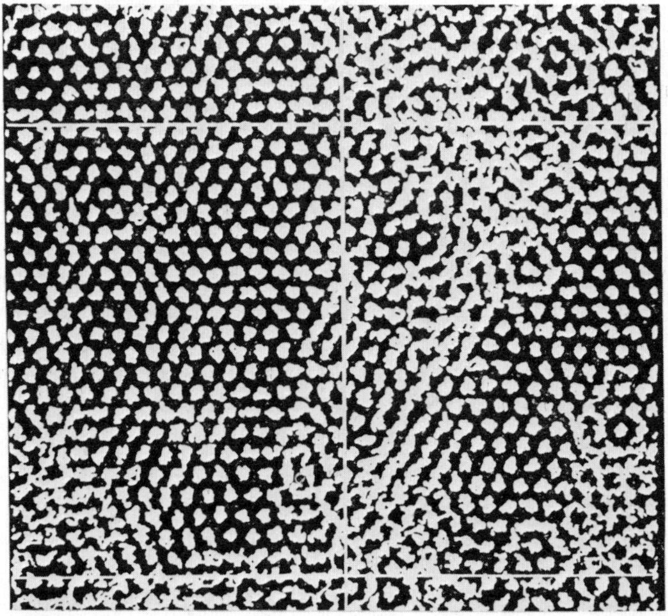

Figure 5.5. A plot in configuration space of the orbits of a number of interacting molecules. In this example, the coexistence of solid and liquid phases is shown (from Alder, B., and Wainwright, T. E., 'Phase Transitions in Electric Disks', *Phys. Rev.*, **127**(2), p. 359, 1962)

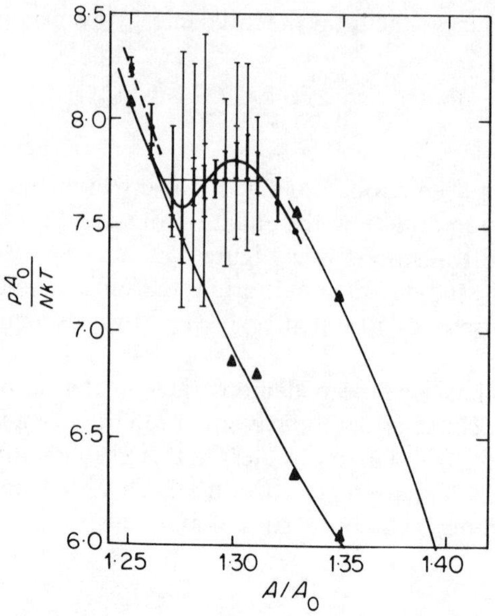

Figure 5.6. An example of a phase diagram obtained by the method of molecular dynamics (from Alder, B., and Wainwright, T. E., 'Phase Transitions in Elastic Disks', *Phys. Rev.*, **127**(2), p. 359, 1962). The curves show points of pressure and area (in three dimensions volume) where phase transitions occur

(*b*) *Ensemble Averaging: the Monte Carlo Method.* Because of the complexity involved in solving the deterministic equations of motion of a many-particle assembly, Gibbsian statistical mechanics has employed the concept of ensemble averaging to replace time averaging. Equally, in computational mechanics, as an alternative approach to obtaining thermodynamic variables of interest, an average over an ensemble of systems can be computed directly (Metropolis *et al.* 1953).

For simplicity, we shall pose the problem in two space dimensions where, as before, any state of a system of N particles may be described as a point in $4N$-dimensional phase space. A 'volume' element in such a space is:

$$\mathrm{d}v^{2N}\,\mathrm{d}x^{2N} = (\mathrm{d}x_1\,\mathrm{d}y_1\,\mathrm{d}x_2\,\mathrm{d}y_2\ldots\mathrm{d}x_N\,\mathrm{d}y_N)$$
$$\cdot\,(\mathrm{d}v_{x1}\,\mathrm{d}v_{y1}\,\mathrm{d}v_{x2}\,\mathrm{d}v_{y2}\ldots\mathrm{d}v_{xN}\,\mathrm{d}v_{yN})$$

where the subscripts refer to each particle belonging to the system. Given the microscopic variable u which is defined at every point in phase space, we are required to determine the equilibrium thermodynamic value $\bar{u}$. We use the canonical ensemble where, according to statistical mechanics, $\bar{u}$ may be determined from the ensemble average,

$$\bar{u} = \int u \exp(-E/kT)\,dx^{2N}\,dv^{2N} \Big/ \int \exp(-E/kT)\,dx^{2N}\,dv^{2N} \qquad (5.23)$$

E is the potential energy of each state in phase space or system belonging to the ensemble.

Effectively, therefore, the problem is to determine a many-thousand-dimensional integral and, patently, this cannot be achieved by some integro-difference technique. On the other hand, mathematically, an estimate of the integral can clearly be achieved by *sampling* the integrand (of equation 5.32) statistically at a set of representative points in phase space. Such an approach is of general application to the evaluation of integrals and is aptly termed the Monte Carlo method.

We may consider the mathematical problem from the alternative perspective of ensemble averaging. If a canonical ensemble can be created computationally then

$$\bar{u} = \frac{1}{M}\sum_{v=1}^{M} u_v \qquad (5.33)$$

where v refers to each system belonging to the ensemble and M, a large number, is the number of systems in the ensemble. However, the ensemble must be so constructed as to satisfy the requirements of statistical mechanics. Adopting the approach of Tolman (1967), the fundamental hypothesis, which must be complied with here in assuming the equivalence of time averaging and ensemble averaging, is that of equal *a priori* probabilities for equal volumes in phase space. And second, the ensemble must be canonical so that the probability of a system belonging to the ensemble is proportional to $\exp(-E/kT)$.

The ensemble may be constructed computationally by the use of a *Markov* process (Figure 5.7) in which the set of systems is evolved in sequence in imaginary time. Given a system which we regard as a point in phase space at step p, a new system belonging to the ensemble is created at step $p + 1$ by moving to a neighbouring point in phase space. The ensemble consists of all the states $p = 1$ to M through which the system moves. For the state at step p, we have a large number of choices of the next state at step $p' = p + 1$. We choose randomly and hence move to one of many states, so that the Markov procedure may be represented as a trace along branches of a growing tree (Figure 5.7). Clearly, after only a very few steps, the ensemble has the possibility of including a very large number of systems or states, though only one particular path is chosen. Providing all the particles in the system are moved, all states in configuration space become accessible, a condition which is an obvious preliminary requirement to satisfy the hypothesis of equal *a priori* probability.

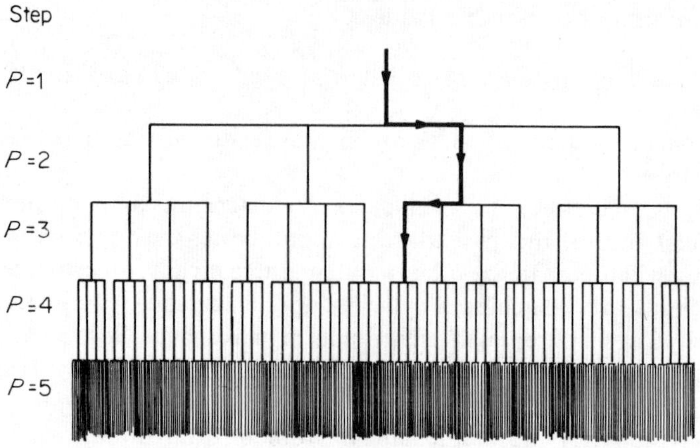

Figure 5.7. A schematic of the Markov process used to generate possible systems belonging to a canonical ensemble for the Monte Carlo method. The number of possible systems which could be obtained multiplies geometrically. One particular path is randomly chosen

We may now quantitatively specify the algorithm to be pursued in constructing the required ensemble. The N particles are distributed over the volume V in any arbitrary initial state. One particular particle is chosen (say $\mu = 1$) and moved randomly to any point inside a square of area a^2,

$$x_\mu^{p+1} = x_\mu^p + a(R_1 - \tfrac{1}{2})$$
$$y_\mu^{p+1} = y_\mu^p + a(R_2 - \tfrac{1}{2}) \tag{5.34}$$

where R_1 and R_2 are random numbers $0 \leqslant R \leqslant 1$. At any step $p + 1$, therefore, a possible new system has been generated in the state $(x_1^p, y_1^p, \ldots, x_\mu^{p+1}, y_\mu^{p+1}, \ldots, x_N^p, y_N^p)$. The new state is obtained in a manner compatible with the requirement of equal *a priori* probability, since the random numbers are not weighted. The move involves a change of energy between the systems ΔE, which is evaluated:

$$\Delta E = \sum_{\substack{\nu=1 \\ \nu \neq \mu}}^{N} \Phi(|\mathbf{x}_\mu^{p+1} - \mathbf{x}_\nu^{p+1}|) - \Phi(|\mathbf{x}_\mu^p - \mathbf{x}_\nu^p|) \tag{5.35}$$

It is our purpose to create a canonical ensemble, so that the frequency of occurrence of a system belonging to the ensemble is proportional to the factor $\exp(-E/kT)$. This distribution in the ensemble is achieved by choosing whether to accept the newly created system or not. If $\Delta E < 0$, so that the move produces a system of lower energy, the system is accepted as a member of the ensemble. On the other hand, if $\Delta E > 0$, so that a system of higher energy has been created, the system is only accepted as a member of the ensemble with a probability $\exp(-\Delta E/kT)$. Choosing a third random number R_3, if $R_3 < \exp(-\Delta E/kT)$, the new system is accepted as a member of the ensemble, while, if $R_3 > \exp(-\Delta E/kT)$, the new system is rejected and the old system is repeated as a member of the ensemble.

The step is now complete and another particle μ' is moved during the next step. Clearly the more systems (M large) included in the ensemble, the better will be the solutions but to satisfy the conditions of equal *a priori* probabilities, at least $M_{min} = NL/a$ systems must be created. The choice of a is otherwise apparently arbitrary, but it is found that the convergence in obtaining thermodynamic quantities is more rapid when a is of the order of the interparticle distance of separation. The thermodynamic variable in which we are interested is simply obtained as the average over the ensemble (equation 5.33).

In general, this method of ensemble averaging using the Monte Carlo technique is considerably more rapid than the direct method of time averaging using the deterministic equations of motion (Section 5.5a). However, the method has less scope, since clearly it cannot be applied to nonequilibrium thermodynamics. Nevertheless, the essential mathematical procedure of evaluating an integral by statistically sampling the integrand is of wide generality.

CHAPTER VI

Particle-Field Calculations

1 The Average Field from Particle Assemblies

The quantitative description of an assembly of N particles by the direct implementation of the two-particle force laws (Chapter 5) is limited since there are of the order of N^2 interactions to be computed. By defining an average force or field due to all the particles in the assembly, the direct computation of each pair of interactions can be avoided and consequently an assembly of a very large number of particles (typically 10^6 particles) may be described computationally. We shall be concerned, in this chapter, with the formulation and application of particle-field models—in particular with those for long-range forces, whereby gravitational galactic or plasma phenomena can be studied computationally. We must therefore examine the conditions for which it is legitimate to define the average force or field of an assembly and the manner in which such an average force should be determined.

(*a*) *Continuous Fields from Long-range Forces.* In terms of the 'exact' action-at-a-distance particle-particle model (Chapter 5), an assembly of N particles is described by $6N$ first-order equations (5.22) for the positions and velocities $(\mathbf{x}, \mathbf{v})$ of each particle:

$$\frac{d\mathbf{x}_\mu}{dt} = \mathbf{v}_\mu$$

$$\frac{d\mathbf{v}_\mu}{dt} = \sum_{\substack{\nu=1 \\ \nu \neq \mu}}^{N} \frac{e_\nu e_\mu(\mathbf{x}_\mu - \mathbf{x}_\nu)}{m_\mu|\mathbf{x}_\mu - \mathbf{x}_\nu|^3} \quad \text{(for all } \mu,\ 1 \leqslant \mu \leqslant N) \tag{6.1}$$

m_μ is the mass of each particle and e_μ is the charge in the electrostatic case and proportional to the mass in the gravitational case $e_\mu = m_\mu\sqrt{-G}$. Thus, for each particle, the right-hand side of the momentum equation is determined by summing over all $N - 1$ forces on the particle.

As an alternative to determining the forces acting on each particle explicitly, and with *no loss of generality*, we may describe the forces by the electric or gravitational field at every point $\mathbf{x}$.

For particle μ, $1 \leqslant \mu \leqslant N$:

$$\frac{d\mathbf{v}_\mu}{dt} = \frac{e_\mu}{m_\mu}\mathbf{E}(\mathbf{x}_\mu) \tag{6.2}$$

and

$$\begin{aligned}\mathbf{E}(\mathbf{x}) &= \sum_{\nu=1}^{N} \mathbf{E}_\nu \\ &= \sum_{\nu=1}^{N} e_\nu \frac{\mathbf{x} - \mathbf{x}_\nu}{|\mathbf{x} - \mathbf{x}_\nu|^3}\end{aligned} \tag{6.3}$$

where it is assumed that the self-field of the particle is not included. The concept of a field applies only to a continuum. To describe the field, all of the configuration space must be divided into a set of small cells. In three dimensions, each cell may be labelled by the indices i, j, k (in a rectangular parallelepiped there are IJK cells).

Rather than defining the vector field on such a mesh, it is preferable to determine the electric or gravitational potential at each cell, so that the equation of motion for each particle μ is:

$$\frac{d\mathbf{v}_\mu}{dt} = -\frac{e_\mu}{m_\mu}\frac{\partial \Phi}{\partial \mathbf{x}_\mu} \tag{6.4}$$

where the notation for the space derivative on the right-hand side is to be interpreted as the gradient operator acting on the potential at the space point $\mathbf{x}_\mu$. The total potential (from equation 6.3) is the sum of the particle potentials:

$$\begin{aligned}\Phi(\mathbf{x}) &= \sum_{\nu=1}^{N} \Phi_\nu \\ &= \sum_{\nu=1}^{N} \frac{e_\nu}{|\mathbf{x} - \mathbf{x}_\nu|}\end{aligned} \tag{6.5}$$

The summation for the total potential on the right-hand side of equation 6.5 is equivalent to, and may be replaced by, Poisson's equation,*

$$\nabla^2 \Phi = -4\pi\rho \tag{6.6}$$

where ρ is the charge or negative mass density of the assembly. Such a charge or mass density is defined on the mesh by assigning, in some manner (Section 6.2), each particle to a cell or cells on the mesh.

If such a particle-field description is to be as precise as the particle-particle description, clearly the mesh must be very fine, so that the mesh step Δ is

* See, for example, Jackson (1963).

very much smaller than the average interparticle distance,

$$\Delta \ll \frac{L}{N^{\frac{1}{3}}} \tag{6.7}$$

where the assembly occupies a volume L^3. Errors arise since the charge density which defines the potential is determined on the mesh by smearing the charge of a particle over the whole area of a local cell or cells. Hence point particles are no longer being described, but rather the assembly is constructed of finite-sized particles or clouds of charge, each typically of volume Δ^3 (see Figures 6.1 and 6.2). At increasing distances from the particle, the cloud of charge approximates to a greater and greater degree a point particle, so that, when condition 6.7 is satisfied, the particle-field assembly becomes quite equivalent to the 'exact' particle-particle assembly. The effect of a distribution or cloud of charge is to maintain a finite potential while, for a point particle, as the distance from the point becomes arbitrarily small, the potential becomes arbitrarily large. At large distances from a particle, the point and cloud potentials become the same (Figure 6.1). When a large distance of separation is achieved in an assembly on the computer (in the sense of condition 6.7), again only a few particles can be represented, since many arithmetic operations are required in the difference solution of Poisson's equation on the mesh.

We may now turn to question the effect, if, contrary to condition 6.7 for the 'exact' simulation of a particle-field assembly, the density of the particles on the potential mesh is increased. Clearly the solutions will be a poor approximation to the exact solutions, since each particle 'sees' a smeared-out

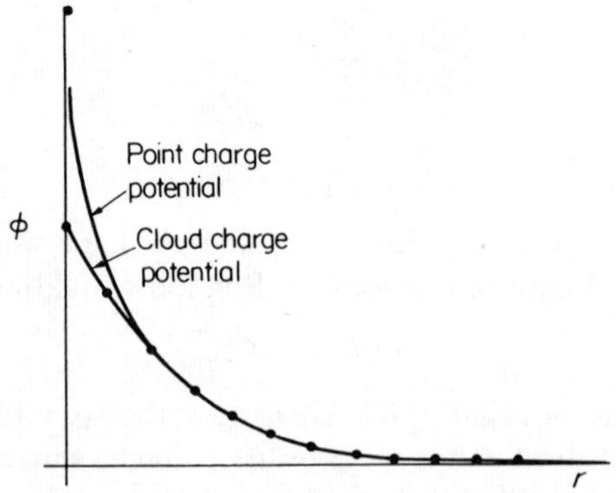

Figure 6.1. Comparison of the potentials due to a point charge in a continuous space and the cloud of charge due to a particle on a difference mesh. At large distances from the point or cloud, the potentials are similar

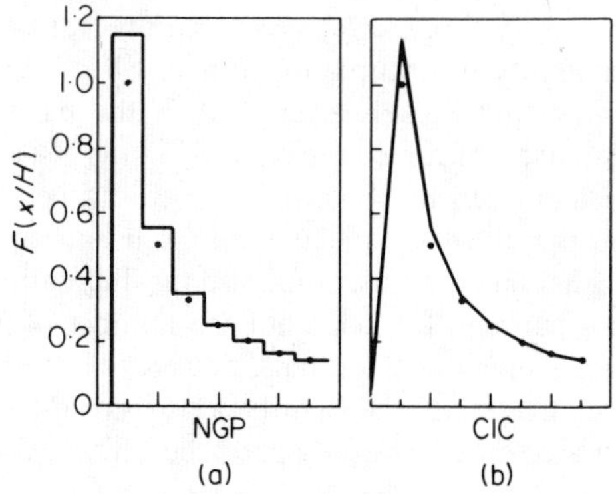

Figure 6.2. The forces experienced by two charged particles on a difference mesh: (a) when the whole of a particle's charge is assigned to the 'nearest grid point'; (b) due to the cloud-in-cell method, when the charge of a particle is interpolated to the four nearest cells (from Hockney, R. W., 'Measurements of Collisions and Heating Times in a Two-Dimensional Thermal Computer Plasma', *J. Comp. Phys.*, **8**(1), p. 19)

or averaged field and not the particular field of local point particles. On the other hand, if we wish to study collective properties of the whole system, rather than a particular exact solution, the generality of averaged fields, rather than particular fields, is of predominant interest. As the number of particles on the mesh is increased, particular information is being lost, but it is possible that the general information is being isolated.

As the density of particles is increased, however, a difficulty arises as small perturbations in the solution give rise to large fluctuations. To illustrate this effect, we shall suppose that the particle p is in a cell $(i, j + 1, k)$ with neighbouring cells (i, j, k) and $(i + 1, j, k)$ (Figure 6.3), and we shall assume that

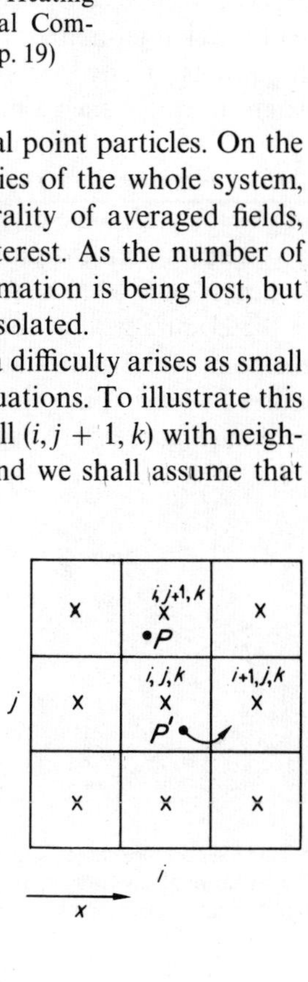

Figure 6.3. Fluctuations arise in the particle-in-cell method. If a small error causes the particle p' to cross the boundary between the cells (i, j, k) and $(i + 1, j, k)$, the force on the particle p can be radically altered

another particle p' is in cell (i, j, k) (since the number of particles has been increased, the probability of occupation of a cell has been increased and neighbouring cells are quite likely occupied). If the particle p' had been slightly displaced so that it occupied the cell $(i + 1, j, k)$ instead, the potential, and consequently field, seen by the particle p would be radically altered. It is clear, therefore, that the noise, or fluctuations, in such a system would be very large. On the other hand, we may diminish the fluctuations by increasing the total number of particles further, until the number of particles in each cell is sufficiently large, such that if one particle were to be removed elsewhere, the density of the cell would only be slightly perturbed, and consequently the field on particle p would only be slightly perturbed. These conditions pertain when

$$\Delta \gg \frac{L}{N^{\frac{1}{3}}} \tag{6.8}$$

Clearly, such a system is a statistical one and the fields, in which a particle is moved, are averaged and not local particular particle fields. The 'averaging' process is achieved by the act of counting the particles over each cell to obtain a mean charge or mass density. We shall question, therefore, the physical conditions, if any, within which such an approach is legitimate, and the extent to which such a model can be applied.

(*b*) *Collisions and the Continuum: the Debye Sphere.* To interpret the physical significance of the statistical particle-field model, we shall consider a system of charged particles distributed over a uniform charge density.* In accordance with the foregoing discussion, we wish to interpret the nature of the electric field at some arbitrary point p in the system. With respect to the point p, it is informative to divide the whole system into two volumes V_1 and V_2, where V_1 is a sphere of radius a centred at p and V_2 is the rest of the system (Figure 6.4). At some sufficiently large radius, which will be denoted

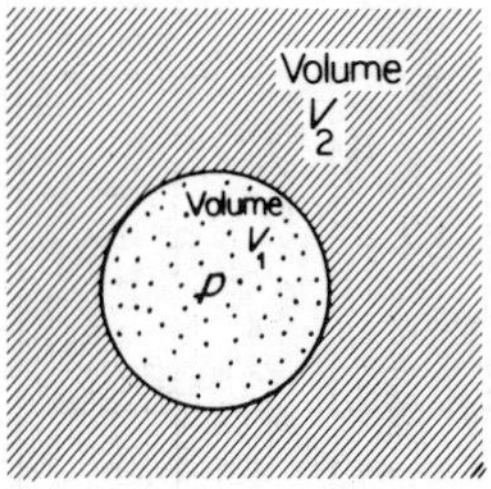

Figure 6.4. The Debye sphere. The particle p interacts with the rest of the assembly, which may be divided into two volumes: V_1 within which the particles appear discrete; V_2 where the rest of the assembly may be regarded as a continuum

* This idea applies directly to an ionized gas. The argument in this section applies equally to a gravitational assembly of stars. For simplicity of notation only, however, the language used will be that appropriate to the electrostatic case.

by $a = \lambda_D$, the charges in the outer volume V_2 appear only as a continuum to the point p and, conversely, the charges within the sphere V_1 will appear discrete to the point p and must therefore be described individually. It is, therefore, useful to think of the electric field at the point p as the sum of two contributions: the local field $\mathbf{E}_1$ due to the particles in the volume V_1 and the average field $\mathbf{E}_2$ due to the apparent continuum of charges in the outer volume V_2:

$$\mathbf{E} = \mathbf{E}_1 + \mathbf{E}_2 \tag{6.9}$$

We shall identify the particular sphere radius $a = \lambda_D$ at which such a division can be made, by noting that if all the charges in the volume V_1 were removed, the consequent potential in the volume V_1 would be:

$$4\pi nea^2$$

where n is the number density of particles. A typical particle of kinetic energy kT entering the sphere would then obtain a potential energy

$$4\pi ne^2a^2$$

If the kinetic energy of the particle, kT, is much larger than this potential energy, the sphere affects its motion only minimally. The Debye length λ_D is therefore defined as:

$$4\pi ne^2\lambda_D^2 = kT$$

or,

$$\lambda_D = \sqrt{\left(\frac{kT}{4\pi ne^2}\right)} \tag{6.10}$$

Over lengths greater than the Debye length the assembly behaves collectively; below the scale length λ_D, microscopic or particle-particle effects are important.

In the statistical particle-field model, we are concerned with collective phenomena (otherwise, to a first approximation, single-particle motion could be followed) and, therefore, in the computational model, macroscopic scale lengths L greater than the Debye length are taken,

$$L > \lambda_D \tag{6.11}$$

The significance of each contribution to the electric field (equation 6.9) at a point p in such an assembly is now apparent, for the field $\mathbf{E}_2$ arising from the continuum V_2 is the averaged field of interest associated with collective phenomena.

The local electric field $\mathbf{E}_1$, due to the particles within the Debye sphere of volume V_1, is difficult to specify quantitatively. As an illustrative example, if we were to suppose that the charges in the sphere were distributed over a regular cubic lattice, it is clear by symmetry (and has been shown by Lorentz) that the local field $\mathbf{E}_1$ would be zero. Alternatively, if in an ensemble of such spheres the charges in each volume V_1 were randomly distributed, the ensemble average of the local field $\mathbf{E}_1$ would sum to zero. It follows that in a gas-like assembly, if the total number of charged particles n_D in the Debye sphere of volume V_1 is large,

$$n_D \gg 1 \tag{6.12}$$

where

$$n_D = \frac{4}{3}\pi\lambda_D^3 n \tag{6.13}$$

then, to a good approximation, the local field $\mathbf{E}_1$ may be taken as zero. It is to be noted that a small local field $\mathbf{E}_1$ will occur depending on the fluctuations in the Debye number n_D and the relative amplitude of the fluctuations will clearly vary as $1/\sqrt{n_D}$. The larger the number of particles in the Debye sphere, the better is the approximation of a zero local electric field.

(*c*) *The Plasma and Gravitational Frequencies.* In accordance with the foregoing concept of dividing the phenomena in an assembly into binary (particle-particle) effects or *collisions*, and *collective* effects, we may associate characteristic time scales with each. A collision time τ_c is defined as the time for a particle to undergo a deflection of 90° due to binary particle-particle collisions within the Debye sphere.

Equally, it is simple to define the characteristic time τ_p for collective oscillations (plasma or gravitational oscillations) to occur. To do so we notice that in rectangular Cartesian coordinates if a $y - z$ plane of particles were all to be displaced a distance x, then, by Gauss's Law, an electric field E_x would be generated:

$$E_x = -4\pi nex \tag{6.14}$$

Consequently, there is a restoring force on each particle,

$$m\ddot{x} = eE_x \tag{6.15}$$

and it follows that there is an equation of motion for the slab of particles,

$$\ddot{x} + \left(\frac{4\pi ne^2}{m}\right)x = 0 \tag{6.16}$$

This equation corresponds to a simple harmonic motion with frequency (the plasma frequency),

$$\omega_p = \sqrt{\left(\frac{4\pi ne^2}{m}\right)} \tag{6.17}$$

or

$$\tau_p = \frac{2\pi}{\omega_p} = \sqrt{\left(\frac{\pi m}{ne^2}\right)} \tag{6.18}$$

The plasma oscillation period is the fundamental time scale associated with collective effects. In a galaxy, the corresponding gravitational oscillation period is the time scale for a star to oscillate freely through the potential well formed by the galaxy.

It is not surprising, and it can be shown rigorously (Birdsall *et al.* 1970), that the particle-particle collision time τ_c is related to the characteristic time for collective oscillations by

$$\frac{\tau_p}{\tau_c} = \frac{K}{n_D} \tag{6.19}$$

where K is approximately constant and of the order of unity. Thus, as the number of particles in the Debye sphere n_D becomes large, the collision time becomes long and the effect of collisions is of secondary importance. Such a system is said to be *collisionless*. In many plasmas, and certainly in galaxies, the number of particles or stars in the Debye sphere is very large (a galaxy in equilibrium is, by definition, a Debye sphere where, typically, $n_D \sim 10^{10}$ stars) so that the collective effects are considerably more important than the collisional effects.

(*d*) *Summary of the Collisionless Model.* It has been argued that in an assembly constituted of particles acting under long-range forces, when the number of particles n_D (equation 6.13) in a Debye sphere is large, the primary phenomena of interest are collective effects rather than binary particle effects. Such a physical system may be described by a particle-field model where the macroscopic scale length L of the computational model is larger than the Debye length,

$$L > \lambda_D$$

To describe the average field E_2 with reasonable completeness, the mesh-step length Δ on which the potential is determined must be taken smaller than the Debye length,

$$\Delta < \lambda_D \tag{6.20}$$

To minimize fluctuations in a cell, of volume Δ^3, the typical number of particles in a cell must be large,

$$n\Delta^3 \gg 1 \tag{6.21}$$

It follows that with these conditions (equations 6.20 and 6.21) satisfied, the number of computational particles in a Debye sphere will be large and, in turn, the computational model will statistically simulate a plasma or galactic system. The model is usually called the collisionless particle-in-cell (PIC) model. The characteristic scale length and scale times of plasma and gravitational assemblies are summarized in Table 6.1.

Table 6.1 Characteristic scale lengths and scale times in assemblies of particles with long-range forces

		Plasma assembly	Gravitational assembly
1. Interparticle separation	a	$\frac{1}{n^{\frac{1}{3}}} =$	$\frac{L}{N^{\frac{1}{3}}}$
2. Collective oscillation period	τ_p	$2\pi\Big/\sqrt{\left(\frac{4\pi ne^2}{m}\right)}$	$2\pi/\sqrt{(4\pi Gmn)}$
3. Debye length	λ_D	$\sqrt{\left(\frac{kT}{4\pi ne^2}\right)}$	$\sqrt{\left(\frac{kT}{4\pi Gm^2 n}\right)}$
4. Particles in a Debye sphere	n_D	$\frac{4}{3}\pi n\lambda_D^3$	
5. Collision time	τ_c	$\frac{n_D}{K}\tau_p$	
6. Collision length	λ_c	$\sqrt{\left(\frac{3kT}{m}\right)}\tau_c$	

where k = Boltzmann's constant; T = temperature of assembly; e = particle charge; n = number density; N = total number of particles in the assembly; G = gravitational constant; m = particle mass; L^3 = volume of assembly; $K \sim 1$.

2 The Collisionless Particle-in-cell Model

(*a*) *Formulation of the PIC Model.* The physical significance of the collisionless particle-in-cell (PIC) model for long-range electrostatic or gravitational forces has been discussed (Section 6.1). It is to be noted that the statistical process of adding the charges or masses in each cell provides averaged potentials and forces and local or microscopic fields are not described ($E_1 = 0$). We shall here consider the formulation, boundary conditions, typical parameters and limitations of the collisionless particle-in-cell model.

For simplicity, a two-dimensional model in rectangular Cartesian coordinates will be described, though the method applies equally to one, two or three space dimensions. Nevertheless, it is of interest to note that, since the field is to be determined by the solution of Poisson's equation, the precise potential law which is operative between particles in one, two or three dimensions is different. In practice, in one space dimension infinite planes or sheets of charge or mass which move perpendicular to their planes are described; in two dimensions rods of charge or mass are described and in three dimensions particles are described.

We shall assume that an area $L \times L$ in the x–y plane is divided into small cells (typically 64×64) by an Eulerian equally spaced mesh (sides $l \times l$).

Using a random number generator and with the aid of appropriate weighting functions, particles are assigned velocity coordinates and are distributed densely according to initial conditions, where it is assumed that every cell contains many particles. If n_0 is the average number of particles in a cell, the total number of particles is $N = n_0 I^2$. Typically, n_0 may be taken between four and one hundred particles, so that the total number of particles which are simulated may approach one million. The coordinates $(x_\mu, y_\mu, v_{x_\mu}, v_{y_\mu})$ of each particle μ, $1 \leqslant \mu \leqslant N$, are stored, typically on external devices such as magnetic disks or tapes.

As before (Chapter 5), the most appropriate integration scheme for the equations of motion of the particles is the leapfrog scheme, so that alternate time levels are used to define the position coordinates (say time step $n - 1$) and velocity coordinates (time step $n - 2$) of all the particles. To integrate the particle coordinates, the cell in which a particle resides is identified, and the electric fields are determined from derivatives of the potential on the mesh. For particle μ in cell (i, j) at step n–1,

$$i = \mathrm{Int}\left(\frac{x_\mu^{n-1}}{\Delta}\right), \qquad j = \mathrm{Int}\left(\frac{y_\mu^{n-1}}{\Delta}\right) \tag{6.22}$$

The notation Int (z) is used to represent the function which truncates the real number z to the corresponding integer. The electric fields across the cell (i, j) which is occupied by particle μ are determined,

$$\begin{aligned} E_{xij}^{n-1} &= -(\Phi_{i+1,j}^{n-1} - \Phi_{i-1,j}^{n-1})/2\Delta \\ E_{yij}^{n-1} &= -(\Phi_{i,j+1}^{n-1} - \Phi_{i,j-1}^{n-1})/2\Delta \end{aligned} \tag{6.23}$$

These electric field components are used in sequence to update the coordinates of the μth particle:

$$\mathbf{v}_\mu^n = \mathbf{v}_\mu^{n-2} + \frac{e_\mu}{m_\mu} 2\,\Delta t \mathbf{E}_{ij}^{n-1} \tag{6.24}$$

$$\mathbf{x}_\mu^{n+1} = \mathbf{x}_\mu^{n-1} + \mathbf{v}_\mu^n 2\,\Delta t \tag{6.25}$$

where Δt is the time step.

As the new coordinates of all the particles are determined, a new charge density (at time step $n + 1$) is defined on the space mesh by the updated distribution of particles. Several methods exist for the determination of the new charge density in space, depending on the manner in which a particle's charge is distributed to local cells (see Section 6.2c and Figure 6.6). In the simplest method, the whole of a particle's charge is assigned to the 'nearest grid point': that is, only to the new cell to which the particle has moved. In this technique, therefore, the charge density is simply determined by adding the particles in each cell:

$$\rho_{ij}^{n+1} = \frac{1}{\Delta^2}\sum_{\mu=1}^{N} e_\mu \delta\left(\mathrm{Int}\left(\frac{x_\mu^{n+1}}{\Delta}\right) - i\right)\delta\left(\mathrm{Int}\left(\frac{y_\mu^{n+1}}{\Delta}\right) - j\right) \tag{6.26}$$

where,

$$\delta(l - m) = \begin{cases} 1 \text{ if } l = m \\ 0 \text{ if } l \neq m \end{cases}$$

In the second stage of each double time step of the calculation, the difference form of Poisson's equation is solved on the space mesh, in order that the updated potential Φ_{ij}^{n+1} at each cell point may be determined,

$$\Phi_{i+1,j}^{n+1} + \Phi_{i-1,j}^{n+1} + \Phi_{i,j+1}^{n+1} + \Phi_{i,j-1}^{n+1} - 4\Phi_{i,j}^{n+1} = -4\pi\Delta^2\rho_{i,j}^{n+1} \tag{6.27}$$

The rapid solution of this sparse matrix equation has been discussed in Section 4.4. The equations (6.22 through 6.27) define the time loop of the particle-in-cell model and are repeated over successive time steps.

It is of value to consider general boundary conditions for such a simulation. Clearly, if a particular physical problem is of interest, boundary conditions consistent with the application at hand must be employed. The particle-in-cell model may be used, however, to study fundamental properties of the plasma or galactic fluid, where it is required to avoid solutions which are dependent on particular boundary conditions. Since an infinite system cannot be represented computationally, the use of periodic boundary conditions, which most closely simulate an infinite system, are of value. The application of periodic boundary conditions to the potential calculation has been considered (Section 4.4). Equally, for the equations of motion for the particles, periodic boundary conditions are simply applied for, if a particle moves outside the domain of interest, the particle is reinserted on the other side of the domain.

It may be noted, however, that with periodic boundary conditions the domain of the calculation is repeated throughout space (Figure 6.5) and, consequently, the phenomena within the domain of the calculation may 'interact' with the repeated images of the domain.

(*b*) *Dimensionless Form of the Equations and the Time Step.* Generally, it is of value to employ dimensionless variables, since in the first instance parameters of importance are then clearly defined. But in particular, in a computational model, the use of equations in dimensionless form is of essential importance since very large or very small numbers are avoided and, in addition, the number of arithmetic operations in the computation can be minimized by removing unnecessary constants. For these reasons and to stress the arithmetic simplicity of the collisionless particle-in-cell model, we shall reformulate the applicable finite equations in dimensionless form.

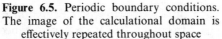

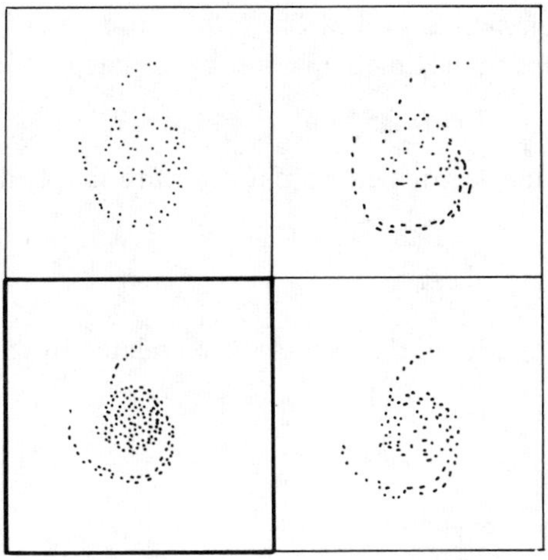

Figure 6.5. Periodic boundary conditions. The image of the calculational domain is effectively repeated throughout space

The space step Δ, and the time step Δt, define the fundamental units of length and time. Denoting the dimensionless variables with a tilde sign, we redefine the particle coordinates as

$$\tilde{\mathbf{x}}_\mu = \mathbf{x}_\mu \frac{1}{\Delta}$$

$$\tilde{\mathbf{v}}_\mu = \mathbf{v}_\mu \frac{2\,\Delta t}{\Delta} \tag{6.28}$$

Equally the field variables may be redefined,

$$\tilde{\mathbf{E}} = \mathbf{E}\frac{4(\Delta t)^2}{\Delta}\frac{e}{m}$$

$$\tilde{\Phi} = \Phi\frac{2(\Delta t)^2}{\Delta^2}\frac{e}{m}$$

$$\tilde{\rho} = \rho\frac{\Delta^2}{4\pi e n_0} \tag{6.29}$$

where, as before, n_0 is the average *number* of particles in a cell.

With such a choice of dimensionless variables, the equations for the particles, which must be applied at each time step to a large number of particles, take their simplest form. The cell coordinates of the particle μ are immediately defined at the time step $n-1$ by the integer truncation function,

$$i = \text{Int}\,(\tilde{x}_\mu^{n-1}), \qquad j = \text{Int}\,(\tilde{y}_\mu^{n-1}) \tag{6.30}$$

and the appropriate electric field components are simply defined,

$$\begin{aligned} \tilde{E}_{xij}^{n-1} &= -\tilde{\Phi}_{i+1,j}^{n-1} + \tilde{\Phi}_{i-1,j}^{n-1} \\ \tilde{E}_{yij}^{n-1} &= -\tilde{\Phi}_{i,j+1}^{n-1} + \tilde{\Phi}_{i,j-1}^{n-1} \end{aligned} \tag{6.31}$$

Equally, the coordinates of the particles are updated by simple additions:

$$\begin{aligned} \tilde{\mathbf{v}}_\mu^n &= \tilde{\mathbf{v}}_\mu^{n-2} + \tilde{\mathbf{E}}_{ij}^{n-1} \\ \tilde{\mathbf{x}}_\mu^{n+1} &= \tilde{\mathbf{x}}_\mu^{n-1} + \tilde{\mathbf{v}}_\mu^n \end{aligned} \tag{6.32}$$

with the charge or mass density defined by:

$$\tilde{\rho}_{ij}^{n+1} = \frac{1}{n_0} \sum_{\mu=1}^{N} \delta(\text{Int}\,(\tilde{x}_\mu^{n+1}) - i)\delta(\text{Int}\,(\tilde{y}_\mu^{n+1}) - j) \tag{6.33}$$

We are left, at each time step, with the calculation of the potential on the space mesh which is to be obtained from the difference form of Poisson's equation. By this suitable choice for the dimensionless variables, all constants in the above equation have been removed. There remains only one dimensionless parameter θ which appears in Poisson's equation:

$$\tilde{\Phi}_{i+1,j}^{n+1} + \tilde{\Phi}_{i-1,j}^{n+1} + \tilde{\Phi}_{i,j+1}^{n+1} + \tilde{\Phi}_{i,j-1}^{n+1} - 4\tilde{\Phi}_{i,j}^{n+1} = -\theta\tilde{\rho}_{i,j}^{n+1} \tag{6.34}$$

From the definition of the dimensionless variables, it is apparent that the dimensionless parameter θ is defined by:

$$\begin{aligned} \theta &= \frac{4\pi n_0 e^2}{\Delta^2 m} 2(\Delta t)^2 \\ &= 2\omega_p^2(\Delta t)^2 \end{aligned} \tag{6.35}$$

since n_0/Δ^2 is the average density on the mesh. Correspondingly for the gravitational assembly,

$$\theta = -2\omega_p^2(\Delta t)^2$$

where, as before, ω_p is the gravitational frequency (Table 6.1). The choice of the parameter θ is yet to be determined, but it is immediately apparent from the dimensionless form of the equations that, if large errors are to be avoided, the magnitude of the parameter θ must certainly be taken of the order or smaller than unity. We may, however, be more specific about the choice of θ, since, as the equations for the coordinates of the particles are integrated by the explicit leapfrog scheme (Section 3.6b), a stability criterion places an upper limit on the time step and certainly, for stability, we must satisfy the condition:

$$|\theta| \leqslant \tfrac{1}{2} \tag{6.36}$$

In practice, to avoid fluctuations caused by particles traversing many space steps during the period of one time step, the parameter θ is generally to be taken much smaller than unity.

(*c*) *Remarks on the Limitations of the PIC Model.* Errors occur in the collisionless particle-in-cell model due to the finite size of both the time step and the space step. The finite size of the space step leads in the first instance to an incomplete description of the Fourier modes in the potential function, so that nonlinear phenomena and, in particular, plasma turbulence, which couples many Fourier modes, may not be adequately described. In addition, it has been noted (Section 5.2) that the leapfrog method does not conserve the energy of a particle identically, so that the finite size of the time step can give rise to pseudo-heating and the nonconservation of total energy.

These errors may be minimized, however, by the choice of small time steps and space steps. The most serious source of errors are those which arise from including in the simulation only a relatively small number of particles (typically 10^6). It has been pointed out (Section 6.1) that the collision time τ_c and the collective interactive time τ_p (for plasma or gravitational oscillations) are related to the number of particles n_D in a Debye sphere (equation 6.19),

$$\frac{\tau_p}{\tau_c} = \frac{K}{n_D} \tag{6.37}$$

Whereas in a physical system the number of particles in a Debye sphere may be very large (in a laboratory plasma typically 10^5; in a galaxy, 10^{10}), such a large number cannot be simulated computationally. Consequently, the particles which are used are pseudo-particles of large mass and charge, so that they each represent many particles of the physical system. As is clear from the single dimensionless parameter θ appearing in the above equations, this of itself is of little concern, but we find that the effect of the model gives rise to pseudo-collisions which occur much more frequently than in physical systems of interest. This effect is apparent from equation 6.37 above, and is due to the relatively small number of simulation particles in a Debye sphere which in turn leads to an enhanced level of fluctuations in the computer assembly.

The essential effect, therefore, of the particle-in-cell method is a greatly reduced collision time which leads to anomalous heating and diffusion in velocity space. Nevertheless, provided the effective collision time is made sufficiently long, by which we mean a time longer than the duration of the calculation, the model is truly collisionless.

The limitations of the model, with respect to relatively small numbers of particles and finite particle sizes, have been investigated particularly by Birdsall and Fuss (1969), Birdsall *et al.* (1970) and Hockney (1971a). Birdsall *et al.* (1970) have reduced the level of noise or fluctuations in the system, at the expense of a more complex algorithm, by using the 'cloud-in-cell' method (Figures 6.2, 6.6), whereby the charge or mass of a particle is interpolated over neighbouring cells according to the precise position of a particle within a cell (Figure 6.6).*

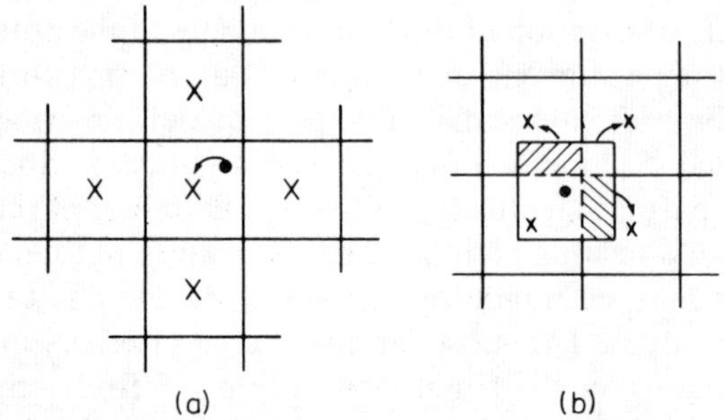

Figure 6.6. Methods of assigning the charge of a particle on a difference mesh for the solution of Poisson's equation: (a) the nearest-grid-point method; (b) the cloud-in-cell method where the charge is distributed to the four nearest cells by area weighting (from Hockney, R. W., 'Measurements of Collision and Heating Times in a Two-Dimensional Thermal Computer Plasma, *J. Comp. Phys.*, **8**(1), p. 19)

3 The Collisionless Particle-in-cell Model Applied to Plasma Simulation

The particle-in-cell model has been of importance in the study of fundamental nonlinear phenomena in plasmas, both in hot ionized gases and in the free-electron 'gases' contained in metallic solids. It is of varied interest too in the simulation of practical devices with application both to fusion and solid-state physics.

We shall illustrate the application of the model to the idealized problem of 'the two-stream instability'. The essential question arises in a collisionless assembly of particles, of the manner in which thermalization of a fluid can

* The collisionless plasma sheet model in one space dimension, which is essentially similar to the finite-size particle-in cell-model discussed here, was introduced by Buneman (1959) and Dawson (1962, 1970).

occur.* In the two-stream problem, an extreme nonthermal situation is represented by the instreaming of two oppositely moving beams. The initial conditions to represent such a simulation give a distribution of particles which is constant in configuration space, but in which particles have an appreciable velocity either in the positive x-direction or in the negative x-directions (Figure 6.7). In the first instance, this is a one-dimensional

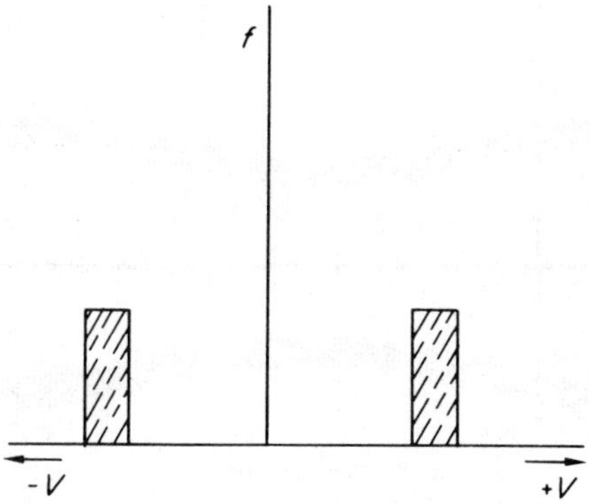

Figure 6.7. A two-stream distribution or density f of particles in velocity space

problem in configuration space, and we may represent the initial conditions by plotting the particles in a two-dimensional phase space (Figure 6.8a). A uniform background charge is assumed to exist from the massive ions which remain stationary.

It is found computationally that such a configuration is unstable. While the electrostatic potential energy in this initial state is zero, it is clear that the directed kinetic energy of the system can be reduced if the two streams thermalize and move towards each other in velocity space (Figure 6.8b). Such an initial growth of the instability can be interpreted with the aid of a linear analysis but, after several plasma frequencies, mixing and 'thermalization' of the two streams occur in a nonlinear configuration (Morse and Nielson 1969).

After long times the nonlinear development of the system demonstrates the persistence of 'holes' in phase space, in which particles are trapped by oscillating with the plasma frequency about a hole (Figure 6.8c). Similar simulations in two or three space dimensions illustrate the final decay of such

* According to classical statistical mechanics, thermalization and thermal equilibrium are produced by collisions: the rate of such a process is given by Boltzmann's H-theorem. The reader is referred for example to Tolman (1967), chapters 6 and 7.

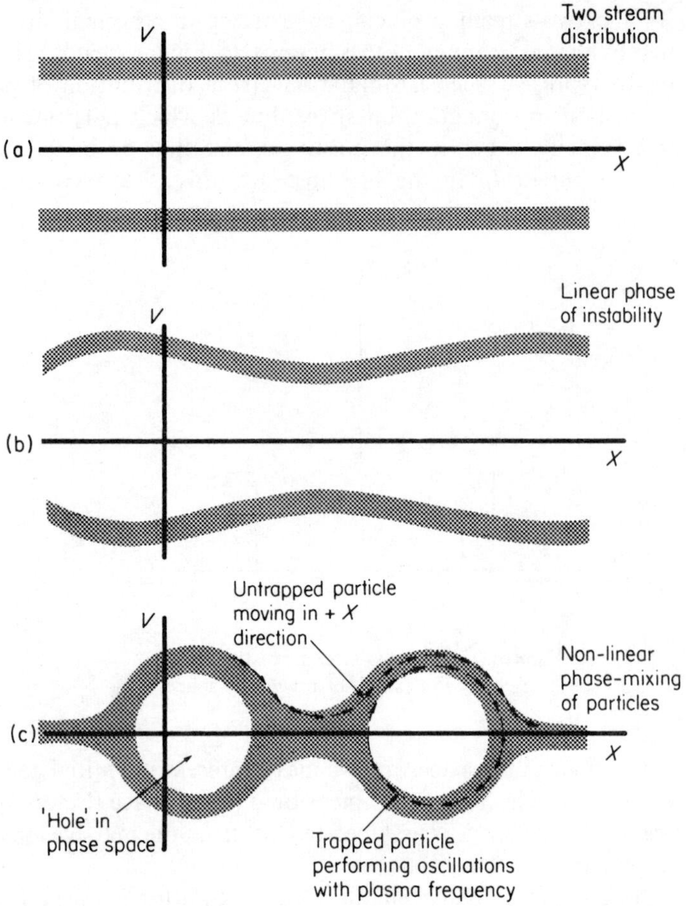

Figure 6.8. The development of the two-stream instability in phase space: (a) two beams of opposite velocity moving through each other in configuration space: (b) linear phase of the instability; (c) nonlinear phase of the instability when holes are formed in phase space around which some particles oscillate with the plasma frequency. Other particles are untrapped and move over each hole. The instability produces a thermalization or mixing in phase space

holes in phase space and the full thermalization, in the absence of collisions, of an initially nonequilibrium state (Morse 1970).

In Figure 6.9, the application of the particle-in-cell model to a solid-state triode system is illustrated. After long times a steady state is reached with the electrons flowing through the substrate (Hockney 1971b).

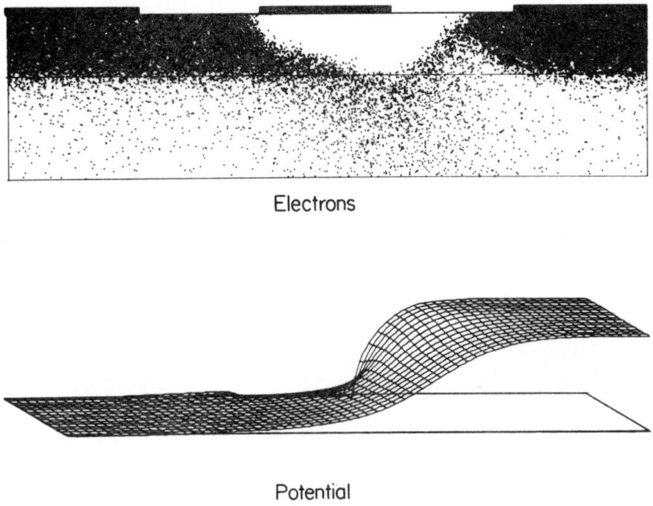

Figure 6.9. Passage of electrons through the substrate of a transistor. The corresponding potential is also shown (from Hockney, 1971b)

4 The Collisionless Particle-in-cell Model Applied to the Simulation of Galaxies

The occurrence, the structure and the evolution of galaxies have been of profound interest to the astrophysicist for many years. But the interpretation of the frequent occurrence of the complex and beautiful spiral structures observed in galaxies poses difficult theoretical problems. Many ideas on the subject have been advanced, but analytic methods are difficult and it is with particle-in-cell simulations particularly that ideas can be examined quantitatively. No attempt will be made here to review the subject,* but a few salient examples to illustrate the simulation of the gravitational problem will be given (Hohl 1970).

Galaxies are observed to be constructed of a three-dimensional central core or nucleus surrounded by a two-dimensional disk of stars of large area. A great proportion of the galaxies in the universe are found to exist in spiral or bar form. A two-dimensional (four-dimensional in phase space) particle-in-cell model is used to describe the galactic plane, whereby either rod stars are described or, by modifying Poisson's equation, point stars can be described (Hohl 1970).

The simulations are usually initiated by considering a uniform rotating disk of stars in radial equilibrium. Hohl (1970) has found that in the absence

* For a discussion on the subject, the reader is referred to Oort (1965).

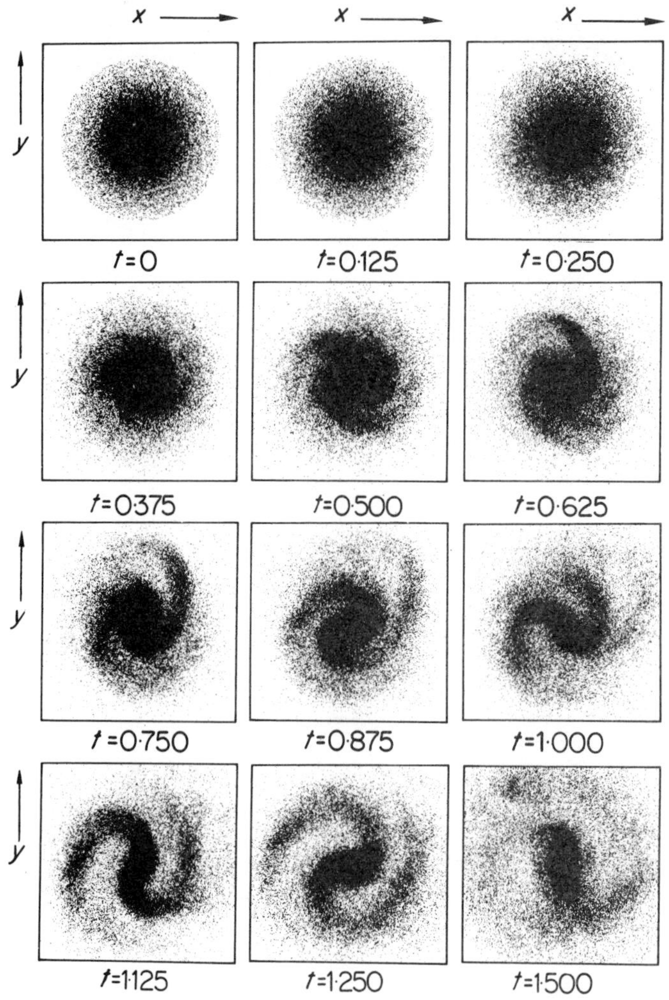

Figure 6.10. The evolution of a disk of 50,000 stars. The time is expressed in rotational periods. In the absence of an additional central force due to a galactic nucleus, the spiral arms which develop do not persist for many rotational periods (from Hohl 1970)

of a spread or dispersion of velocities, such a rigidly rotating disk of stars is unstable, so that nonuniformity quickly develops with both radial and azimuthal modes. Spiral structures occur, but do not persist, though bar-structured galaxies are frequently found with such simulations.

On the other hand, when an additional central force to simulate the core or nucleus of the galaxy is included (corresponding to 50 per cent or more of

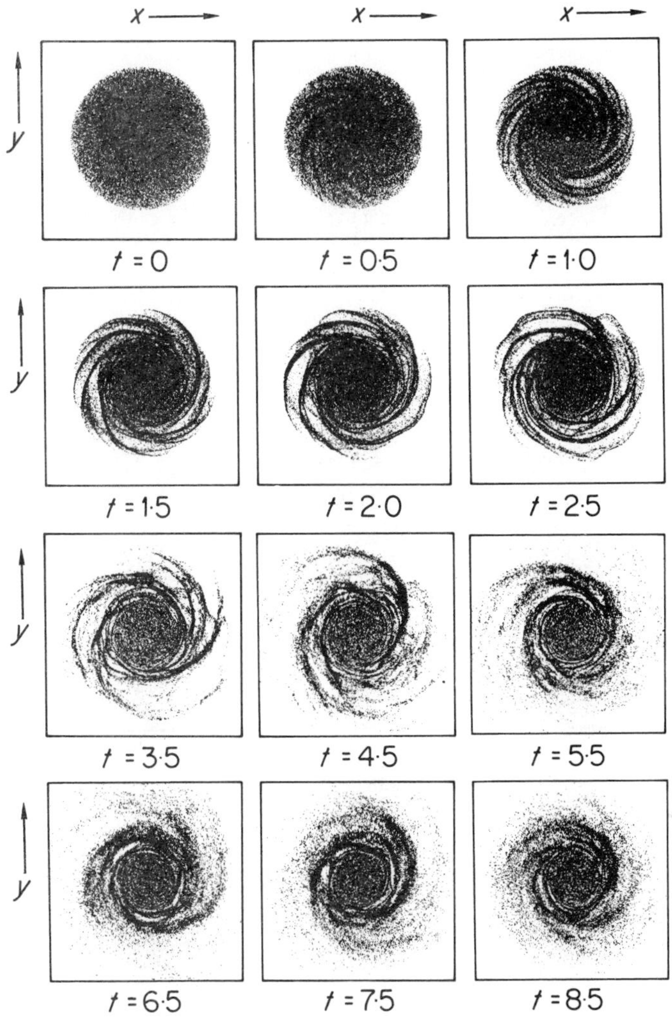

Figure 6.11. The evolution of a disk of 50,000 stars under the influence of a central force due to an assumed galactic nucleus. The disk of stars contains 10 per cent of the total mass of the system. Time is measured in galactic rotational periods and it is evident that the spiral arms persist for many periods (from Hohl 1970)

the galactic mass), spiral structures occur and persist for many rotations of the galaxy (Figures 6.10, 6.11, 6.12; Hohl 1970, Hockney and Hohl 1969). In Figures 6.13 and 6.14, a comparison (due to Hohl 1971) is made between some observed galaxies and those found in simulations.

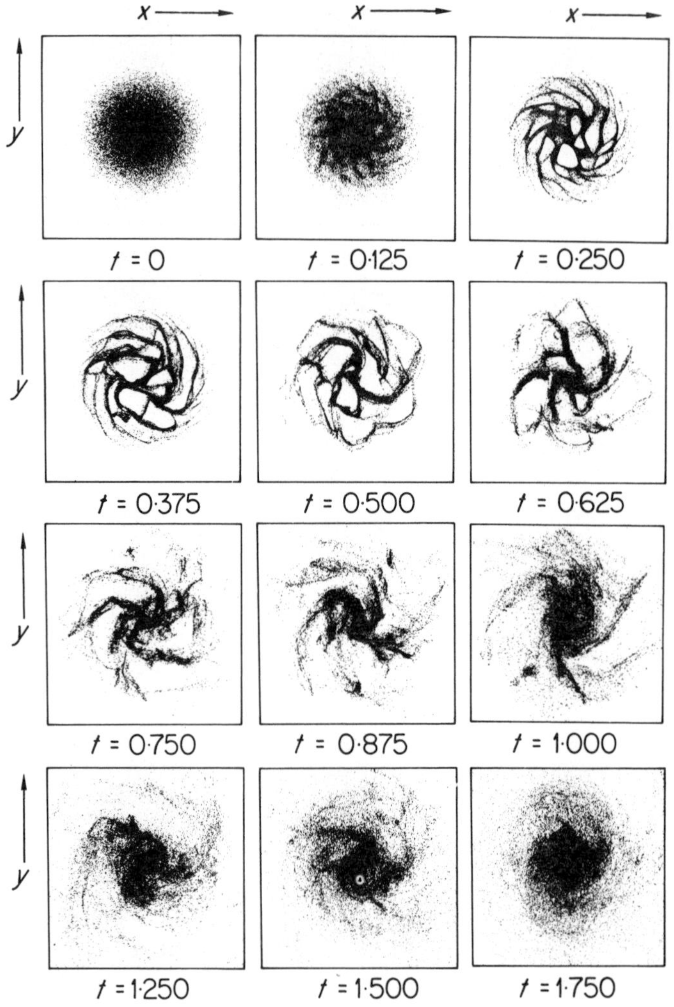

Figure 6.12. The evolution of an initially uniform rotating disk of stars. The simulation shows how the Rayleigh–Jeans instability contributes to the breaking up and formation of spiral arms from an initially uniform distribution (from Hohl 1970)

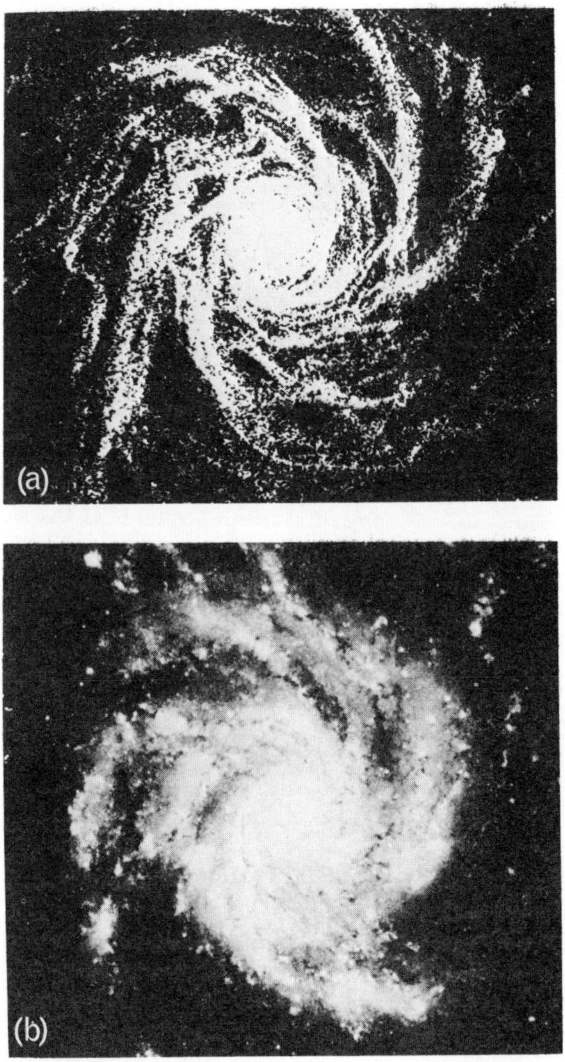

Figure 6.13. A comparison of the spiral structure obtained by the particle-in-cell model (a) with the observed galaxy M-101, Sc-type spiral galaxy (b) (from Hohl 1971)

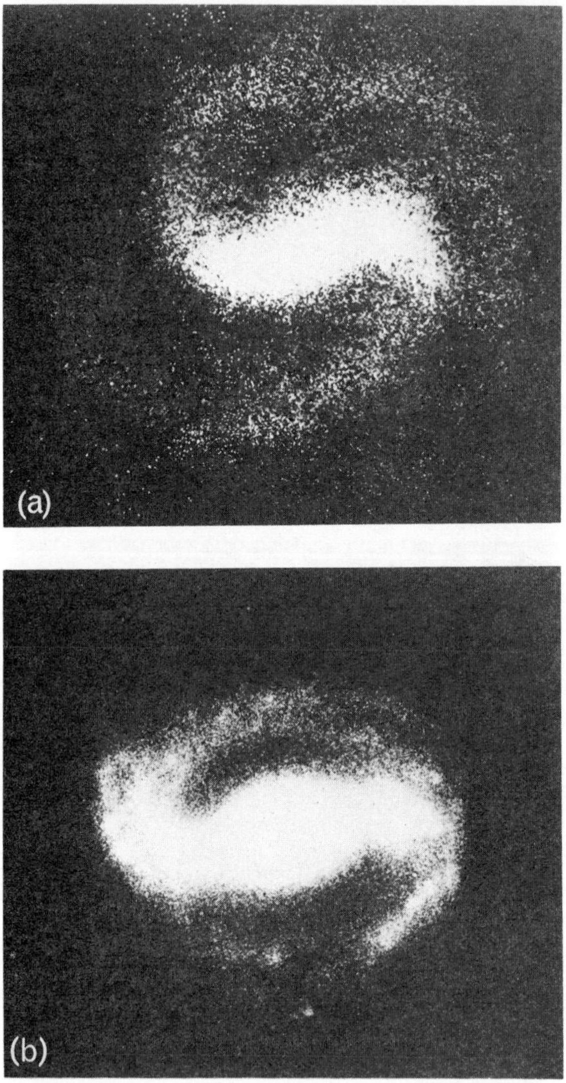

Figure 6.14. A comparison of the bar type structure frequently found in simulations by the particle-in-cell method (a) with the observed galaxy WGC-175 (b) (from Hohl 1971)

5 The Collision-Dominated PIC Model in Hydrodynamics

In the preceding sections of this chapter, we have seen the usefulness of describing a plasma or galaxy by a large number of individual particles which are constrained to move in the average field of the assembly. The particles describe the motion of the assembly and the determining fields are calculated on Eulerian (fixed in configuration space) cells. Since in hydrodynamics, the gaseous or liquid flows which we wish to describe occur in reality as a consequence of molecular motion, it is tempting to describe the motion of a fluid by a similar model of many particles moving on an Eulerian lattice. However, in questioning what form the 'equations of motion' of such particles would take, it is immediately apparent that quite different microscopic properties pertain.

In contrast to the particles in a galaxy or plasma and in the time scales of interest, the molecules in, say, the atmosphere, or a typical laboratory fluid, are perpetually undergoing many collisions with neighbouring molecules. The momentum and energy of such gas-like particles are therefore not retained, but are rather shared with surrounding particles. Expressed in terms of statistical mechanics, the molecules of a gas within some small local region establish a thermal equilibrium over a short time scale and, in velocity space, the molecules are distributed according to the Maxwell–Boltzmann law.

Based on this fundamental realization of the molecular motion of typical fluid flows, we shall here develop a simple heuristic model to describe the 'ideal' hydrodynamic equations.

(*a*) *Lagrangian Motion and the Hydrodynamic Equations.* In developing the collision-dominated particle-in-cell model (the original method was introduced by Harlow 1964), the algorithm for the motion of the 'particles' must be linked with the time-dependent equations of a continuous fluid. Equations for the density ρ and momentum density $\rho\mathbf{v}$ of a simple fluid have been obtained (Section 3.1) by considering the conservation of mass and the conservation of momentum:

$$\frac{\partial \rho}{\partial t} + \nabla \cdot \rho \mathbf{v} = 0 \tag{6.38}$$

$$\frac{\partial \rho \mathbf{v}}{\partial t} + \nabla \cdot (\rho \mathbf{v}\mathbf{v}) = -\nabla p \tag{6.39}$$

where p is the pressure in the fluid. Here, the density, velocity and pressure are fluid variables, defined only at every point $(\mathbf{x}, t)$ in configuration space and time. It is instructive to rewrite these equations with respect to the changes that occur in a local fluid element as it moves in configuration space. To do this the divergence term in the continuity equation (6.38) may be expanded,

$$\frac{\partial \rho}{\partial t} + \mathbf{v} . \boldsymbol{\nabla} \rho + \rho \boldsymbol{\nabla} . \mathbf{v} = 0 \tag{6.40}$$

or

$$\frac{\mathrm{d}\rho}{\mathrm{d}t} = -\rho \boldsymbol{\nabla} . \mathbf{v} \tag{6.41}$$

where

$$\frac{\mathrm{d}}{\mathrm{d}t} = \frac{\partial}{\partial t} + \mathbf{v} . \boldsymbol{\nabla} \tag{6.42}$$

The operator $\mathrm{d}/\mathrm{d}t$ (equation 6.42) is the total time derivative with respect to a point moving with the fluid, so that equation 6.41 is said to be in Lagrangian form. We have earlier considered (Chapter 3) the Lagrangian time-derivative, when it arises from the advection of fluid properties by the centre-of-mass motion of a fluid.

Expressed in Lagrangian form, the equation for the density of the fluid (equation 6.41) is a quantitative statement to the effect that the increase in density of the local fluid is simply produced by the compression ($\boldsymbol{\nabla} . \mathbf{v}$) of the fluid. Similarly, we may expand the momentum equations (6.39),

$$\frac{\partial \rho}{\partial t}\mathbf{v} + \rho \frac{\partial \mathbf{v}}{\partial t} + (\boldsymbol{\nabla} . \rho\mathbf{v})\mathbf{v} + (\rho\mathbf{v} . \boldsymbol{\nabla})\mathbf{v} = -\boldsymbol{\nabla} p \tag{6.43}$$

The first and third terms cancel according to the equation for the conservation of mass (6.38) and the other two terms on the left-hand side of this equation form the Lagrangian time derivative:

$$\rho \frac{\mathrm{d}\mathbf{v}}{\mathrm{d}t} = -\boldsymbol{\nabla} p \tag{6.44}$$

From the equation for the conservation of momentum we have therefore obtained an equation for the local acceleration of the fluid, which accordingly is only due to the pressure of the fluid. Essentially, it is this equation of motion which may be used to define the velocities of the 'particles' in our required algorithm.

To obtain a closed set of equations, it is necessary to determine the pressure in the fluid. For an ideal polytropic fluid, the adiabatic law furnishes such an equation for the pressure, since, for any fluid element,

$$\frac{p}{\rho^{\gamma}} = \text{constant} \tag{6.45}$$

where γ is the ratio of specific heats. Accordingly, the Lagrangian derivative of this quantity vanishes:

$$\frac{\mathrm{d}}{\mathrm{d}t}\left(\frac{p}{\rho^{\gamma}}\right) = 0 \tag{6.46}$$

It is to be noted that this equation is only valid for an ideal gas in the absence of entropy production (more complex situations are considered in the later Chapters, 9 and 10).

The hydrodynamic equations, when expressed in Lagrangian form (equations 6.41, 6.44, 6.46), have particular simplicity. Clearly, this is a consequence of molecular motion, which transports fluid properties and it provides the justification for using a particle-in-cell method, since the simulation particles will transport mass, momentum and energy in a Lagrangian manner.

Finally it is of value to express the invariant p/ρ^{γ}, for an ideal gas, in conservation form. Using the equation for the conservation of mass, we form the sum (from equations 6.38 and 6.46):

$$\frac{p}{\rho^{\gamma}}\left\{\frac{\partial \rho}{\partial t} + \nabla \,.\, \rho \mathbf{v}\right\} + \rho \frac{\mathrm{d}}{\mathrm{d}t}\left(\frac{p}{\rho^{\gamma}}\right) = 0$$

$$\frac{\partial}{\partial t}\left(\rho \frac{p}{\rho^{\gamma}}\right) + \nabla \,.\, \left(\rho \frac{p}{\rho^{\gamma}} \mathbf{v}\right) = 0 \tag{6.47}$$

(*b*) *The PIC Method for an Ideal Gas.* We may summarize the fluid equations for an ideal gas, which have been discussed above, in terms of conservation by particle convection (equations 6.38, 6.39, 6.47):

$$\frac{\partial (nm)}{\partial t} + \nabla \,.\, (nm\mathbf{v}) = 0 \tag{6.48}$$

$$\frac{\partial (nm\mathbf{v})}{\partial t} + \nabla \,.\, (nm\mathbf{v}\mathbf{v}) = -\nabla p \tag{6.49}$$

$$\frac{\partial}{\partial t}\left(n \frac{p}{\rho^{\gamma}}\right) + \nabla \,.\, \left(n \frac{p}{\rho^{\gamma}} \mathbf{v}\right) = 0 \tag{6.50}$$

where n is the number density of particles at any point and time $(\mathbf{x}, t)$. It is sufficient to illustrate the particle-in-cell algorithm in two space dimensions, though the method applies equally to one, two or three space dimensions. The region of interest is divided by an Eulerian lattice of cells on which a large number of simulation particles are distributed so that, in general and to minimize fluctuations, the number of particles per cell is large. At time t^n, the particle μ has coordinates $(x_{\mu}^{n}, y_{\mu}^{n})$ in configuration space.

If, to each particle, a discrete mass, momentum and the invariant $e = p/\rho^{\gamma}$ are assigned then, according to equations 6.48, 6.49 and 6.50, conservation of particles will ensure the conservation of the fluid mass, momentum and the invariant p/ρ^γ. At the time step n, therefore, when all the particles have taken up their new positions, six variables are associated with each particle:

$$\mathbf{u}_\mu^n = (x_\mu^n, y_\mu^n, m_\mu, mv_{x\mu}^n, mv_{y\mu}^n, e_\mu^n) \qquad (\text{particles } 1 \leqslant \mu \leqslant N) \qquad (6.51)$$

Therefore, to determine cell-like quantities, we assume that within a cell the particles experience many collisions and equilibrate their momentum and the energy-related variable e_μ. Hence cell variables are average variables determined by adding over the particles within a cell:

$$n_{ij}^n = \frac{1}{\Delta^2}\sum_\mu \delta\left(\mathrm{Int}\left(\frac{x_\mu^n}{\Delta}\right) - i\right)\delta\left(\mathrm{Int}\left(\frac{y_\mu^n}{\Delta}\right) - j\right)$$

$$\rho_{ij}^n = \frac{1}{\Delta^2}\sum_\mu m_\mu \delta\left(\mathrm{Int}\left(\frac{x_\mu^n}{\Delta}\right) - i\right)\delta\left(\mathrm{Int}\left(\frac{y_\mu^n}{\Delta}\right) - j\right)$$

$$(\rho\mathbf{v})_{ij}^n = \frac{1}{\Delta^2}\sum_\mu m_\mu \mathbf{v}_\mu^n \delta\left(\mathrm{Int}\left(\frac{x_\mu^n}{\Delta}\right) - i\right)\delta\left(\mathrm{Int}\left(\frac{y_\mu^n}{\Delta}\right) - j\right)$$

$$n_{ij}^n \frac{p_{ij}^n}{(\rho_{ij}^n)^\gamma} = \frac{1}{\Delta^2}\sum_\mu e_\mu \delta\left(\mathrm{Int}\left(\frac{x_\mu^n}{\Delta}\right) - i\right)\delta\left(\mathrm{Int}\left(\frac{y_\mu^n}{\Delta}\right) - j\right) \qquad (6.52)$$

where the notation is as before (Section 6.2), and the cell variables of density, pressure and velocities are derived. The Eulerian part of the calculation is to be carried out on the mesh: namely, the effect of the pressure gradient in the acceleration equation to determine temporary (before the Lagrangian motion) values of the cell velocities. Accordingly, we choose a time step Δt and obtain:

$$\tilde{v}_{xij}^{n+1} = v_{xij}^n - \frac{\Delta t}{2\Delta\rho_{ij}^n}(p_{i+1,j}^n - p_{i-1,j}^n)$$

$$\tilde{v}_{yij}^{n+1} = v_{yij}^n - \frac{\Delta t}{2\Delta\rho_{ij}^n}(p_{i,j-1}^n - p_{i,j-1}^n) \qquad (6.53)$$

It is necessary to ensure that the time step related to the newly obtained velocities satisfies the Courant–Friedrichs–Lewy condition (Chapter 3, equation 3.79). If this is not the case, a reduced time step is taken, and the velocities are recalculated.

Consequently, new particle variables may be defined with a view to the Lagrangian step of the simulation procedure. The essential point of the algorithm here is that the new momenta and energy of the particles are obtained from the cell variables, rather than from the previous particle values, on the assumption that particles within a cell equilibrate. Hence, if the particle μ is in the cell (i, j), it is assigned cell variables of momentum and energy:

$$\mathbf{u}_\mu^{n+1} = (x_\mu^n, y_\mu^n, m_\mu, m_\mu v_{xij}^{n+1}, m_\mu v_{yij}^{n+1}, \tilde{e}_{ij}^{n+1}) \tag{6.54}$$

where

$$\tilde{e}_{ij}^{n+1} = \frac{p_{ij}^n}{(\rho_{ij}^n)^\gamma} \tag{6.55}$$

It remains, therefore, to move the particles to new coordinates according to the equations:

$$\frac{dx_\mu}{dt} = \bar{v}_{x\mu}, \qquad \frac{dy_\mu}{dt} = \bar{v}_{y\mu} \tag{6.56}$$

It is to be noted that in this integration scheme, the particles are moved with a velocity $\bar{\mathbf{v}}_\mu$, which is not equivalent to the momentum $m\mathbf{v}_\mu^{n+1}$ which is transported by the particle. Rather, it is necessary to interpolate $\bar{\mathbf{v}}_\mu$ in both time and space so that for the particle in the cell (i, j) (Figure 6.15):

$$\bar{\mathbf{v}}_\mu = \frac{1}{\Delta^2}(a_{i+1,j}\bar{\mathbf{v}}_{i+1,j} + a_{i+1,j+1}\bar{\mathbf{v}}_{i+1,j+1} + a_{i,j+1}\bar{\mathbf{v}}_{i,j+1} + a_{i,j}\bar{\mathbf{v}}_{i,j}) \tag{6.57}$$

where the cell variables of the velocity are interpolated in time,

$$\bar{\mathbf{v}}_{ij} = \tfrac{1}{2}(\tilde{\mathbf{v}}_{ij}^{n+1} + \mathbf{v}_{ij}^n) \tag{6.58}$$

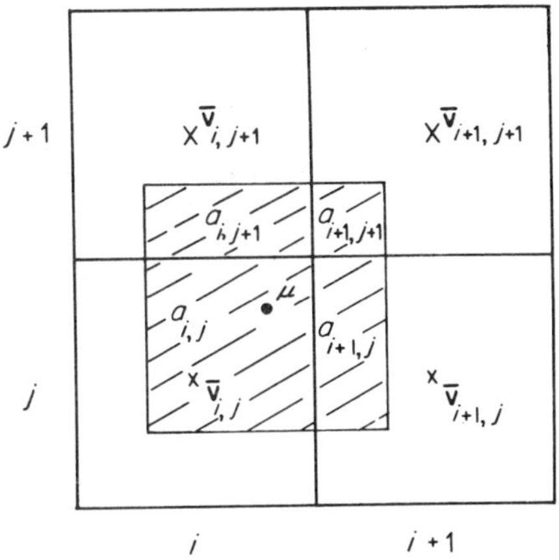

Figure 6.15. Interpolation for a 'particle' velocity, in the collision-dominated PIC method, from the velocities associated with the four nearest cells

and the coefficients a_{ij} are the area overlap of a particle over four surrounding cells (the area-weighting method, Figure 6.15). Hence new particle positions are determined:

$$x_\mu^{n+1} = x_\mu^n + \Delta t\, \bar{v}_{x\mu}$$
$$y_\mu^{n+1} = y_\mu^n + \Delta t\, \bar{v}_{y\mu} \qquad (6.59)$$

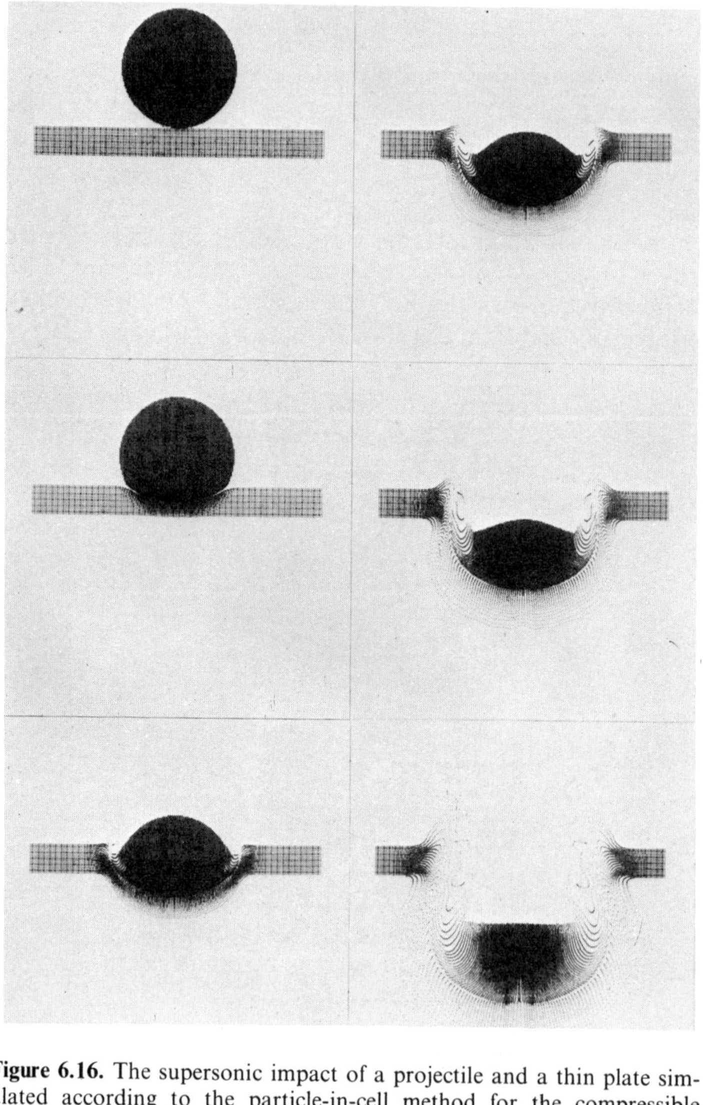

Figure 6.16. The supersonic impact of a projectile and a thin plate simulated according to the particle-in-cell method for the compressible hydrodynamic equations (from Amsden 1966). Figure provided by Group T-3, Los Alamos Scientific Laboratory, Los Alamos, U.S.A.

The time step is completed once all the particle variables are updated, u_μ^{n+1}. It is clear from the algorithm that if, according to equation 6.59, a particle crosses a cell boundary, the mass, momentum, and energy of the new cell are incremented. It is readily verified from the algorithm, that the mass, momentum and adiabatic invariant p/ρ^γ are identically conserved on the mesh.

The collision-dominated particle-in-cell method is patently simple to apply, and has the added advantage of great flexibility: for example, a system comprised of several fluids with internal boundaries may be described with ease by the use of different types of particles. On the other hand, the PIC model uses considerably more storage than is necessary since, in addition to cell variables, all the particle variables at any time step are required to be retained.

Most seriously, however, the PIC method is not precisely centred in time and stability is achieved through diffusion which arises from the calculation of the intermediate velocities (equations 6.53) and which, in subsonic flows, occurs particularly through the area-weighting interpolation. Consequently, momentum and energy tend to be diffused with a diffusion coefficient:

$$D \sim \left\langle v^2 + \frac{\gamma p}{\rho} \right\rangle \Delta t \tag{6.60}$$

Due to the limitations imposed by numerical diffusion, the collision-dominated PIC model has its greatest application in supersonic hydrodynamic problems. The results from a PIC simulation of the impact of a supersonic projectile on a thin plate, are included in Figure 6.16 (Amsden 1966). The effects of compression can clearly be seen as the projectile breaks through the plate. In Figure 6.17, the splash of a drop of compressible fluid from a shallow fluid is illustrated (Amsden 1966). This figure may be compared with the markedly different solutions obtained for an incompressible fluid (Section 9.4).

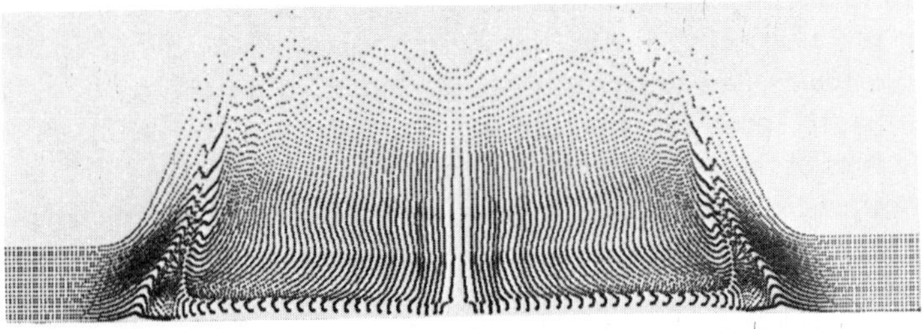

Figure 6.17. The particle-in-cell method used to simulate the splash of a drop into a shallow compressible fluid (cf. Figure 9.12: the splash of an incompressible fluid). The positions of the particles used in the method are plotted and illustrate the physical solution (from Harlow, F. H., and Shannon, J. P., 'The Splash of a Liquid Drop', *J. Appl. Phys.*, **38**, No. 10, pp. 3855–3866, 1967)

CHAPTER VII

Particles in Self-consistent Fields: Atoms and Solids

1 Self-Consistent Fields in the Quantum Theory of Particle Assemblies

(*a*) *Preliminary Discussion.* In applying quantum theory to fundamentally important many-particle assemblies, such as the electrons in an atom, a molecule or a crystal lattice, we may obtain particle-field models which bear a very close resemblance to the classical case described in Chapter 6.* We shall here be concerned only with the time-independent problem and consequently with systems which are in some state of minimum energy. By applying the variational principle to minimize the energy of the whole assembly, we obtain the so-called Hartree–Fock equations which describe the state of each particle of the assembly separately. There are, therefore, N equations for the N interacting independent particles. The approach is very similar to the classical problem in that we attempt to describe the state of one electron, say, and in turn we determine the field which this state produces. The aggregate field of all electrons in the assembly will be used to evaluate improved states of the single electrons. The problem may, therefore, be formulated as one in which the states of individual particles must be determined in the 'self-consistent' fields of all other particles in the assembly.

As in the classical problem, the Hamiltonians which define the particle states are themselves dependent on the states of all the particles, and it follows that the problem is nonlinear. We might say that the Hamiltonians are functions of the 'distributions' of the particles.

In contrast to the classical problem, however, the equations defining the particle states in quantum theory are algebraically more complex and, even computationally, only relatively simple systems have at present been described. We shall be concerned in this chapter with assemblies of particles which interact with long-range forces and, in particular, with the electrostatic forces between electrons and nuclei. It is the electrostatic forces which

* It is assumed that the reader possesses an understanding of the fundamentals of quantum mechanics. See, for example, Dicke and Wittke (1960) or Matthews (1968).

usually dominate the structure of atoms, molecules and crystal lattices. Furthermore, as a first approximation, the massive nuclei may be regarded as fixed, in the sense of only contributing a positive electrostatic potential from fixed coordinates.

As in the classical problem, we shall distinguish between few-particle assemblies and many-particle assemblies. In few-particle assemblies such as individual atoms, ions or simple molecules, the effect of correlations between particles, or 'binary interactions', are in general to be described. The situation is then similar to the 'exact' particle-particle models described in Chapter 5 in that there are $\frac{1}{2}N(N-1)$ interactions. On the other hand, in many-particle assemblies such as the electrons in a crystal lattice, we cannot hope to describe each particle-particle interaction and, as in the classical problem, an averaged field defined by a simple 'exchange term' (Section 7.2) must be used.

(*b*) *Variational Principle and the Hartree Equations.* We consider a system of N electrons, each of mass m, which interact through the Coulomb electrostatic force. According to quantum theory, if relativistic effects and spin-orbit interactions are neglected, the state of the whole system is to be described by a wave function ψ which depends on the coordinates $\mathbf{x}_1, \mathbf{x}_2, \ldots, \mathbf{x}_N$ of all particles in the system:

$$\psi = \psi(\mathbf{x}_1, \mathbf{x}_2, \ldots, \mathbf{x}_N) \tag{7.1}$$

The solutions ψ, describing the whole system, satisfy Schrödinger's equation:

$$\mathrm{H}\psi = E\psi \tag{7.2}$$

The eigenvalue E is the total energy of the system while H is the total Hamiltonian, being the sum of the potential and kinetic energy operators of all the electrons:

$$\mathrm{H} = \sum_{\mu=1}^{N} \mathrm{H}_\mu + \frac{1}{2}\sum_{\mu=1}^{N}\sum_{\substack{\nu=1\\ \nu\neq\mu}}^{N} \frac{e^2}{|\mathbf{x}_\mu - \mathbf{x}_\nu|} \tag{7.3}$$

The first summation on the right-hand side of this equation is over a separate Hamiltonian for each electron which consists of the kinetic energy of each electron and the applied potential energy due to a fixed distribution of nuclear charges. The second double summation describes the electrostatic repulsion between the $\frac{1}{2}N(N-1)$ pairs of electrons. The Hamiltonian H_μ for the μth electron in the potential, V_{nuc}, of the nuclei is:

$$\mathrm{H}_\mu = -\frac{\hbar^2}{2m}\nabla_\mu^2 + V_{\text{nuc}}(\mathbf{x}_\mu) \tag{7.4}$$

$\hbar$ is Planck's constant divided by 2π and, for example, for an atom where the positive nucleus is taken as the origin, the potential V_{nuc} has the form,

$$V_{\text{nuc}} = -\frac{Ne^2}{|\mathbf{x}_\mu|} \tag{7.5}$$

The problem is to determine ψ as an eigenfunction of the equation (7.2) with eigenvalue E. Ignoring the spin coordinates, the function ψ is a function of $3N$ independent variables and therefore to map ψ for only ten values of each independent variable would require 10^{3N} values. Clearly, whenever the number of interacting independent particles is greater than two or three, the need for an approximation, or a reduced description of the system, is essential.

We shall illustrate how a reduced description of any many-particle system may be obtained. The manner by which such a reduced description is obtained is included here in order to stress the essential approximations involved. We are concerned with time-independent problems and with the ground states of the whole system where the total energy E,

$$E = \int \psi^* \mathrm{H} \psi \, \mathrm{d}\tau \tag{7.6}$$

is to be minimized. The integration here is to be carried out over all coordinates of the system $\mathrm{d}\tau = \mathbf{dx}_1, \mathbf{dx}_2, \ldots, \mathbf{dx}_N$. By varying the wave function ψ, we seek those particular solutions which minimize E,

$$\delta E = 0 = \delta \int \psi^* \mathrm{H} \psi \, \mathrm{d}\tau \tag{7.7}$$

The second limitation which is imposed in obtaining a viable description of a many-particle assembly is to look for particular solutions of the separable-variable type. That is, we shall attempt to describe the system by a set of N independent wave functions u_μ for each of the electrons $1 \leqslant \mu \leqslant N$ of the assembly. The problem can therefore be formulated in an analogous manner to the classical N-body problem. This approximation is introduced by writing the total wave function ψ as a product of the individual wave functions for each of the electrons:

$$\psi = u_1(\mathbf{x}_1)\, u_2(\mathbf{x}_2) \cdots u_N(\mathbf{x}_N) \tag{7.8}$$

It is to be noted that this wave function will not satisfy the symmetry requirements of a quantum assembly under the permutations of pairs of particles, and in section (*c*) we shall modify the form of ψ to take such symmetry requirements into account. A wave function of this form (equation 7.8) is an approximation since it does not describe the correlations between pairs of electrons: that is, it is here assumed that each electron moves independently of every other electron.

We may now attempt to obtain N separate equations describing the wave functions u of each electron in the assembly, where each electron will now move in the potential of the fixed nuclei and the potential of all other electrons.

The total Hamiltonian (equation 7.3) operating on such a wave function (equation 7.8) yields:

$$\mathrm{H}\psi = \sum_{\mu=1}^{N} (u_1 \cdots u_{\mu-1}u_{\mu+1} \cdots u_N)\mathrm{H}_\mu u_\mu + \frac{1}{2}\sum_{\mu=1}^{N}\sum_{\substack{\nu=1\\ \mu\neq\nu}}^{N} (u_1 \cdots)\frac{e^2}{x_{\mu\nu}}u_\mu u_\nu \tag{7.9}$$

where we have written $x_{\mu\nu} = |\mathbf{x}_\mu - \mathbf{x}_\nu|$, and each contribution to the Hamiltonian H_μ operates only on the coordinates $\mathbf{x}_\mu$. If this form for $\mathrm{H}\psi$ is included in the energy integral, the integrals over each coordinate in $d\tau$ may be carried out separately except for those variables appearing on the right-hand side of the Hamiltonian operators. We obtain, therefore, for the energy integral:

$$E = \sum_{\mu=1}^{N} \int u_\mu^* \left(\mathrm{H}_\mu + \sum_{\substack{\nu=1\\ \nu\neq\mu}}^{N} \frac{1}{2} \int u_\nu^* \frac{e^2}{x_{\mu\nu}} u_\nu \, \mathbf{dx}_\nu \right) u_\mu \, \mathbf{dx}_\mu \tag{7.10}$$

It has been assumed that each of the separate electron wave functions u_λ are normalized,

$$\int u_\lambda^* u_\lambda \, \mathbf{dx}_\lambda = 1 \tag{7.11}$$

The energy E of the system can now be minimized (equation 7.7) according to the variational principle. We are interested in the set of one-electron wave functions u_μ that minimize E and we vary each wave function δu_μ until the energy is minimized. The variations are to be carried out independently so that each variation is orthogonal to the wave function, namely:

$$\int \delta u_\mu^* u_\mu \, \mathbf{dx}_\mu = 0 \tag{7.12}$$

and the energy integral is minimized:

$$\sum_\mu \int \delta u_\mu^* \left\{ \mathrm{H}_\mu + \sum_{\substack{\nu=1\\ \nu\neq\mu}}^{N} \int u_\nu^* \frac{e^2}{x_{\mu\nu}} u_\nu \, \mathbf{dx}_\nu \right\} u_\mu \, \mathbf{dx}_\mu = 0 \tag{7.13}$$

In order to carry out each variation independently subject to the conditions (equation 7.12), we multiply conditions 7.12 by Lagrange multipliers ϵ_μ and subtract each of the conditions from equation 7.13,

$$\sum_\mu \int \delta u_\mu^* \left\{ \mathrm{H}_\mu + \sum_{\substack{\nu=1\\ \nu\neq\mu}}^{N} \int u_\nu^* \frac{e^2}{x_{\mu\nu}} u_\nu \, \mathbf{dx}_\nu - \epsilon_\mu \right\} u_\mu \, \mathbf{dx}_\mu = 0 \tag{7.14}$$

The use of Lagrange multipliers allows the variations to be carried out separately and it follows that the coefficient of each variation δu_μ^* must separately be equal to zero. Consequently we arrive at N separate equations for each of the wave functions u_μ:

$$\left(\mathrm{H}_\mu + \sum_{\substack{\nu=1\\ \nu\neq\mu}}^{N} \int u_\nu^* \frac{e^2}{x_{\mu\nu}} u_\nu \, \mathbf{dx}_\nu \right) u_\mu = \epsilon_\mu u_\mu \quad (1 \leqslant \mu \leqslant N) \tag{7.15}$$

These equations, known as the Hartree equations, could have been written down heuristically, since they may be regarded as a set (N) of Schrödinger equations for each of the N electrons in the system. The total effective Hamiltonian operating on each one-electron wave function consists of the kinetic energy and nuclear potential energy operators (H_μ) and the sum of the 'expectation potential' $V_{\nu\mu}$ from each of the other $N - 1$ electrons ν for $\nu \neq \mu$:

$$V_{\nu\mu} = e^2 \int \frac{u_\nu^* u_\nu}{|\mathbf{x}_\mu - \mathbf{x}_\nu|} \mathbf{dx}_\nu \tag{7.16}$$

These terms describe the electrostatic repulsion between all pairs of electrons, and thereby the self-consistent field of the assembly.

The set of Hartree equations to be solved in describing a many-particle assembly are a set of N coupled integro-differential equations and, as in the classical case, they are nonlinear since the Hamiltonians operating on each electron wave function depend on the states of all other electrons in the system. It is instructive to proceed further and interpret the electron-electron potential energy operators $V_{\nu\mu}$ (equation 7.16) in terms of the electrostatic potential Φ_ν produced by the electron ν acting on the electron μ at the point $\mathbf{x}_\mu$,

$$\Phi_\nu(\mathbf{x}_\mu) = e \int \frac{u_\nu^* u_\nu}{|\mathbf{x}_\mu - \mathbf{x}_\nu|} \mathbf{dx}_\nu \tag{7.17}$$

This equation is just the integral form of Poisson's equation (cf. Section 6.1) with a charge density $\rho_\nu = e u_\nu^* u_\nu$:

$$\nabla^2 \Phi_\nu = -4\pi e u_\nu^* u_\nu \tag{7.18}$$

In terms of the electrostatic self-consistent potentials Φ_ν, the Hartree equations may be written as:

$$\left\{ -\frac{\hbar^2}{2m} \nabla_\mu^2 + V_{\mathrm{nuc}}(\mathbf{x}_\mu) + \sum_{\substack{\nu=1\\ \nu\neq\mu}}^{N} e\Phi_\nu(\mathbf{x}_\mu) \right\} u_\mu = \epsilon_\mu u_\mu \quad (1 \leqslant \mu \leqslant N) \tag{7.19}$$

where the self-consistent potentials Φ_ν are to be obtained from a set (N) of Poisson's equations:

$$\nabla^2 \Phi_\nu = -4\pi e u_\nu^* u_\nu \quad (1 \leqslant \nu \leqslant N) \tag{7.20}$$

It is to be noted that, when there is a very large number of electrons in the assembly (N large), we might ignore the effect of the self-field of the particle in the summation over the potentials (equation 7.19). Thus, instead of describing the separate potentials Φ_ν produced by each electron, we might use the potential Φ produced by all the electrons. The N separate Poisson's equations (equations 7.20) would in turn be replaced by one equation for the total potential:

$$\nabla^2 \Phi = -4\pi e \sum_{\nu=1}^{N} u_\nu^* u_\nu \tag{7.21}$$

It is clear that such an *average* field would simplify the problem considerably. However, before defining such an average field it is necessary to consider the symmetry properties of the total wave function under permutations of the particles which, in the Hartree equations, have been ignored.

2 Indistinguishability of the Particles and the Exchange Potential

We have obtained the Hartree equations above by assuming that the total wave function of the system could be written as a simple product of one-electron wave functions (equation 7.8). However, the electrons in an assembly are indistinguishable, in the sense that if we interchange one electron with another, the system is in no manner altered. There is no way of labelling the electrons. It is clear that the total Hamiltonian describing the whole system (equation 7.3) is invariant under a permutation or exchange of the individual particle coordinates. We shall show, therefore, that the simple wave function (equation 7.8) is inconsistent with the requirement of indistinguishability of the electrons in an assembly. And we must consider how the Hartree equations are to be altered to take the indistinguishability of the electrons into account.

It is sufficient here to consider a system of two electrons. It is assumed that electron 1 is in state u_α and electron 2 is in state u_β. According to our previous assumption of the separability of the wave function (equation 7.8), the wave function for the whole system is:

$$\psi = u_\alpha(1)u_\beta(2) \tag{7.22}$$

However, since the electrons in the system are indistinguishable, the wave function ψ' in which electron 1 is in state u_β and electron 2 is in state u_α is entirely equivalent to equation 7.22,

$$\psi' = u_\beta(1)u_\alpha(2) \tag{7.23}$$

If these wave functions are equivalent in describing the same system, then, according to the superposition principle, any linear combination is an acceptable wave function

$$\psi'' = c_1 u_\alpha(1)u_\beta(2) + c_2 u_\beta(1)u_\alpha(2)$$

where c_1 and c_2 are constants subject to the requirement that ψ'' be normalized, $|c_1|^2 + |c_2|^2 = 1$. On the other hand, since the particles are indistinguishable there is no reason why either state ψ' (equation 7.23) or ψ (equation 7.22) is more likely and it follows that we should take:

$$|c_1|^2 = |c_2|^2$$

Consequently for such a two-particle system, the correct form for the total wave function, as constructed from one-particle functions and consistent with the requirement of indistinguishability, is:

$$\psi = \frac{1}{\sqrt{2}}\{u_\alpha(1)u_\beta(2) \pm u_\beta(1)u_\alpha(2)\} \tag{7.24}$$

The question arises as to which of the two possible signs in the wave function ψ (equation 7.24) is to be taken. For electrons, according to experiment, the antisymmetric wave function (negative sign) must be taken. All particles, which, in an assembly of identical particles, are to be described by antisymmetric wave functions are said to be fermions.

It is not difficult to extend this idea to an assembly of N indistinguishable fermions. According to experiment, we require that the wave function ψ describing the assembly is antisymmetric under the permutation of any pair of particles. It follows that ψ may be written as a determinant of one-particle wave functions:

$$\psi = \frac{1}{\sqrt{N!}}\begin{vmatrix} u_\alpha(1) & u_\alpha(2) & \cdots & u_\alpha(N) \\ & & & \vdots \\ u_\beta(1) & u_\beta(2) & & \vdots \\ \vdots & \vdots & & \vdots \\ u_\zeta(1) & u_\zeta(2) & \cdots & u_\zeta(N) \end{vmatrix} \tag{7.25}$$

This is the Slater determinant and it is clear that it is consistent with the requirement of the antisymmetry of the wave function for, if we interchange any pair of columns, the sign of the determinant changes. It is noteworthy too that the Pauli exclusion principle follows immediately from the Slater determinant. According to the Pauli principle, a system of identical fermions cannot be in a state described by a wave function constructed of two one-particle wave functions which are the same. Evidently the Slater determinant vanishes if any pair of one-particle wave functions are the same.

It is to be noted that the independent coordinates for one particle consist of three space coordinates and the spin coordinate. In general, therefore, in constructing the Slater determinant (equation 7.25), account is to be taken in the one-particle wave functions of different states as specified by the spin coordinate. We shall not be concerned here with the spin interaction and the reader is referred to the literature (Dicke and Wittke 1960).

We may now return to the central problem and consider how the Hartree equations are to be modified when the effect of the indistinguishability of the electrons in an assembly is included. The variational principle may be applied in precisely the same way as in Section 7.1 but now with a total wave function ψ as specified by the Slater determinant (equation 7.25). Thus, taking the antisymmetric requirement of the wave function into account, it is evident that the Hartree equations (7.15) will now be modified as:

$$\mathrm{H}_\mu u_\mu(\mathbf{x}_\mu) + \sum_{\substack{\nu=1 \\ \nu\neq\mu}}^{N} \int u_\nu^*(\mathbf{x}_\nu)\frac{e^2}{x_{\mu\nu}}\{u_\nu(\mathbf{x}_\nu)u_\mu(\mathbf{x}_\mu) - u_\nu(\mathbf{x}_\mu)u_\mu(\mathbf{x}_\nu)\}\,\mathbf{dx}_\nu = \epsilon_\mu u_\mu(\mathbf{x}_\mu)$$

$$(1 \leqslant \mu \leqslant N) \quad (7.26)$$

The integral terms under the summation sign are identified as the electrostatic repulsive potential between an electron in the state μ and every other electron state. The two-electron wave function appears in the integrals since the self-potential of an electron must be excluded. It is to be noted that the summation in equation 7.26 is over $N - 1$ states but with no loss of generality we may extend the summation over all N states, including the case $\nu = \mu$, since then the two-electron wave function vanishes. The set of N equations (7.26) may therefore be rewritten as:

$$\mathrm{H}_\mu u_\mu + \left(\sum_{\nu=1}^{N}\int u_\nu^*\frac{e^2}{x_{\mu\nu}}u_\nu\,\mathbf{dx}_\nu\right)u_\mu - \sum_{\nu=1}^{N}\left(\int u_\nu^*\frac{e^2}{x_{\mu\nu}}u_\mu\,\mathbf{dx}_\nu\right)u_\nu = \epsilon_\mu u_\mu$$

$$(1 \leqslant \mu \leqslant N) \quad (7.27)$$

This set of N equations in the N one-electron states are the Hartree–Fock equations. The set of terms under the second summation are the exchange potential terms which arise since the electron does not act on itself. The set of terms under the first summation expresses the total electrostatic potential of the assembly of electrons. As in Section 7.1, it is informative to express the integrals in terms of the particle potentials which may be obtained as solutions to Poisson's equation. The first summation is the total potential of the electron assembly:

$$e\Phi(\mathbf{x}) = \sum_\nu \int u_\nu^*(\mathbf{s})\frac{e^2}{|\mathbf{x} - \mathbf{s}|}u_\nu(\mathbf{s})\,\mathbf{ds} \quad (7.28)$$

which again is the integral form of Poisson's equation:

$$\nabla^2\Phi = -4\pi e\sum_{\nu=1}^{N} u_\nu^* u_\nu \quad (7.29)$$

The second set of terms must be expressed separately as the exchange potentials $\Phi_{\mu\nu}$ between any pair of states:

$$e\Phi_{\mu\nu}(\mathbf{x}) = \int u_\nu^*(\mathbf{s})\frac{e^2}{|\mathbf{x} - \mathbf{s}|}u_\mu(\mathbf{s})\,\mathbf{ds} \quad (7.30)$$

where, in differential form, $\Phi_{\mu\nu}$ satisfies Poisson's equation

$$\nabla^2 \Phi_{\mu\nu} = -4\pi e u_\nu^* u_\mu \tag{7.31}$$

There are $\frac{1}{2}N(N-1)$ such equations describing all pairs of states. The Hartree–Fock equations are now simply expressed in terms of these potentials:

$$\left\{-\frac{\hbar^2}{2m}\nabla^2 + V_{\text{nuc}} + e\Phi\right\}u_\mu - \sum_{\nu=1}^{N} e\Phi_{\mu\nu}u_\nu = \epsilon_\mu u_\mu \quad (\text{all } \mu, 1 \leqslant \mu \leqslant N) \tag{7.32}$$

Several points are to be noted with regard to these equations, where it is useful to draw an analogy with the classical assembly of particles interacting with long-range forces. The Hartree–Fock equations (7.32) are a set of N coupled nonlinear equations describing the N one-particle states in the assembly.* The equations are nonlinear and coupled through the self-consistent potentials (equations 7.29 and 7.31). There are $\frac{1}{2}N(N-1)$ interactions between all pairs of states as described by the exchange potential terms $\Phi_{\mu\nu}$, and we may regard this description as 'exact' in the sense that all pairs of interactions are described in analogy to the 'exact' classical case (Chapter 5). And again, as in the classical case, it is clear that only a limited number of one-particle states may be described computationally in this model since there are $\frac{1}{2}N(N-1)$ interactions to be described by the $\frac{1}{2}N(N-1)$ Poisson's equations for the exchange potentials (equations 7.31). The full Hartree–Fock equations are, therefore, only of value for a 'few-electron' system such as an atom or a simple molecule. On the other hand, for a 'many-electron' system (for example, the electrons in a crystal lattice) an averaged self-consistent potential must be introduced, in the same spirit as an averaged field was introduced for the many-particle plasma or gravitational assembly (Chapter 6).

As a first approximation for the *many-particle* assembly interacting with the long-range electrostatic force, the binary interactions described by the exchange potentials may be regarded as small in contrast to the collective potential Φ (Chapter 6). More correctly, however, we may average the exchange potential to take into account the effect of Debye screening, so that the effect of the average exchange potential occurs only over short distances (Bohm and Pines 1950; 1953). Slater (1953, 1968) has therefore suggested the replacement of the exact exchange potential terms by an averaging exchange potential of the form:

$$\Phi_{\text{exc}} = 3|e|\left\{\frac{3}{4\pi}n_{\text{e}}\right\}^{\frac{1}{3}} \tag{7.33}$$

* According to Koopman's theorem, the Hartree–Fock equations (7.32) will, in addition to the ground state, describe the excited states of a many-electron assembly.

where n_e is the density of electrons per unit volume. Consequently the Hartree–Fock equations for a many-electron system describe states in the averaged field:

$$\left\{-\frac{\hbar^2}{2m}\nabla^2 + V_{\text{nuc}} + e\Phi - e\Phi_{\text{exc}}\right\} u_\mu = \epsilon_\mu u_\mu \tag{7.34}$$

With such an averaged field, the self-consistent many-electron problem is greatly simplified, since the $\frac{1}{2}N(N-1)$ pairs of interactions are not directly described. There remains only one field equation.

We shall consider these two limiting cases of assemblies of quantum particles. In analysing the structure of atoms, the assembly of electrons is to be regarded as a few-electron system where the exchange terms for the two-particle states (equation 7.32) are to be included. On the other hand, the electrons in a crystal lattice are to be regarded as a many-particle assembly and the Hartree–Fock equations with the averaged field (equation 7.34) are to be used.

3 The Atom as a Few-Electron Assembly

In applying the Hartree–Fock equations (7.31, 7.32) to the analysis of the structure of atoms, the applied nuclear electrostatic potential has the particularly simple central form. We shall, therefore, assume that the nucleus is fixed at the origin and spherical coordinates (r, θ, ϕ) (Figure 7.1) will be used. The electron potential energy from the nucleus of an atom of atomic number N has the simple form:

$$V_{\text{nuc}} = -\frac{Ne^2}{r} \tag{7.35}$$

It is useful too, to introduce dimensionless variables where the unit of length is the Bohr radius $\hbar^2/me^2$ and the unit of energy is $me^4/\hbar^2$. The states u_μ for the N electrons in the atom are then to be described by the N coupled Hartree–Fock equations:

$$(\tfrac{1}{2}\nabla^2 - \Phi + \epsilon_\mu)u_\mu + \sum_{\nu=1}^{N} \Phi_{\mu\nu}u_\nu = 0 \quad (1 \leqslant \mu \leqslant N) \tag{7.36}$$

Φ is the total potential due to the nuclear charge and all electron states,

$$\nabla^2\Phi = -4\pi N\delta(r) + 4\pi \sum_\mu u_\mu^* u_\mu \tag{7.37}$$

and the exchange potentials between pairs of electron states are defined by the set of $\frac{1}{2}N(N-1)$ Poisson's equations:

$$\nabla^2\Phi_{\mu\nu} = 4\pi u_\nu^* u_\mu \tag{7.38}$$

The effective Hamiltonians acting on each of the electron states u_μ (equation 7.36) are in this form spatial differential operators. As a general

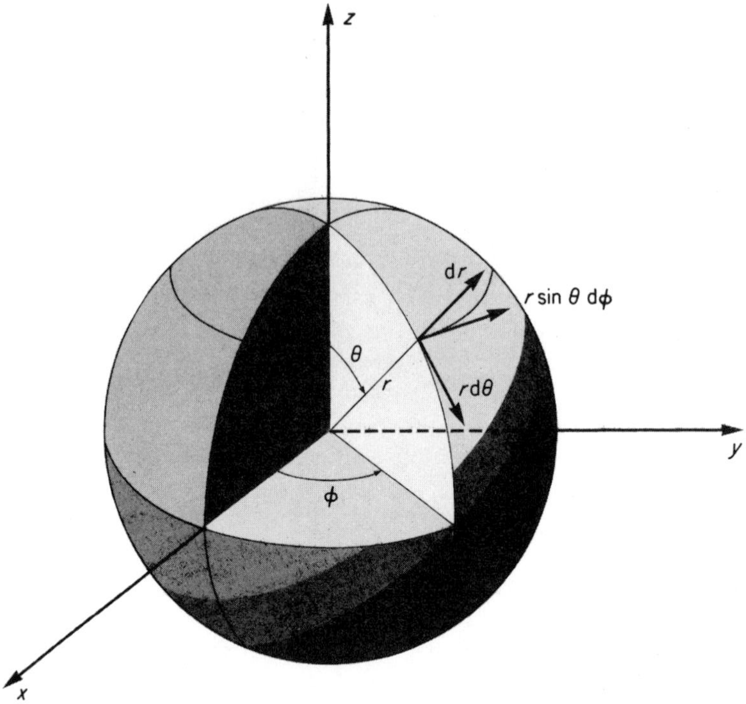

Figure 7.1. Spherical coordinates

approach to the solution of these equations, therefore, we might express the equations on a three-dimensional difference mesh when, according to the methods in Chapter 4, the Hamiltonians become heptadiagonal matrices. The eigenvalues and eigenvectors of such matrices may correspondingly be determined (Chapter 4). However, in obtaining a reasonable resolution along each coordinate (r, θ, ϕ), the Hamiltonian difference matrices become very large and the resulting computational problem is severe. Because of the dominant nuclear central potential, an alternative and successful approach is to expand the solutions in the polar angles (θ, ϕ), while using the difference method in the radial direction.

The Laplacian operator which occurs both in Poisson's equation and in the kinetic energy contributions to the Hamiltonians is expressed in spherical coordinates:

$$\nabla^2 = \frac{1}{r}\frac{\partial^2}{\partial r^2}\cdot r + \frac{1}{r^2 \sin\theta}\cdot\frac{\partial}{\partial\theta}\sin\theta\frac{\partial}{\partial\theta} + \frac{1}{r^2\sin^2\theta}\cdot\frac{\partial^2}{\partial\phi^2}$$

$$= \frac{1}{r}\frac{\partial^2}{\partial r^2}\cdot r + \frac{1}{r^2}\Lambda^2 \tag{7.39}$$

It is evident that the eigenfunctions s_μ of the Laplacian operator may be expressed as functions of the separable-variable type:

$$s_\mu = R_l(r) P_l^m(\theta)\, e^{im\phi} \tag{7.40}$$

where the functions P_l^m satisfy the generalized Legendre equation:

$$\frac{1}{\sin\theta}\cdot\frac{d}{d\theta}\sin\theta\frac{d}{d\theta}P_l^m + \left\{l(l+1) - \frac{m^2}{\sin^2\theta}\right\}P_l^m = 0 \tag{7:41}$$

l and m are integers. The associated radial functions R_l satisfy:

$$\frac{1}{r}\frac{d^2}{dr^2}rR_l - \frac{l(l+1)}{r^2}R_l = \lambda_l R_l \tag{7.42}$$

where λ_l is the related eigenvalue. The generalized Legendre equation has been much studied and the desired solutions (the associated Legendre polynomials (Table 7.1) may be obtained analytically in terms of a power series (Courant

Table 7.1 Spherical harmonics. The associated Legendre polynomials $P_l^m(\cos\theta)$ (solutions to equation 7.41) are related to the normalized spherical harmonics $Y_{lm}(\theta, \phi)$:

$$Y_{lm}(\theta,\phi) = \sqrt{\left\{\frac{2l+1}{4\pi}\cdot\frac{(l-m)!}{(l+m)!}\right\}}P_l^m(\cos\theta)\, e^{im\phi}$$

$l = 0$	$Y_{00} = \dfrac{1}{\sqrt{(4\pi)}}$
$l = 0$	$Y_{10} = \sqrt{\left(\dfrac{3}{4\pi}\right)}\cos\theta$
	$Y_{11} = -\sqrt{\left(\dfrac{3}{8\pi}\right)}(\sin\theta)\, e^{i\phi}$
$l = 2$	$Y_{20} = \sqrt{\left(\dfrac{5}{4\pi}\right)}\left(\dfrac{3}{2}\cos\theta - \dfrac{1}{2}\right)$
	$Y_{21} = -\sqrt{\left(\dfrac{15}{8\pi}\right)}(\sin\theta\cos\theta)\, e^{i\phi}$
	$Y_{22} = \dfrac{1}{4}\sqrt{\left(\dfrac{15}{2\pi}\right)}(\sin^2\theta)\, e^{i2\phi}$
$l = 3$	$Y_{30} = \sqrt{\left(\dfrac{7}{4\pi}\right)}\left(\dfrac{5}{2}\cos^3\theta - \dfrac{3}{2}\cos\theta\right)$
	$Y_{31} = -\dfrac{1}{4}\sqrt{\left(\dfrac{21}{4\pi}\right)}\sin\theta\,(5\cos^2\theta - 1)\, e^{i\phi}$
	$Y_{32} = \dfrac{1}{4}\sqrt{\left(\dfrac{105}{2\pi}\right)}\sin^2\theta\cos\theta\, e^{i2\phi}$
	$Y_{33} = -\dfrac{1}{4}\sqrt{\left(\dfrac{35}{4\pi}\right)}\sin^3\theta\, e^{i3\phi}$

and Hilbert 1953). We may therefore use the known eigenfunctions of the angular part of the Laplacian operators $(1/r^2)\Lambda^2$ to express the atomic solutions of the Hartree–Fock equations (7.36).

(*a*) *The Central-field Approximation.* The central-field approximation is introduced to simplify the set of three-dimensional Hartree–Fock equations so that they may be expressed as differential equations in the one radial dimension (Hartree 1957). Certainly for the inner electron states, the dominant potential is the central nuclear potential. It is therefore to be expected that the total effective potential due to the positive nucleus and the negative electrons in the Hartree–Fock equations depends strongly on the radius and only weakly on the polar angles (θ, ϕ). If the exchange potentials were to be ignored, and if it were assumed that the total effective potential Φ was spherically symmetric, say $\tilde{\Phi}(r)$, then each solution to the Hartree–Fock equations (7.36) may be expressed as:

$$u_\mu = R_{nl}(r)P_l^m(\theta)\,\mathrm{e}^{\mathrm{i}m_l\phi} \tag{7.43}$$

P_l^m are the associated Legendre polynomials as above and the radial wave functions satisfy the ordinary differential equation:

$$\frac{\mathrm{d}^2}{r\,\mathrm{d}r^2}(rR_{nl}) - \left\{2\tilde{\Phi}(r) + \frac{l(l+1)}{r^2} - 2\epsilon_{nl}\right\}R_{nl} = 0 \tag{7.44}$$

Before a more exact self-consistent solution is considered, several points are to be noted with regard to such solutions. An electron wave function of the simple product form (equation 7.43) is said to be of the central-field type and is specified by the three parameters $\mu = (n, l, m_l)$, related to (r, θ, ϕ) respectively. n is the principal quantum number, while the quantum number l is related to the total angular momentum of the electron state and m_l is termed the magnetic quantum number. In the special case of the hydrogen atom, the central potential energy is just the nuclear potential,

$$\Phi(r) = -\frac{1}{r} \tag{7.45}$$

and it may readily be shown that the energy eigenvalues depend simply on the principal quantum number,*

$$\epsilon_n = -\frac{1}{2n^2} \tag{7.46}$$

For each value of n, l can take $n - 1$ values,

$$l + 1 \leqslant n \tag{7.47}$$

* See, for example, Dicke and Wittke (1960).

More generally, however, (equation 7.44) the energy eigenvalue ϵ_{nl} depends on both the quantum numbers n and l, and the set of electron states with the same quantum numbers n and l are said to form an 'nl-group'.

Within each 'nl-group' there are $2l + 1$ possible values for the magnetic quantum number:

$$-l \leqslant m_l \leqslant l \tag{7.48}$$

If the two possible spin states are taken into account, each 'nl-group' consists of $2(2l + 1)$ electron states. According to the Pauli exclusion principle discussed in Section 2 with reference to the Slater determinant, no two electrons may occupy the same state. It follows that we may assign electrons accordingly to the N eigenfunctions of lowest energy.

We may now turn to the central problem here of obtaining self-consistent solutions to the Hartree–Fock equations (7.36, 7.37 and 7.38). In the central-field approximation and according to the discussion above, it is assumed that all the wave functions are of the central-field type (equation 7.43). We are required to determine the self-consistent potentials and exchange potentials defined by such wave functions (equation 7.38):

$$\left\{\frac{1}{r}\frac{\partial^2}{\partial r^2} + \frac{1}{r^3}\Lambda^2\right\} r\Phi_{nl,n'l'} = 4\pi R_{nl} R_{n'l'} Y^*_{l'm'} Y_{l_m} \tag{7.49}$$

$Y_{lm}(\theta, \phi)$ are the associated Legendre polynomials (Table 7.1). It follows that the total potential and exchange potentials are functions of the polar angles (θ, ϕ). However, it is evident that the potentials are greatly simplified when use is made of the *addition theorem* for spherical harmonics (Courant and Hilbert 1953):

$$P_l(\cos\gamma) = \frac{4\pi}{2l + 1} \sum_{m=-l}^{l} Y^*_{lm}(\theta', \phi') Y_{lm}(\theta, \phi) \tag{7.50}$$

where $\cos\gamma = \cos\theta\cos\theta' + \sin\theta\sin\theta'\cos(\phi - \phi')$. It immediately follows that the exchange potentials due to all the electrons within a complete nl-group are independent of the azimuthal angle, ϕ. The exchange potentials between electron states in complete groups may be expanded in terms of ordinary Legendre polynomials P_k,

$$\Phi_{nl,n'l'} = \sum_{k=|l-l'|}^{l+l'} \hat{\Phi}_{nl,n'l',k}(r) P_k \tag{7.51}$$

Thus for configurations of *complete groups*, the coupled Hartree–Fock equations and the associated Poisson's equations for the potentials are:

$$\left\{\frac{1}{r}\cdot\frac{d^2}{dr^2}\cdot r - \frac{k(k+1)}{r^2}\right\}\hat{\Phi}_{nl,n'l',k}(r) = (2k+1)R_{nl}(r)R_{n'l'}(r)$$

$$\text{(for all } nl, n'l', \text{ and } k, 0 \leqslant k \leqslant l + l') \tag{7.52}$$

Correspondingly there are m Hartree–Fock equations for the m radial wave functions related to each nl-group:

$$\left[\frac{1}{r}\cdot\frac{\mathrm{d}^2}{\mathrm{d}r^2}\cdot r - \frac{l(l+1)}{r^2} - 2\left\{\frac{N}{r^2} - \sum_{n'l'} 2(2l'+1)\hat{\Phi}_{n'l',n'l',0}\right\} + 2\epsilon_{nl}\right] R_{nl}(r)$$

$$= 2\sum_{n'l'}\sum_{k=0}^{l+l'} \beta_{ll'k}\hat{\Phi}_{nl,n'l',k}(r)R_{n'l'}(r) \tag{7.53}$$

The various terms in this equation are readily identified. The Hamiltonian on the left-hand side includes the effect of the radial kinetic energy, the orbital kinetic energy, the potential due to the positive charge on the nucleus, and the total potential due to the $2(2l'+1)$ electrons in each $n'l'$-group. The terms on the right-hand side of the equation are the effective exchange potential terms, where the constants $\beta_{ll'k}$ arise from the coupling of Legendre polynomials and are specified by the expansion of the product of polynomials $P_l, P_k, (0 \leqslant k \leqslant l + l')$,

$$P_l P_k = \sum_{l'} \beta_{ll'k} P_{l'} \tag{7.54}$$

The coefficients $\beta_{ll'k}$ are tabulated in Table 7.2, where for finite $\beta_{ll'k}$, k may take the values $|l - l'| \leqslant k \leqslant l + l'$.

Table 7.2 Coupling coefficients of the harmonics of the exchange potential for complete groups

$\beta_{ll'k}$

	$l' = 0$	$l' = 1$	$l' = 2$
$l = 0$	$\beta_{000} = 1$	$\beta_{011} = 1$	$\beta_{022} = 1$
$l = 1$	$\beta_{101} = \frac{1}{3}$	$\beta_{110} = 1$	$\beta_{121} = \frac{2}{3}$
		$\beta_{112} = \frac{2}{5}$	$\beta_{123} = \frac{3}{7}$
$l = 2$	$\beta_{202} = \frac{1}{5}$	$\beta_{211} = \frac{2}{5}$	$\beta_{220} = 1$
		$\beta_{213} = \frac{9}{35}$	$\beta_{222} = \frac{2}{7}$
			$\beta_{224} = \frac{2}{7}$

(*b*) *Successive Solution for the Potentials and Radial Wave Functions.* With the aid of the central-field approximation, the set (N) of Hartree–Fock equations for the wave functions of N electron states in three space dimensions have been reduced to a set of coupled ordinary differential equations in the independent variable r (equation 7.53). All the electrons within a complete nl-group possess the same radial wave function, R_μ, where the index μ refers to a particular nl-group in the atom. We apply the difference method (Chapter 2) to discretize the radial dimension as r_j, $1 \leqslant j \leqslant J$, so that the coupled equations (7.52, 7.53), are written as the matrix equations (Chapter 4):

$$(\mathrm{H}_\mu + 2\epsilon_\mu)R_\mu = \sum_\nu \mathrm{h}_{\mu\nu} R_\nu \quad (\text{all } \mu, 1 \leqslant \mu \leqslant m) \tag{7.55}$$

$$\mathrm{L}_k \hat{\Phi}_{\mu\nu k} = R_\mu R_\nu \quad (\text{all } \mu, \nu, k) \tag{7.56}$$

m is the total number of nl-groups in the atom. In differential form, the operators H_μ, L_k, and $\mathrm{h}_{\mu\nu}$ are:

$$\mathrm{H}_\mu = \frac{1}{r} \cdot \frac{\mathrm{d}^2}{\mathrm{d}r^2} \cdot r - \frac{l(l+1)}{r^2} - 2\left\{\frac{N}{r^2} - \sum_{\nu=1}^{m} 2(2l'+1)\hat{\Phi}_{\nu\nu 0}\right\} \tag{7.57}$$

$$\mathrm{L}_k = \left\{\frac{1}{r} \cdot \frac{\mathrm{d}^2}{\mathrm{d}r^2} \cdot r - \frac{k(k+1)}{r^2}\right\} \frac{1}{2k+1} \tag{7.58}$$

$$\mathrm{h}_{\mu\nu} = \sum_{k=0}^{l+l'} \beta_{ll'k} \hat{\Phi}_{\mu\nu k} \tag{7.59}$$

Correspondingly, in difference form, both H_μ and L_k are *tridiagonal* matrices, while $\mathrm{h}_{\mu\nu}$ are diagonal matrices on the mesh r_j. The problem is a nonlinear one since the tridiagonal matrices H_μ are functions of the potentials, $\mathrm{H}_\mu(\Phi_{\nu\nu})$.

We are therefore required to solve for the radial wave functions R_μ with the radial two-point boundary conditions:

$$R_\mu(0) = 0, \qquad R_\mu(r) \to 0 \quad \text{as } r \to \infty \tag{7.60}$$

subject to the condition that the radial wave functions are normalized:

$$\int_0^\infty R_\mu^2(r) r^2 \,\mathrm{d}r = 1 \tag{7.61}$$

The boundary conditions on the potentials are:

$$\begin{aligned} \Phi_{\mu\nu}(r) &\to 0, \qquad r \to \infty \\ \frac{\mathrm{d}\Phi_{\mu\nu}}{\mathrm{d}r} &= 0, \qquad r = 0 \end{aligned} \tag{7.62}$$

These two-point boundary conditions (equations 7.60, 7.62) may be included in difference form in the tridiagonal matrices H_μ and L_k respectively, according to the method of Section 4.2. It then follows that the normalization conditions (equation 7.61) will define the eigenvalues in the inhomogeneous equations (7.55). The methods for resolving tridiagonal matrix equations are rapid and have been discussed in Sections 4.2, 4.3 and 4.7.

The main difficulty in resolving these coupled equations occurs through nonlinearity, since the Hamiltonians are functions of the potentials. An iterative method is therefore to be used in obtaining self-consistent solutions.

It is to be noted that while there are m equations for the radial wave functions there is a very much larger number of equations for the harmonics of the potentials and exchange potentials. There are $\frac{1}{2}m(m-1)$ pairs of groups and for each pair of groups there are of the order $k = |l - l'|$ harmonics.

The simplest iterative method is to 'guess' an appropriate set of radial wave functions, from which the potentials may be determined. These potentials are then used to construct the Hamiltonians and exchange potential terms in the equations for the radial wave functions. Improved solutions to the radial wave functions are in turn determined, and the procedure is continued until convergence is achieved.

At the iteration step p, the radial functions $R_\mu^{(p)}$ and associated eigenvalues $\epsilon_\mu^{(p)}$ are known. The potentials and exchange potentials are accordingly determined:

$$\mathrm{L}_k \hat{\Phi}_{\mu\nu k}^{(p)} = R_\mu^{(p)} R_\nu^{(p)} \tag{7.63}$$

Since each of the matrices L_k are tridiagonal and linear, these equations are readily solved by tridiagonal inversion (Section 4.3). From the potentials, the Hamiltonians and exchange operators are constructed:

$$\mathrm{H}_\mu^{(p)} = \mathrm{H}_\mu(\hat{\Phi}_{\mu\nu 0}^{(p)}), \qquad \mathrm{h}_{\mu\nu}^{(p)} = \mathrm{h}_{\mu\nu}(\hat{\Phi}_{\mu\nu k}^{(p)}) \tag{7.64}$$

Consequently the new radial wave functions are evaluated from the m 'Schrödinger' equations:

$$\{\mathrm{H}_\mu^{(p)} + 2\epsilon_\mu^{(p+1)}\} R_\mu^{(p+1)} = \sum_\nu \mathrm{h}_{\mu\nu}^{(p)} R_\nu^{(p)} \tag{7.65}$$

subject to the normalization conditions:

$$\sum_j (R_{\mu j}^{(p+1)})^2 r_j^2 \, \Delta r_j = 1 \tag{7.66}$$

The normalization conditions define the eigenvalues $\epsilon_\mu^{(p+1)}$ in the inhomogeneous equations (7.65). The simplest procedure in solving these equations at each step p is to iterate by assuming a value for the eigenvalue $\epsilon_\mu^{(p+1)}$ (say $\epsilon_\mu^{(p)}$). Equation 7.65 is then a simple tridiagonal matrix equation which is inverted (Section 4.3). The resulting radial wave function will not satisfy the normalization condition (equation 7.66) so that the eigenvalue is varied accordingly and equation 7.65 solved again until consistent solutions to equations 7.65 and 7.66 are obtained.

The essential method of solution for the atomic wave functions outlined here has successfully been applied to most of the atoms in the periodic table.* We have only discussed the solution for atoms or ions with complete nl-groups, since, strictly, the radial wave functions are then independent of the magnetic quantum number m_l. When an atom or ion contains an incomplete nl-group (for example, say, the potassium atom) the assumption

* For a detailed discussion of these methods the reader is referred to Hartree (1957).

that the wave functions are of the central-field type (equation 7.43), or of the separable-variable type, is not strictly valid. However, in general, very few electrons in the atom will be in incomplete *nl*-groups, so that even for atoms with incomplete groups, the central-field approximation is a very good approximation. For atoms with incomplete groups the radial wave functions will no longer be strictly independent and to preserve the orthogonality of the radial wave functions, off-diagonal Lagrange multipliers must be introduced in the approximate Hartree–Fock equations (7.53) (Hartree 1957).

An alternative approach to employing difference equations in the radial dimension is to expand the solutions in terms of known radial functions. Indeed, we may regard the expansion method as a general approach to the solution of atomic and molecular wave functions (Slater 1963). For a discussion of the expansion method the reader is referred to Roothan (1951) and Roothan and Bagus (1963).

4 Solids as an Example of Many-Electron Systems

The electronic structure of solids is a classic example of a many-body assembly to be described by quantum theory. As is clear from the previous discussion in this chapter, the algebraic complexity of describing a many-particle assembly according to quantum theory is severe and for a solid only a limited computational description has as yet been achieved. In the description of solids many phenomena should be incorporated, including the self-consistent interactions between ions and electrons, between 'phonons' and electrons, and between imperfections in the solid and the electrons.

In order to reduce the problem to manageable proportions many approximations are made. In determining the electronic structure of solids, the most fundamental first approximation is to assume that the nuclei are cold and fixed rigidly at a set of lattice points. That is, it is assumed that the nuclei are stationary and make up a perfect crystal. We may then define a unit cell in the crystal which is repeated and which can be defined by three vectors $(\mathbf{a}_1, \mathbf{a}_2, \mathbf{a}_3)$. It is sufficient here to consider a crystal composed of one type of nucleus with one nucleus per unit cell. The location of each nucleus in the crystal is specified by a lattice vector $\mathbf{R}_i$:

$$\mathbf{R}_i = i\mathbf{a}_1 + j\mathbf{a}_2 + k\mathbf{a}_3 \tag{7.67}$$

where i, j, k are integers. From the unit lattice vectors $\mathbf{a}_i$ we shall define reciprocal lattice vectors $\mathbf{b}_j$ such that,

$$\mathbf{a}_i \,.\, \mathbf{b}_j = \delta_{ij} \tag{7.68}$$

We may simplify the many-electron problem in the crystal by ignoring the surfaces of the solid and by assuming that the unit cell is repeated over all space. It follows that attention can be confined to the unit cell of the lattice with the assumption of periodic boundary conditions, so that functions which may be assumed to be periodic over the unit cell can always be expanded over an infinite set of reciprocal lattice vectors $\mathbf{K}_i$:

$$\mathbf{K}_i = 2\pi(i_1\mathbf{b}_1 + i_2\mathbf{b}_2 + i_3\mathbf{b}_3) \tag{7.69}$$

where i_1, i_2, i_3 are integers. For example, the electrostatic potential of the nuclei Φ_{nuc} may be described by the set of Fourier amplitudes $\hat{\Phi}(\mathbf{K}_i)$:

$$\Phi_{\text{nuc}}(\mathbf{r}) = \sum_{\mathbf{K}_i} \hat{\Phi}(\mathbf{K}_i)\, e^{i\mathbf{K}_i \cdot \mathbf{r}} \tag{7.70}$$

We are therefore required to solve the Hartree–Fock equations for the N (a large number) electrons in the crystal:

$$\left\{ -\frac{\hbar^2}{2m}\nabla^2 + e\Phi_{\text{nuc}} + e\Phi - e\bar{\Phi}_{\text{exc}} \right\} u_\mu = \epsilon_\mu u_\mu \quad (1 \leqslant \mu \leqslant N) \tag{7.71}$$

for, in general, the N lowest energy states ϵ_μ. As before, Φ is the potential of the electrons which in principle is to be determined self-consistently from Poisson's equation (7.29), and, according to the discussion of Section 7.2, we shall assume an averaged exchange potential $\bar{\Phi}_{\text{exc}}$ given by equation 7.33.

However, even if we knew the potentials in the Hartree–Fock equations (7.71), the difficulty of solving the then linear equations for a sufficient number of electrons remains severe, and generally in the study of the electronic configurations of solids, little attempt has yet been made at solving the self-consistent problem. *Ab initio*, therefore, a given potential due to the nuclei, electrons and exchange terms is assumed: say, the total potential energy of the crystal is V,

$$V = e\Phi_{\text{nuc}} + e\Phi - e\bar{\Phi}_{\text{exc}} \tag{7.72}$$

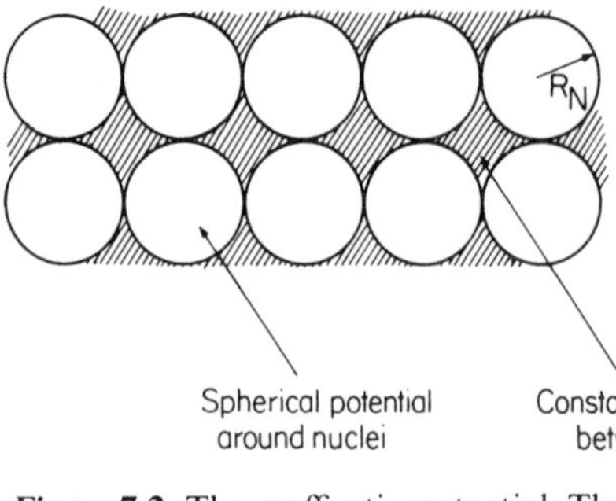

Figure 7.2. The muffin-tin potential. The potential within the touching spheres centred on each nucleus is assumed to be spherically symmetric. Between the spheres, the potential is assumed constant

An example of such an artificial assumed potential which is used is the so-called 'muffin-tin' potential (Figure 7.2). Each nucleus is assumed to be surrounded by a sphere, where any pair of neighbouring spheres just touch. Within each sphere the crystal potential is assumed to be spherically symmetric, while the potential in the interstices between spheres is assumed to be constant and may be regarded as zero by normalizing all energies with respect to it.

With these approximations (which are clearly severe), we may turn to consider how the linear Hartree–Fock equations,

$$\left\{-\frac{\hbar^2}{2m}\nabla^2 + V(\mathbf{r})\right\}u_\mu = \epsilon_\mu u_\mu \quad (\text{all } \mu, 1 \leqslant \mu \leqslant N) \tag{7.73}$$

are to be resolved. While the potential $V(\mathbf{r})$ is to be considered periodic over the unit cell, it is clear that the wave functions u_μ may not be so regarded. If the potential were constant, the solutions to the 'Schrödinger' equations (7.73) are well known as plane waves of wavenumber $\mathbf{k}$,

$$u(\mathbf{r}) = e^{\pm i\mathbf{k}.\mathbf{r}} \tag{7.74}$$

where $\hbar k/m$ is the momentum associated with the particle. Correspondingly in a periodic potential, an important theorem, the Bloch theorem (Floquet's theorem in mathematics), specifies that the solutions to equation 7.73 must have the form:

$$u_\mathbf{k}(\mathbf{r}) = e^{\pm i\mathbf{k}.\mathbf{r}}\pi_\mathbf{k}(r) \tag{7.75}$$

where $\pi_\mathbf{k}$ is a periodic function over the unit cell,*

$$\pi_\mathbf{k}(\mathbf{r}) = \pi_\mathbf{k}(\mathbf{r} + \mathbf{R}_i) \tag{7.76}$$

Since $\pi_\mathbf{k}$ is periodic over the unit cell, we can expand it as an infinite Fourier series over the reciprocal lattice vectors $\mathbf{K}_i$,

$$\pi_\mathbf{k}(\mathbf{r}) = \sum_{\mathbf{K}_i} C(\mathbf{K}_i)\, e^{i\mathbf{K}_i.\mathbf{r}} \tag{7.77}$$

Alternatively we may write the wave function itself (equation 7.75) as a set of plane waves:

$$u_\mathbf{k}(\mathbf{r}) = \sum_{\mathbf{K}_i} C(\mathbf{K}_i)\, e^{i(\mathbf{k}+\mathbf{K}_i).\mathbf{r}} \tag{7.78}$$

Several points are to be noted with regard to such a wave function. If the solid were of infinite extent the parameter $\mathbf{k}$ could take all values. However, because of the properties of Bloch waves, values of $\mathbf{k}$ which are incremented

* For an elementary discussion of Bloch waves and their simple properties, the reader is referred, for example, to Dekker (1957).

by the reciprocal lattice vectors are related: for $\mathbf{k}' = \mathbf{k} + \mathbf{K}_j$,

$$\begin{aligned} u_{\mathbf{k}}(\mathbf{r}) &= \mathrm{e}^{\mathrm{i}(\mathbf{k}' - \mathbf{K}_j).\mathbf{r}} \pi_{\mathbf{k}}(\mathbf{r}) \\ &= \mathrm{e}^{\mathrm{i}\mathbf{k}'.\mathbf{r}} \pi_{\mathbf{k}}(\mathbf{r})\, \mathrm{e}^{-\mathrm{i}\mathbf{K}_j.\mathbf{r}} \end{aligned}$$

But since $\pi_{\mathbf{k}}$ is periodic over the unit cell, we obtain

$$\begin{aligned} u_{\mathbf{k}}(\mathbf{r}) &= \mathrm{e}^{\mathrm{i}\mathbf{k}'.\mathbf{r}} \pi_{\mathbf{k}'}(\mathbf{r}) \\ &= u_{\mathbf{k}'}(\mathbf{r}) \end{aligned} \tag{7.79}$$

$u_{\mathbf{k}'}(\mathbf{r})$ is also a satisfactory wave function of Bloch form. Consequently we may in particular identify those long wavelength values of $\mathbf{k}$ between zero and the reciprocal lattice vector of smallest magnitude. This region of $\mathbf{k}$-space is termed the first Brillouin zone, and the wave functions within the first Brillouin zone have lowest energy and are said to belong to the first energy band. Successively higher Brillouin zones contain wave functions associated with higher energy bands.

In practice, for finite crystals, the boundary conditions on the wave function demand discrete values of $\mathbf{k}$, though they are very close in $\mathbf{k}$-space. If there are N unit cells in the crystal, $\mathbf{k}$ may take $2N$ values in the first Brillouin zone. It follows that, according to the Pauli exclusion principle, $2N$ electrons may be assigned to the first energy band, followed by $2N$ electrons to the next energy band etc.

Accordingly, we may express the problem of determining the electronic configuration of a solid, as requiring solutions to the Hartree–Fock equations (7.73) for various values of $\mathbf{k}$, $0 \leqslant |\mathbf{k}| \leqslant |\mathbf{K}_1|$, and for a sufficient number of bands n to describe the ground states and the first excited states. We wish to map the energy eigenvalue $\epsilon_n(\mathbf{k})$ as a function of $\mathbf{k}$ for the lower bands n.

5 Expansion of the Hartree–Fock Equations for Bloch Waves

By a suitable choice of basis functions, the Hamiltonian operator in the Hartree–Fock equations may be expressed as an infinite matrix operator for each value of $\mathbf{k}$ (Section 7.4). The determination of the lowest eigenvalues of the matrix then provides values of the electron energy in the crystal for the lowest bands.

(*a*) *Plane Waves*. To illustrate the procedure we shall first expand the Hamiltonian operator in terms of plane waves. As has been discussed above, the given muffin-tin potential is expanded over the unit cell as an infinite Fourier series, in accordance with the assumption of periodicity over the unit cell:

$$V(\mathbf{r}) = \sum_l \hat{V}(\mathbf{K}_l)\, \mathrm{e}^{\mathrm{i}\mathbf{K}_l.\mathbf{r}} \tag{7.80}$$

$\hat{V}(\mathbf{K}_l)$ is the amplitude of the Fourier mode with wavenumber $\mathbf{K}_l$, a reciprocal lattice vector. The amplitudes of the Fourier coefficients can be determined by the inverse Fourier integral (Chapter 2). The required wave function must have the form of a Bloch function,

$$u_{\mathbf{k}}(\mathbf{r}) = \sum_i C(\mathbf{K}_i)\, e^{i(\mathbf{k}+\mathbf{K}_i).\mathbf{r}}$$

which we may regard as an infinite set of plane waves with wavenumbers $\mathbf{k} + \mathbf{K}_i$. We examine the Hartree–Fock Hamiltonian operator (equation 7.73) operating on such a function:

$$\left\{-\frac{\hbar^2}{2m}\nabla^2 + \sum_l \hat{V}(\mathbf{K}_l)\, e^{i\mathbf{K}_l.\mathbf{r}}\right\} \sum_i C(\mathbf{K}_i)\, e^{i(\mathbf{k}+\mathbf{K}_i).\mathbf{r}}$$
$$= \epsilon(\mathbf{k}) \sum_i C(\mathbf{K}_i)\, e^{i(\mathbf{k}+\mathbf{K}_i).\mathbf{r}} \qquad (7.81)$$

The plane waves are eigenfunctions of the kinetic energy operator so that this equation may be re-expressed as:

$$\sum_i \left\{\frac{\hbar^2}{2m}(\mathbf{k} + \mathbf{K}_i)^2\, e^{i(\mathbf{k}+\mathbf{K}_i).\mathbf{r}} + \sum_l \hat{V}(\mathbf{K}_l)\, e^{i(\mathbf{k}+\mathbf{K}_l+\mathbf{K}_i).\mathbf{r}}\right\} C(\mathbf{K}_i)$$
$$= \epsilon(\mathbf{k}) \sum_i C(\mathbf{K}_i)\, e^{i(\mathbf{k}+\mathbf{K}_i).\mathbf{r}} \qquad (7.82)$$

If this equation is multiplied by a plane wave $e^{-i(\mathbf{k}+\mathbf{K}_j)}$ and integrated over the volume of the unit cell, then due to the orthogonality properties of such waves, we obtain:

$$\frac{\hbar^2}{2m}(\mathbf{k} + \mathbf{K}_j)^2 C(\mathbf{K}_j) + \sum_i \hat{V}(\mathbf{K}_j - \mathbf{K}_i) C(\mathbf{K}_i) = \epsilon(\mathbf{k}) C(\mathbf{K}_j) \qquad (7.83)$$

This is an infinite set of linear homogeneous simultaneous equations for the coefficients $C(\mathbf{K}_1), C(\mathbf{K}_2), \ldots$ of the Bloch waves which form the wave functions u. It is convenient to interpret the equation as a matrix equation where each element of the Hamiltonian matrix is:

$$H_{ji} = \frac{\hbar^2}{2m}(\mathbf{k} + \mathbf{K}_j)^2 \delta_{ij} + \hat{V}(\mathbf{K}_j - \mathbf{K}_i) \qquad (7.84)$$

We notice that the diagonal elements depend on the wavenumber $\mathbf{k}$ in the first Brillouin zone. Equations 7.83 are an eigenvector–eigenvalue problem for each value of $\mathbf{k}$,

$$H(\mathbf{k})_{ji} C(\mathbf{K}_i) = \epsilon(\mathbf{k}) C(\mathbf{K}_j) \qquad (7.85)$$

For each value of $\mathbf{k}$, the matrix H_{ji} is infinite, and each eigenvalue $\epsilon_n(\mathbf{k})$ is associated with a particular energy band. The corresponding eigenvectors $C_n(\mathbf{K}_i)$ define the related wave function when the Bloch sum is formed.

Computationally we cannot, of course, solve an infinite matrix equation, but if a sufficiently large finite matrix is taken (small $\mathbf{K}_i$ values), then it is to be assumed that the low eigenvalues $\epsilon_n(\mathbf{k})$ (small n) approximate the required exact eigenvalues. And we are only concerned with obtaining the energies in the low energy bands. The solutions must be repeated, however, for various values of $\mathbf{k}$ in the first Brillouin zone.

The computational solution of such a matrix equation has been discussed in Chapter 4 where, for a matrix with nonzero elements everywhere (equation 7.84), the problem is severe. To obtain solutions, therefore, the truncation of the Hamiltonian matrix to a matrix of relatively small dimension is required. The accuracy of the solutions for the energy $\epsilon_n(\mathbf{k})$ then depends on the choice of basis functions which have been used to expand the Hamiltonian (equation 7.84). Here we have used simple plane waves, and it is clear that to describe the core electrons, namely those electrons closely tied to a nucleus, many plane waves would be required since they are patently not a good approximation to the core or nearly atomic electrons. It follows that such an expansion after the Hamiltonian has been truncated is not an effective method to the solution of the problem.

We may obtain more viable methods of solution by using, as basis functions, functions which more closely approximate the expected wave functions in the crystal. We shall briefly discuss the principles of such an improved approach to the problem, but for a more general discussion the reader is referred to Slater (1965).

(b) Orthogonalized Plane Waves. It has been pointed out in the previous section that, while the use of plane waves is a possibly good description of the valence or nearly free electrons in the crystal, many such plane waves would be required to describe the atomic or nearly atomic orbitals. The opposite extreme of a plane wave which represents a free electron, is to consider a wave function constructed from the atomic orbitals a_α of the free atom (Section 7.3). However, the wave functions in the crystal must be Bloch functions, which can be constructed by taking a sum of atomic wave functions at different lattice sites in the crystal $\mathbf{R}_l$,

$$\phi_{\alpha\mathbf{k}}(\mathbf{r}) = \sum_l \mathrm{e}^{i\mathbf{k}.\mathbf{R}_l} a_\alpha(\mathbf{r} - \mathbf{R}_l) \tag{7.86}$$

It is easily seen that such a wave function satisfies the requirements of a Bloch function (equation 7.75). If we were only to use such functions as the basis functions in the expansion of the Hamiltonian, the method is described as the tightly bound approximation. On the other hand, such a simplified description would not adequately describe the valence electrons.

In the method of orthogonalized plane waves (Herring 1940, Slater 1965), a composite set of basis functions is taken. This consists of a limited number of Bloch sums of atomic orbitals ϕ_α and, to describe the valence electrons, a number of plane waves:

$$\zeta_{i\mathbf{k}} = \mathrm{e}^{\mathrm{i}(\mathbf{k}+\mathbf{K}_i).\mathbf{r}} \tag{7.87}$$

In general, the set of basis functions (equations 7.86, 7.87) will not be orthogonal. Although the atomic orbitals a_α at the same lattice site will be orthogonal, the Bloch sums ϕ_α will not be orthogonal to each other. However, the overlap integrals of ϕ_α will only be finite due to the very small overlap between neighbouring atoms. If this effect is ignored, the Bloch sums ϕ_α may be regarded orthogonal. Alternatively they can be pre-orthogonalized by, say, the Schmidt procedure.

We can improve the set of basis functions (equations 7.86, 7.87) by also requiring that the plane waves $\zeta_{i\mathbf{k}}$ be orthogonal to the Bloch sums $\phi_{\alpha\mathbf{k}}$. To do this, the plane waves are to be modified by subtracting the overlap with each sum of atomic wave functions,

$$\psi_{i\mathbf{k}}(\mathbf{r}) = \mathrm{e}^{\mathrm{i}(\mathbf{k}+\mathbf{K}_i).\mathbf{r}} - \sum_\alpha A_{\alpha i}(\mathbf{k})\phi_{\alpha\mathbf{k}}(\mathbf{r}) \tag{7.88}$$

where the sum is over the atomic orbitals α and the coefficients $A_{\alpha i}$ are the overlap integrals:

$$A_{\alpha i} = \int \phi^*_{\alpha\mathbf{k}}\, \mathrm{e}^{\mathrm{i}(\mathbf{k}+\mathbf{K}_i).\mathbf{r}}\, \mathrm{d}\tau \tag{7.89}$$

$\phi_{\alpha\mathbf{k}}$ is assumed to be normalized. It is apparent that the functions (equation 7.88) will now be orthogonal to each of the Bloch sums $\phi_{\alpha\mathbf{k}}$ and the functions $\psi_{i\mathbf{k}}(\mathbf{r})$ are said to be orthogonalized plane waves.

The procedure is then to use the composite set of basis functions (equations 7.86 and 7.88) to determine the matrix elements of the Hamiltonian as before. It is stressed that again the solution is not a self-consistent one in that the crystal potential $V(\mathbf{r})$ is assumed to be specified. The procedure of using orthogonalized plane waves allows a more accurate calculation when the resulting Hamiltonian is truncated.

Based on a heuristic approach developed from the idea of orthogonalized plane waves, an effective method has been developed in which a *pseudo-potential* is used so that the Hamiltonian can be expanded in terms of simple plane waves (Cohen and Heine 1961, Brust 1968).

Another procedure which has been used in obtaining solutions to the linear Hartree–Fock equations is the method of augmented plane waves (Slater 1965). Assuming a muffin-tin potential (Figure 7.2) it is apparent that in the interstices between spheres, the wave functions will be plane waves. Equally, solutions are assumed to satisfy the radial Schrödinger equation within the spheres. In the augmented-plane-wave method, the problem is reduced to matching the solutions and their derivatives at the boundaries of the spheres (Slater 1965).

CHAPTER VIII

Phase Fluids

1 The Density of Particles in Phase Space and the Vlasov Equation

In earlier chapters, the computational mechanics has been formulated in terms of particles and systems of interacting particles. The description of physical systems by the direct simulation of the motion of the constituent particles is limited, since the number of particles which may be followed is relatively small. In Chapter 5 the 'exact' solution of a system of N particles interacting through two-particle force laws was discussed but since there are $\frac{1}{2}N(N-1)$ interactions, only $N \sim 10^3$ particles could be considered. By assuming that particle motion was governed by averaged fields, systems with $N \sim 10^6$ particles could be followed in the statistical particle-in-cell model (Chapter 6), but to extend the computational mechanics to very large assemblies of particles it is necessary to employ the concept of a fluid. Continuous functions to describe particle assemblies, or fluid functions, may be defined in phase space or real space and we shall introduce the discussion in this chapter by considering phase fluids.

It is to be assumed that the system of interest is composed of so many particles that, effectively, the number N may be considered infinite and, in lieu of the coordinates of all particles, the system is described by a *density* or distribution f of particles in phase space. The distribution function defining the state of the phase fluid is a continuous function, $f = f(\mathbf{x}, \mathbf{v}, t)$, of the one-particle position and velocity coordinates $\mathbf{x}, \mathbf{v}$. We define f at each point $(\mathbf{x}, \mathbf{v})$ as the number of particles $\mathrm{d}n$ in the 'volume' $\mathbf{dx}\,\mathbf{dv}$,

$$\mathrm{d}n = f(\mathbf{x}, \mathbf{v}, t)\,\mathbf{dx}\,\mathbf{dv} \tag{8.1}$$

By applying the Hamiltonian equations of motion in the limit of many particles, an integro-differential equation for the density function f may be obtained (Chapman and Cowling 1953). In general, in three space dimensions, the distribution function is a function of three space coordinates, three momentum coordinates and time, but for simplicity we shall restrict the description to a two-dimensional phase space (x, v). Our purpose here is to obtain a deterministic equation for f and to illustrate some important properties of phase fluids with a view to employing such properties to good effect in computational models.

In the special instance of collisionless assemblies, appropriate for systems of particles which interact with long-range forces (Chapter 6) as in gravitational or plasma problems, the differential equation for the density function f takes a particularly simple form. We shall consider a fixed 'volume' V of the fluid in phase space (Figure 8.1). According to the definition of f, the total

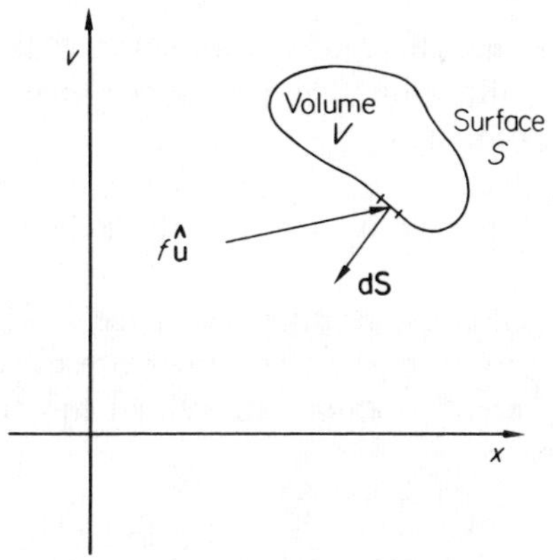

Figure 8.1. Conservation of particles and momentum in the absence of collisions is applied to a finite 'volume', V, in phase space. An equation for the time development of the distribution in phase space is obtained

number N_V of particles in the volume V is the integral over the volume of $f\,\mathrm{d}\tau$, where $\mathrm{d}\tau$ is the volume element, $\mathrm{d}\tau = \mathrm{d}x\,\mathrm{d}v$:

$$N_V = \int_V f\,\mathrm{d}x\,\mathrm{d}v \tag{8.2}$$

Our concern is to establish the change in the number N_V of particles in the volume with respect to time. In the first instance, it is clear that if collisions between the particles occur, then, at the time of a collision, a particle disappears from one point of velocity space and reappears at another, since in the event of a collision a particle's momentum alters discontinuously. On the other hand in the absence of collisions, we may employ the principles of conservation of particles (mass) and conservation of momentum to state that particles may enter or leave the volume V only by crossing the 'surface' S of the volume V in phase space. It follows that the rate of change in the number of particles in the volume V must be equated with the flux of particles crossing the surface S,

$$\frac{\partial}{\partial t}\int_V f \,\mathrm{d}x\,\mathrm{d}v = -\oint_S f\hat{\mathbf{u}}\,.\,\mathbf{dS} \tag{8.3}$$

The flux $f\hat{\mathbf{u}}$ is the flux of particles in phase space, where $\hat{\mathbf{u}}$ is the vector velocity of the phase fluid:

$$\hat{\mathbf{u}} = (v, a) \tag{8.4}$$

v is the velocity of the phase fluid in the x-direction, while a, the acceleration, is the velocity in the v-direction. The divergence theorem is applied to the surface integral (equation 8.3):

$$\frac{\partial}{\partial t}\int_V f\,\mathrm{d}x\,\mathrm{d}v + \int_V \boldsymbol{\nabla}'\,.\,(f\hat{\mathbf{u}})\,\mathrm{d}x\,\mathrm{d}v = 0 \tag{8.5}$$

where $\boldsymbol{\nabla}'$ is the divergence operator in phase space. The integral equation (8.5) is appropriate for any fixed volume V and, therefore, by equating the integrand separately to zero, we may obtain a differential equation for the density of the fluid in phase space:

$$\frac{\partial f}{\partial t} + \boldsymbol{\nabla}'.f\hat{\mathbf{u}} = 0$$

or

$$\frac{\partial f}{\partial t} + \frac{\partial}{\partial x}(fv) + \frac{\partial}{\partial v}(fa) = 0 \tag{8.6}$$

We may proceed further in the situation of an assembly interacting through *conservative* forces, by noting that the partial derivatives $(\partial v/\partial x)_v$ and $(\partial a/\partial v)_v$ are each zero:

$$\frac{\partial f}{\partial t} + v\frac{\partial f}{\partial x} + a\frac{\partial f}{\partial v} = 0 \tag{8.7}$$

The phase density f therefore satisfies the simple advective equation in phase space. This is the *Vlasov equation* which describes a collisionless system of many particles which interact through conservative forces.

2 Remarks and Some Examples of the Vlasov Equation

It is of value to consider several properties of phase fluids in general and the Vlasov equation in particular. It is to be noted that an explicit knowledge of the distribution function defines the state of the fluid in configuration space for, by integrating the distribution function over velocity space, the spatial fluid properties of density $n(x, t)$, centre-of-mass velocity $u(x, t)$, or temperature $T(x, t)$ are readily obtained. These fluid properties, defined in configuration space, are obtained as moments of the distribution function and it is clear that there exists an infinite set of such moments. As examples, the fluid density and centre-of-mass velocity are defined by the moments:

$$n(x, t) = \int_{-\infty}^{\infty} f(v, x, t)\, dv \tag{8.8}$$

$$nu(x, t) = \int_{-\infty}^{\infty} vf(v, x, t)\, dv \tag{8.9}$$

Higher moments, such as temperature, or heat flow, might or might not be of value as variables, depending on whether a thermal equilibrium is established by collisions or possibly by some other mechanism.

The Vlasov equation (8.7) is homogeneous and describes a perfectly collisionless system. By inspection of equation 8.7 it is clear that f is conserved and that all powers of f are conserved. In particular, it is evident that the quantity $f \ln f$ is conserved:

$$\frac{\partial}{\partial t} \int f \ln f\, dx\, dv = 0 \tag{8.10}$$

The quantity $f \ln f$ is proportional to the entropy of the system (Tolman 1967) and it follows that the Vlasov equation, in describing a collisionless system, does not produce entropy.

It is to be noted that the Vlasov equation is of wider applicability than might be inferred from Section 8.1. A general and informative way of expressing the Vlasov equation is in terms of the Hamiltonian, $H = H(q, p)$ of the system (Goldstein 1962), where q and p are the generalized space and momentum coordinates. Then Hamilton's equations are:

$$\dot{q} = \frac{\partial H}{\partial p}$$

$$\dot{p} = -\frac{\partial H}{\partial q} \tag{8.11}$$

and correspondingly the Vlasov equation takes the form

$$\frac{\partial f}{\partial t} + \frac{\partial H}{\partial p}\frac{\partial f}{\partial q} - \frac{\partial H}{\partial q}\frac{\partial f}{\partial q} = 0$$

or

$$\frac{\partial f}{\partial t} + \frac{\partial(f, H)}{\partial(q, p)} = 0 \tag{8.12}$$

where the second term is expressed as a Jacobian. In an equilibrium configuration, it follows that the distribution f is a function of the Hamiltonian only, $f = f(H)$.

To apply the Vlasov equation it is necessary to define the Hamiltonian for the system of interest and, in self-consistent problems, the Hamiltonian is in general determined by the distribution function. To obtain a closed set of equations, therefore, the Vlasov equation (8.7) must be supplemented by a 'force law', namely an equation to determine the Hamiltonian from the distribution function. We shall illustrate the formulation by defining the two particular systems of a gravitational and a plasma assembly.

(*a*) *Collisionless Gravitational System.* In one space dimension and one velocity dimension a self-gravitating phase fluid is described by the Vlasov equation:

$$\frac{\partial f}{\partial t} + v\frac{\partial f}{\partial x} + g\frac{\partial f}{\partial v} = 0 \tag{8.13}$$

where $g(x, t)$ is the local gravitational acceleration. We define g by employing Poisson's equation which describes the gravitational potential,

$$\frac{\mathrm{d}^2\Phi}{\mathrm{d}x^2} = 4\pi G m n(x) \tag{8.14}$$

Here, G is the gravitational constant, m is the mass of a particle and $n(x)$ is the number density determined in configuration space from f,

$$n(x, t) = \int_{-\infty}^{\infty} f(v, x, t)\,\mathrm{d}v \tag{8.15}$$

If Poisson's equation is solved to determine the potential Φ, then the acceleration g in Vlasov's equation is given by

$$g = -\frac{\partial \Phi}{\partial x} \tag{8.16}$$

It is to be noted that, in this instance, the one-dimensional Poisson's equation describes not a set of point stars but rather the potential from a set of parallel infinite gravitational 'sheets' moving perpendicular to their planes.

(*b*) *Collisionless One-species Plasma.* The equations describing a one-species plasma (the electrons) in a background uniform positive charge of ions are essentially similar to the gravitational problem but the acceleration in the Vlasov equation is to be determined from the electric field E,

$$\frac{\partial f}{\partial t} + v\frac{\partial f}{\partial x} + \frac{e}{m}E\frac{\partial f}{\partial v} = 0 \tag{8.17}$$

where e is the charge (negative for the electrons) and m is the particle mass.

The electrostatic potential is similarly determined from Poisson's equation,

$$\frac{d^2\Phi}{dx^2} = -4\pi\sigma \tag{8.18}$$

σ is the charge density defined from the distribution function,

$$\sigma(x, t) = |e|\left(n_0 - \int_{-\infty}^{\infty} f \, dv\right) \tag{8.19}$$

and n_0 is the uniform background density of ions.

3 The Difference Solution of the Vlasov Equation

The simplicity of the Vlasov equation to describe the collisionless 'infinite-body' problem is appealing in that the system is described by one time-dependent partial differential equation in two 'space-like' dimensions or, in general, six dimensions. In effect, the equation to be solved is an example of the advective equation (Section 3.1), which may clearly be resolved by direct solution on a difference mesh.

We shall divide the phase space of interest by an Eulerian lattice with space step Δx and velocity step Δv, and we shall then regard the phase space as constructed of a large number of small cells defined by the indices (i, j) (Figure 8.2). The distribution function f is to be defined at every cell so that the number of particles in the cell (i, j) at time t^n will be $f_{ij}^n \, \Delta x \, \Delta v$. Any of the integration methods for the solution of the advective equation, which

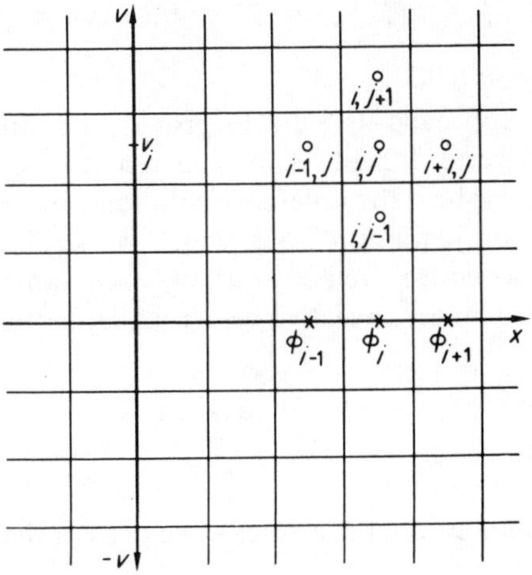

Figure 8.2. The Vlasov equation may be integrated on a difference mesh in phase space

have been described in Section 3.7, may be employed though we shall formulate the difference solution here in terms of the Lax method. According to the algorithm (equation 3.58) the density f_{ij}^{n+1} at the new time level t^{n+1} is defined over the time step Δt as:

$$f_{ij}^{n+1} = \tfrac{1}{4}(f_{i+1,j}^{n} + f_{i-1,j}^{n} + f_{i,j+1}^{n} + f_{i,j-1}^{n}) - \frac{\Delta t\, v_j}{2\,\Delta x}(f_{i+1,j}^{n} - f_{i-1,j}^{n})$$

$$+ \frac{\Delta t}{2\,\Delta v}\frac{(\Phi_{i+1}^{n} - \Phi_{i-1}^{n})}{\Delta x}(f_{i,j+1}^{n} - f_{i,j-1}^{n}) \tag{8.20}$$

It is to be noted that, although this equation is not strictly in conservative difference form (see equation 3.80), the quantity f is nevertheless identically conserved on the difference mesh, since the coefficients v_j are independent of the index i (or the x coordinate) and similarly the potential Φ is independent of the velocity coordinate. The equation is expressed for the gravitational case, but it may be formulated with equal simplicity for alternative examples. At the completion of each time step, the mass density in each space cell is to be determined according to the integral (equation 8.15):

$$m_i^{n+1} = m_0 \sum_{j=-J}^{J} f_{i,j}^{n+1}\,\Delta v \tag{8.21}$$

m_0 is the mass of a star and the boundary points, J and $-J$, in the velocity dimension, are taken sufficiently large to ensure that the distribution function f at J and $-J$ is zero. It remains to determine the potential, at the time step t^{n+1}, from the difference form of Poisson's equation:

$$\Phi_{i+1}^{n+1} - 2\Phi_i^{n+1} + \Phi_{i-1}^{n+1} = 4\pi G(\Delta x)^2 m_i^{n+1} \tag{8.22}$$

which is readily solved (Chapter 4).

In principle this approach appears to provide a simple solution to the Vlasov equation. Several difficulties are inherent to the method however. Since the method is explicit, the difference solution is to be obtained subject to a stability criterion (equation 3.62), which we shall express separately according to the advective 'velocity' along each dimension. Along the x-dimension, the maximum advective velocity is v_J so that the time step is limited as,

$$\Delta t \leqslant \frac{\Delta x}{v_J} \tag{8.23}$$

Along the velocity dimension, the advective velocity is the acceleration,

$$\Delta t \leqslant \frac{\Delta v}{|g|_{\max}} \tag{8.24}$$

where

$$|g|_{\max} = \max \frac{\{\Phi_{i+1}^{n+1} - \Phi_{i-1}^{n+1}\}}{2\,\Delta x}$$

This second condition on the time step (equation 8.24) is related to the gravitational frequency (Section 6.1c) since, using Gauss's law, the acceleration g is related to the mass density:

$$g \sim 4\pi G m n\,\Delta x$$

$$(\Delta t)^2 \lesssim \frac{\Delta v}{v_J}\frac{1}{\omega_G^2} \tag{8.25}$$

The stability criteria impose a severe limitation on the allowable time step.

In addition, and of particular relevance to the solution of the Vlasov equation, a difference solution can give rise to an anomalous numerical diffusion (Section 3.5). This is particularly true of the Lax difference method and to attempt to minimize, though not to eradicate, the effect, a higher-order difference method such as the Lax–Wendroff or leapfrog method could be used (Section 3.7). Numerical diffusion is a severe effect in a Vlasov solution since the simulation will be entropy producing in that the quantity $f \ln f$ will not be conserved in the difference solution: the effect of numerical diffusion can therefore be shown to be equivalent to the occurrence of anomalous collisions.

4 Incompressibility of the Phase Fluid

It has been shown that the density of a fluid in phase space varies according to the process of advection. We may acquire an alternative perspective on the properties of phase fluids, and at the same time devise alternative finite methods, by considering changes in the phase density from a Lagrangian point of view.

It is useful, therefore, to consider the properties of a phase 'volume' V in (x, v) phase space which moves with the phase fluid (Figure 8.3). In the absence of collisions, the number of particles, say N_V, inside the volume V moving with the phase fluid must remain constant. To clarify this point, we may suppose that a particle approaches the boundary of V from either side: it must then move with the boundary, since the boundary moves everywhere with the particle coordinates and it follows that no particle can ever cross the boundary of the moving volume. The number of particles in the volume V is therefore constant.

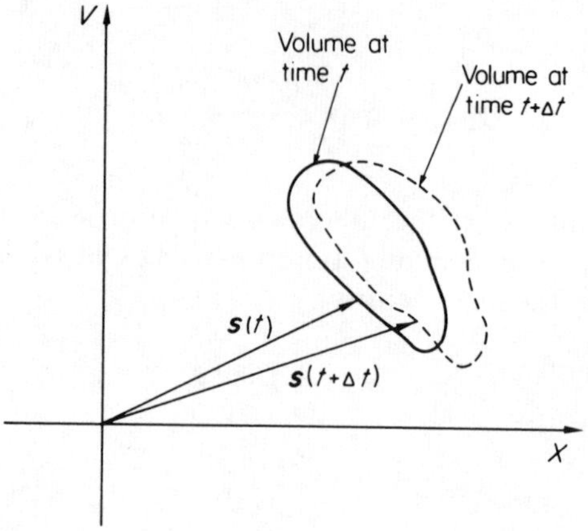

Figure 8.3. In the absence of collisions, a phase fluid is incompressible. The diagram shows the motion of an arbitrary volume fixed in the frame of the fluid

In addition, the magnitude of the moving volume at any time is defined as:

$$V = \int \mathrm{d}x \, \mathrm{d}v$$

$$= \tfrac{1}{2} \oint \mathbf{s} \wedge \mathbf{dl} \tag{8.26}$$

where $\mathbf{s} = (x, v)$ is the vector position in phase space of any point on the boundary of V and $\mathbf{dl}$ is an element of 'length' in phase space, parallel to the boundary. Over a small time Δt, therefore, the change in the moving volume is:

$$\Delta V = \tfrac{1}{2} \oint \mathbf{s}_{t+\Delta t} \wedge \mathbf{dl} - \tfrac{1}{2} \oint \mathbf{s}_t \wedge \mathbf{dl} \tag{8.27}$$

Through the definition of the moving volume, the point $\mathbf{s}$ at the time $t + \Delta t$ is related to the point $\mathbf{s}$ at time t, according to the 'velocity' of the phase fluid,

$$\mathbf{s}_{t+\Delta t} = (x + v\,\Delta t, v + a\,\Delta t)$$

$$= \mathbf{s}_t + (v, a)\,\Delta t \tag{8.28}$$

It follows from equation 8.28 that:

$$\frac{\Delta V}{\Delta t} = \tfrac{1}{2} \oint (v, a) \wedge \mathbf{dl}$$

$$= \tfrac{1}{2} \oint \{v \, \mathrm{d}v - a \, \mathrm{d}x\} \tag{8.29}$$

If the forces on the particles are conservative, the acceleration of a particle a may be written in terms of a potential Φ,

$$a = -\frac{1}{m}\frac{\partial \Phi}{\partial x}$$

and it follows that the change in the moving volume with respect to time may be expressed as a perfect integral:

$$\frac{\Delta V}{\Delta t} = \frac{1}{2m}\oint \mathrm{d}\left(\frac{mv^2}{2} + \Phi\right)$$
$$= 0 \tag{8.30}$$

The magnitude of any volume moving with the phase fluid is an *invariant* of the motion, so that the phase fluid may be regarded as *incompressible.* We shall employ this concept in devising a finite model to describe the Vlasov system (the waterbag model), but it is to be noted here that the Vlasov equation also follows from the arguments in this section. For, if an infinitesimally small moving volume dV is considered, the volume of the fluid element is constant and equally the number of particles inside the volume is constant. It follows that the phase space density f must remain constant with respect to the moving fluid:

$$\frac{\mathrm{d}f}{\mathrm{d}t} = 0 \tag{8.31}$$

This is the advective equation, where the Lagrangian time derivative in phase space is

$$\frac{\mathrm{d}}{\mathrm{d}t} = \frac{\partial}{\partial t} + v\frac{\partial}{\partial x} + a\frac{\partial}{\partial v}. \tag{8.32}$$

5 The Waterbag Method

Some of the properties of a Vlasov system have been pointed out and of particular importance is the incompressibility of the phase fluid with the consequence that the density f is only subject to advection. This property permits a particularly simple description of the system for, instead of describing the distribution function f as a function of the coordinates (x, v, t) everywhere, the function f may be represented by a set of contours (lines of constant density f) in two-dimensional phase space (Figure 8.4). Since the fluid is incompressible, the area inside each contour is an invariant of the motion but, as time proceeds, the contours move in the plane and describe the evolution of the system. Clearly the contours will become distorted with time, but they maintain their identity and no two contours can ever cross.

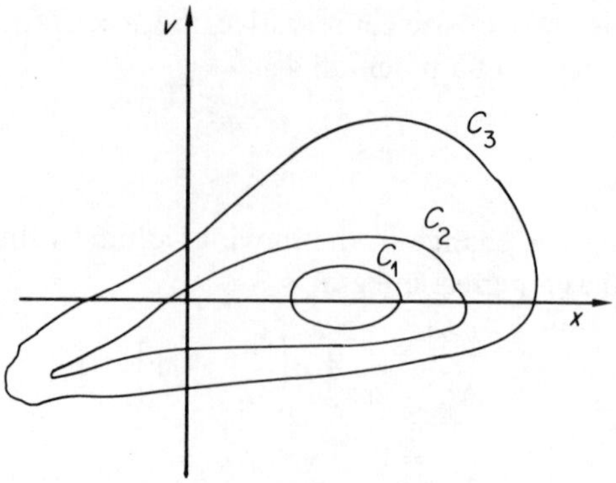

Figure 8.4. The waterbag model. The distribution function is represented by a finite set of contours in phase space

A computational model, the waterbag model (Berk and Roberts 1970), can be formulated to describe the evolution in time of such a system. Step functions are used to describe the distribution function f (Figure 8.5) so that, between the contours, the density f is constant. We shall label each of the J contours, C_j ($1 \leqslant j \leqslant J$), and each contour is to be defined by a finite set of points (x_i, v_i) distributed along the contour (Figure 8.6). These points may be regarded as points in phase space and consequently they move

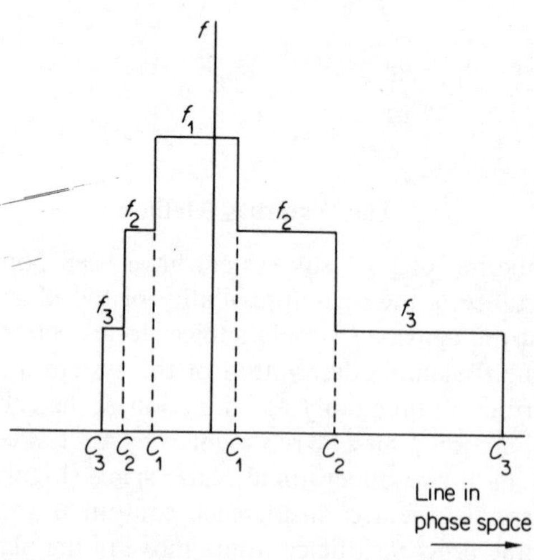

Figure 8.5. The density of particles between contours in phase space is assumed constant

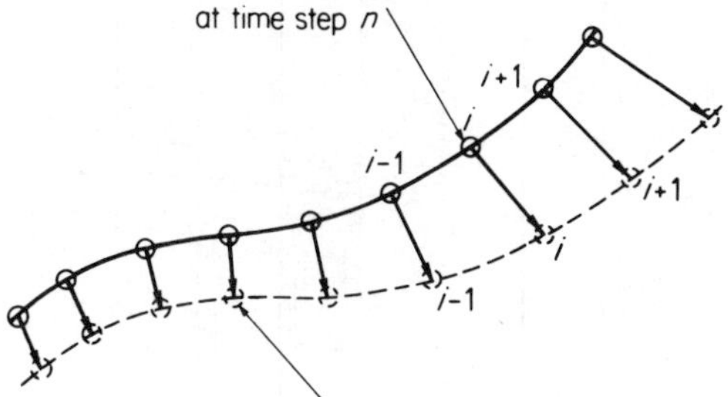

Figure 8.6. Each contour in the waterbag model is defined by a set of points along the contour. Each point moves with the Hamiltonian equations of motion

according to the Hamiltonian equations of motion. For the point (x_i, v_i) on the contour C_j the equations of motion are:

$$\frac{dx_i}{dt} = v_i$$

$$\frac{dv_i}{dt} = g(x_i) \tag{8.33}$$

where g is the acceleration for a gravitational system.

The integration in time is accomplished by moving the contours over small time steps where the new position of each contour is defined by the integrated point coordinates:

$$v_i^n = v_i^{n-2} + 2\,\Delta t\, g(x_i^{n-1}) \tag{8.34}$$

$$x_i^n = x_i^{n-2} + 2\,\Delta t\, v_i^{n-1} \tag{8.35}$$

For the purpose of second-order accuracy, we have used the leapfrog method here with both position and velocity coordinates defined at all times and it follows that decoupling between the time-levels can occur (Section 2.6b). To avoid serious errors which might arise from this effect, at infrequent intervals (typically every 100 time steps) the variables at an intermediate time level must be reconstructed by interpolation from adjacent time levels.

To determine the mass density (or charge density in the electrostatic problem) in configuration space at the time step n, the x-dimension is divided into a set of Eulerian cells (Figure 8.7). Each contour may, or may not, cross a particular Eulerian cell a number of times, but, in general, the positions in velocity space u_j^n at which each Eulerian cell is crossed by a contour are

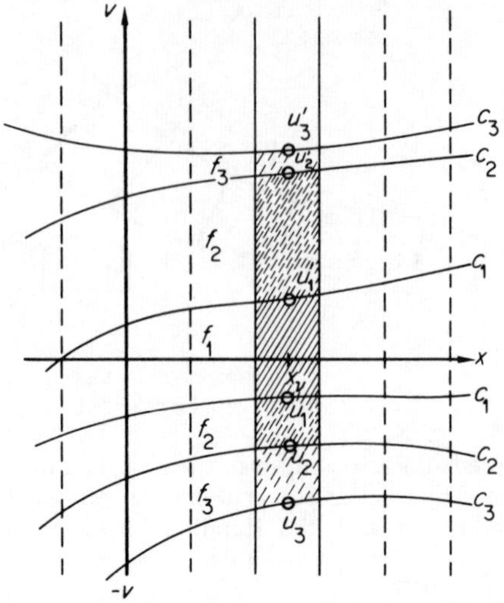

Figure 8.7. The integration of the distribution function in the waterbag model to define the density at each point in configuration space

determined. The mass in each cell is to be determined according to the equation:

$$m_\nu^n = m_0 \int_{-\infty}^{\infty} f \, \mathrm{d}v \tag{8.36}$$

$$= m_0 \left\{ fv|_{-\infty}^{\infty} - \int v \, \mathrm{d}f \right\}, \tag{8.37}$$

$$m_\nu^n \simeq m_0 \sum_j u_{\nu j}^n \Delta f_j \tag{8.38}$$

where we have integrated by parts and used the condition that $f = 0$, at $v = \pm\infty$. In equation 8.38, Δf_j is the absolute change in f across each contour (Figure 8.5) and, by assigning a clockwise direction to each contour, the sign of Δf_j is defined according to whether the contour crosses the point x_ν in the positive or negative direction. Consequently, using equation 8.38, the determination of the mass density at each time level is simply accomplished.

Again, with the mass density defined, Poisson's equation may be solved to determine the potential,

$$\Phi_{\nu+1}^n - 2\Phi_\nu^n + \Phi_{\nu-1}^n = 4\pi \Delta^2 G m_\nu^n \tag{8.39}$$

We illustrate the waterbag model by its application to the nonthermal equilibrium problem of a 'bump-on-tail' distribution function (Figure 8.8)

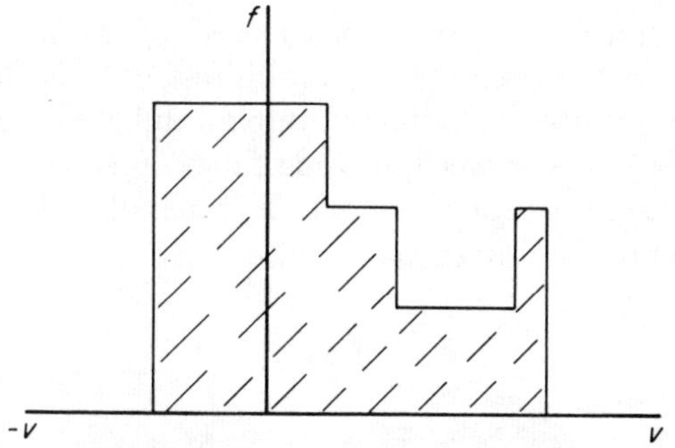

Figure 8.8. The 'bump-on-tail' distribution

(Berk and Roberts 1970). The 'bump-on-tail' effect occurs frequently in plasma physics where hot electrons in the absence of collisions are preferentially accelerated by an electric field. The distribution is unstable through the two-stream instability (Section 6.3) and, in the nonlinear stages, the tail tends to be thermalized by vortices in phase space (Figure 8.9).

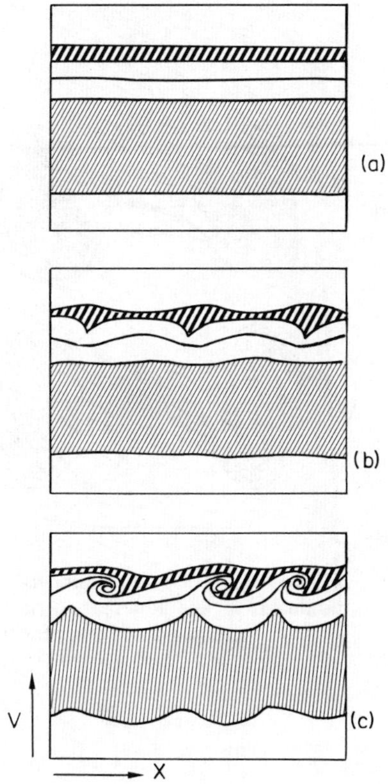

Figure 8.9. The evolution of a 'bump-on-tail' distribution in $(v - x)$ phase space according to the waterbag method. The shaded area shows the high-density region in phase space associated with the bump-on-tail. The distribution (a) is unstable by the two-stream mechanism and produces eddies (c) and mixing in phase space (from Berk and Roberts, 1970)

The waterbag model is free from the effect of diffusion which, in velocity space, is related to the occurrence of collisions. On the other hand, the contours tend to become increasingly complex and elongated (Figure 8.10) and, if the calculation is pursued for long times, it is necessary to simplify and reduce the length of each contour. This has the effect of arbitrarily adding a diffusion to the simulation.

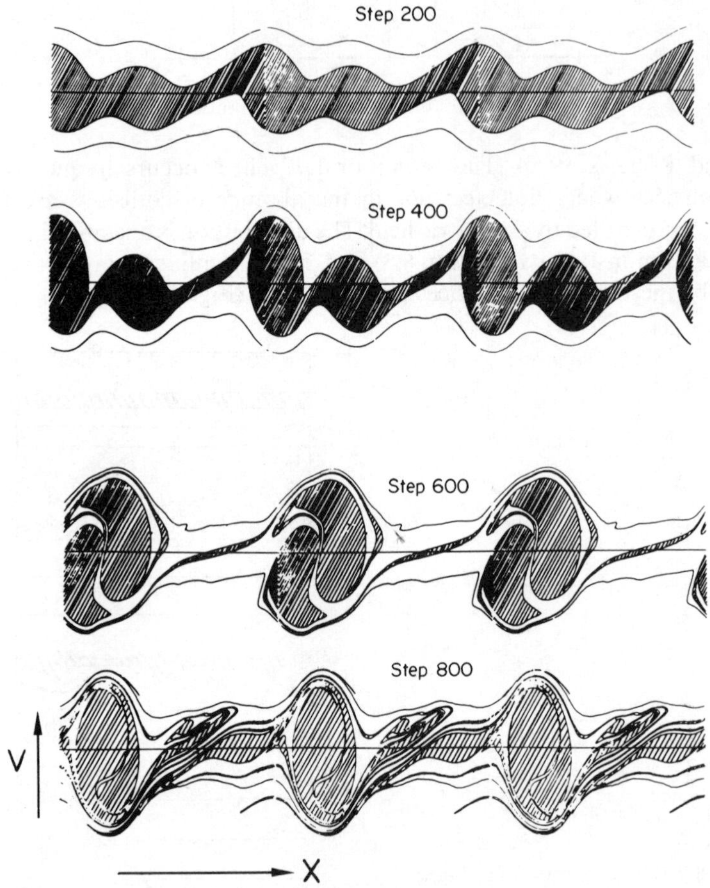

Figure 8.10. The two-stream instability by the waterbag method. In the nonlinear regime 'holes' are formed in phase space. The contours become elongated and complex (from Berk and Roberts, 1970)

CHAPTER IX

Classical Fluid Dynamics

1 Introductory Remarks on the Hydrodynamic Equations

Classical hydrodynamics attempts to describe the dynamics of fluids with which we are familiar in the daily environment: phenomena such as tides, waves on the ocean, the winds, eddies and motion of the atmosphere, waterfalls, the rising smoke from a chimney, sound waves and the occurrence of thunder. These are but a few examples of fluid effects, the frequent occurrence of which provides us with a natural intuition of and interest in hydrodynamics. On the other hand the complexity and variety of fluid phenomena is mirrored in the complexity of the hydrodynamic equations which, since a fluid is an example of the many-body problem, are nonlinear and difficult to resolve.

By using essential principles of conservation (Section 3.1), we have established the form of the hydrodynamics equations from a macroscopic point of view, without regard to the structure of the 'continuous fluid medium'. It is to be appreciated, however, that we are attempting to describe a many-body assembly and it is of value, therefore, to consider the equations from the molecular point of view (Chapman and Cowling 1953).

The principle of describing the many-body problem by a continuous function has been discussed in terms of the Vlasov equation. The collisionless Vlasov equation may describe an arbitrary configuration of the particles in velocity space. On the other hand, in the familiar fluids of interest, as for example in meteorology, in oceanography and in aerodynamics, molecular collisions occur very frequently over the time scales of our interest. This property permits the use of equations with fewer degrees of freedom and, therefore, equations which are easier to resolve. Because molecular collisions occur so frequently, the essential approximation that all particles at any local space point remain in a quasi-thermal equilibrium may be made. By a local thermal equilibrium in this context, we mean that the particles at any local point are distributed in velocity space as a Maxwellian distribution, or at least in a near Maxwellian distribution and it follows that we can describe such a distribution by a *centre-of-mass velocity* v and a temperature T, or square of the *width* of the distribution. Therefore, in place of the distribution function in phase space, the collisional many-particle system can now be described by the fluid variables in configuration space of density ρ, velocity $\mathbf{v}$, and temperature T (or pressure p, or internal energy ϵ).

While the significance from a molecular point of view of the hydrodynamic equation is stressed, it is sufficient, here, to summarize the form and some fundamental properties of the equations.*

(*a*) *Conservative, Eulerian and Lagrangian formulations.* The ideal hydrodynamic equations have been introduced in Sections 3.1 and 6.5. In general, in three space dimensions and time $(\mathbf{x}, t)$, there are five equations for the density ρ, velocity $\mathbf{v}$ and internal energy density ϵ related to the temperature. We express the equations in Eulerian form, namely in differential form with respect to some fixed frame of reference:

mass:
$$\frac{\partial \rho}{\partial t} + \nabla . \rho \mathbf{v} = 0 \tag{9.1}$$

momentum:
$$\frac{\partial \rho \mathbf{v}}{\partial t} + \nabla . \rho \mathbf{v}\mathbf{v} = - \nabla p \tag{9.2}$$

internal energy:
$$\frac{\partial \rho \epsilon}{\partial t} + p \nabla . \mathbf{v} + \nabla . (\rho \epsilon \mathbf{v}) = 0 \tag{9.3}$$

These basic equations must be supplemented by an 'equation of state' according to the particular thermodynamic properties of the fluid of interest. The equation of state relates the local specific internal energy to the density and pressure of the fluid,

$$\epsilon = \epsilon(p, \rho). \tag{9.4}$$

A common example is the equation of state for an ideal gas:

$$\epsilon = \frac{kT}{m(\gamma - 1)} = \frac{p}{\rho(\gamma - 1)} \tag{9.5}$$

where γ is the ratio of specific heats ($\gamma = \frac{5}{3}$ for a monatomic gas), m is the molecular mass, k is Boltzmann's constant and T is the temperature of the fluid.

We have noted previously that one of the essential processes described by the hydrodynamic equation is advection where the properties of the fluid are transported by the fluid motion (Sections 3.1 and 3.5). If we express the equations in terms of the time derivative with respect to the moving fluid, the Lagrangian derivative $\mathrm{d}/\mathrm{d}t = \partial/\partial t + \mathbf{v} . \nabla$, the equations take a partic-

* For the derivation of the equations of hydrodynamics from the point of view of a molecular distribution function, the reader is referred to Chapman and Cowling (1953).

ularly simple form. By differentiating the divergence terms in equations 9.1, 9.2 and 9.3, we obtain the Lagrangian form of the equations.

mass:
$$\frac{d\rho}{dt} = -\rho\nabla\cdot\mathbf{v} \tag{9.6}$$

acceleration:
$$\rho\frac{d\mathbf{v}}{dt} = -\nabla p \tag{9.7}$$

internal energy:
$$\rho\frac{d\epsilon}{dt} = -p\nabla\cdot\mathbf{v} \tag{9.8}$$

For the special case of an ideal gas (equation 9.5), equation 9.8 simplifies further to an expression for the adiabatic law. It is to be noted that the Lagrangian time derivative includes the effect of advection, while the terms on the right-hand side of the equations are each associated with the compression of the fluid.

The properties of the conservation of physical variables are to be stressed, both from the point of view of physical processes (Section 3.1) and from the point of view of a finite difference simulation (Section 3.6). The Eulerian equations (9.1 and 9.2) for the mass and momentum of the fluid are expressed in conservative form. Similarly, we can obtain an equation for the conservation of energy to replace the non-conservative equation (9.3) for the internal energy, by observing that the total energy density in the fluid is the sum of the internal energy density $\rho\epsilon$ and the centre-of-mass kinetic-energy density $\frac{1}{2}\rho v^2$. Therefore, if we add to the equation for the internal energy density (equation 9.3) the dot product of the momentum equation with $\mathbf{v}$, and minus $\frac{1}{2}v^2$ multiplied by the continuity equation, a conservative equation for the energy is obtained:

$$\frac{\partial}{\partial t}(\rho\epsilon + \tfrac{1}{2}\rho v^2) + \nabla\cdot\{(\tfrac{1}{2}\rho v^2 + \rho\epsilon + p)\mathbf{v}\} = 0 \tag{9.9}$$

The energy flux under the divergence operator represents the flux of centre-of-mass kinetic energy, of internal energy and the work done by material pressure respectively.

(*b*) *Diffusion Processes.* The hydrodynamic equations which have been introduced above are hyperbolic in form and describe the advection and compression, associated with sound waves, of the fluid. These equations are applicable when the properties of the fluid are localized by particle-particle collisions and the distribution of the particles in phase space has a Maxwellian form. Properties of the fluid are localized, when the mean free path λ over which a particle travels between collisions is small. More specifically, the equations are valid where λ is much smaller than the characteristic scale length L over which the fluid properties of density, velocity and energy vary.

When the mean free path λ can no longer be regarded as small, we must take into account the transport of momentum and energy through the fluid, by the diffusion of particles in configuration space. We say the fluid becomes viscous, since momentum is diffused through the fluid by microscopic particle motion. The effect is described in the momentum equations (9.2) by the inclusion of a traceless viscous tensor, $\mathbf{V}$,

$$\frac{\partial}{\partial t}\rho\mathbf{v} + \nabla . (\rho\mathbf{v}\mathbf{v} + p\mathbf{I} + \mathbf{V}) = 0 \tag{9.10}$$

where $\mathbf{I}$ is the unit tensor.

Similarly, internal energy is transported through the fluid by a heat conduction flux $\mathbf{q}$ (in a gas, more correctly heat diffusion),

$$\frac{\partial}{\partial t}\rho\epsilon + (p\mathbf{I} + \mathbf{V}) : \nabla\mathbf{v} + \nabla . (\rho\epsilon\mathbf{v} + \mathbf{q}) = 0 \tag{9.11}$$

In addition to the heat conduction in the internal energy equation, we have included the effect of viscous heating on the fluid. Clearly, if the viscous tensor is included in the momentum equations, to conserve energy account must be taken of the work done on the gas by viscous forces. This is included in the second term of the internal energy equation.

We relate the diffusion fluxes, of heat conduction and of viscosity, to the temperature gradients or velocity gradients in the fluid. As in a solid (Section 3.1) heat will tend to flow down a temperature gradient, so that the heat conduction flux takes the form:

$$\mathbf{q} = -\kappa\nabla T \tag{9.12}$$

where κ is the coefficient of thermal conductivity.

Similarly, the viscous tensor expresses the transport of momentum between both sheared motion and compressible motion. The viscous tensor $\mathbf{V}$ is related to the Navier–Stokes tensor $\mathbf{U}$ (Landau and Lifshitz 1959):

$$\mathbf{V} = -\mu\mathbf{U} \tag{9.13}$$

where

$$\mathbf{U} = \nabla\mathbf{v} + \widetilde{\nabla\mathbf{v}} - \tfrac{2}{3}(\nabla . \mathbf{v})\mathbf{I} \tag{9.14}$$

μ is the viscosity of the fluid and $\widetilde{\nabla\mathbf{v}}$ is the transpose of $\nabla\mathbf{v}$.

Both the conductivity κ and viscosity μ depend on the rate at which particles experience collisions, and κ and μ can depend on the local temperature and density of the fluid, so that the transport terms may be nonlinear (Chapman and Cowling 1953).

The inclusion of these diffusion phenomena makes the hydrodynamic equations parabolic, and their effect is to diffuse or smooth the momentum and temperature in the fluid.

(*c*) *Compressible and Incompressible Formulations.* The hydrodynamic equations as outlined above are applicable to a very broad range of fluids and associated phenomena: effects as diverse as those occurring in aerodynamics or in a waterfall. We shall distinguish, however, between two broad classes of problems, depending on whether the fluid is compressible or incompressible.

The general situation is described by the compressible hydrodynamic equations of which in three dimensions there are five time-dependent equations for the density ρ, three components of momentum density $\rho\mathbf{v}$ and internal energy density $\rho\epsilon$. The effects of viscosity and heat conduction may or may not be important.

The energy density of the fluid consists of two components,

$$\rho\epsilon + \tfrac{1}{2}\rho v^2$$

namely, the internal energy or thermal energy and the centre-of-mass kinetic energy. In compressible fluids an interchange can occur between the thermal density and the kinetic energy density, so that in some regions of the fluid most of the local energy can reside in the form of heat, whereas elsewhere, it is the energy of flow which is dominant. On the other hand, in many fluids of interest, such as the ocean or the atmosphere, the energy of flow is small compared to the internal or thermal energy in the fluid. The flow can then not compress the fluid since an interchange of energies can not occur and we say the fluid is incompressible.

We can view the comparison between compressible and incompressible fluids from a different perspective. In the absence of viscosity and heat conduction, the hydrodynamic equations are hyperbolic and contain two characteristic speeds: the centre-of-mass speed v; and the sound speed, $v_s = \sqrt{(\gamma p/\rho)}$, associated with the thermal velocity. Clearly when solving these equations with an explicit difference solution, to satisfy stability criteria a time step Δt is to be chosen such that:

$$\Delta t \leqslant \frac{\Delta}{|\mathbf{v}| + v_s} \tag{9.15}$$

where Δ is a mesh-step length (Chapter 3). We can distinguish between classes of problems, depending on the relative magnitudes of the flow speed and sound speed. Where the flow speed is small in this sense, to avoid a restriction on the time step imposed by a large sound speed, we make the approximation that the fluid is incompressible:

$$\frac{\mathrm{d}\rho}{\mathrm{d}t} = 0 \tag{9.16}$$

Using the continuity equation (9.6), this condition for incompressibility is equivalent to:

$$\nabla . \mathbf{v} = 0 \tag{9.17}$$

This assumption simplifies the form of the equations considerably, and has the effect of defining an infinite sound speed.

Therefore, in the incompressible formulation, we eliminate the need for an energy equation since the three velocity equations (9.7) and the condition that the velocity field be divergence-free define the four dependent variables of pressure and velocity components.

In the subsequent sections of this chapter, we shall first consider the solution and description by finite methods of the incompressible equations. By including moving surfaces and the effect of buoyancy due to gravity on the surface of the earth, the incompressible equations may be applied to such phenomena as splashes and waterfalls and to tides in the ocean. Second, we shall consider the difference solution of the compressible hydrodynamic equations, which are applicable to such phenomena as shock waves and the compressible flows which occur in aerodynamics.

2 Difference Solution of the Incompressible Equations

The motion of an incompressible fluid is defined by the three equations for the acceleration of the fluid with the condition that the velocity field be divergence-free (Section 9.1):

$$\frac{\partial \mathbf{v}}{\partial t} + \mathbf{v} . \nabla \mathbf{v} = -\frac{1}{\rho} \nabla p + \nu \nabla^2 \mathbf{v} \tag{9.18}$$

$$\nabla . \mathbf{v} = 0 \tag{9.19}$$

ν is the kinematic viscosity, $\nu = \mu/\rho$, and for a one-component fluid the density ρ is assumed constant. These four equations define the four dependent variables of three velocity components and scalar pressure.

There are two approaches to the solution of the incompressible equations. We may eliminate the pressure *ab initio* by taking the curl of the acceleration equations and thereby obtain relations for the vorticity of the fluid. By using the divergence condition (equation 9.19) we may define a stream function which is related to the vorticity through a Poisson's equation. In the second approach, the divergence of the velocity equations (9.18) is considered, and again a Poisson's equation is obtained which defines the pressure, directly from the centrifugal and Coriolis forces.

(*a*) *Vorticity and the Stream Function.* Incompressible motions of a fluid are of necessity rotational in form and consequently it is expedient to define the vorticity $\boldsymbol{\xi}(\mathbf{x}, t)$ of the fluid:

$$\boldsymbol{\xi} = \nabla \wedge \mathbf{v} \tag{9.20}$$

We shall eliminate the pressure variable from the equations defining the motion of the fluid and obtain time-dependent equations for the vorticity of the fluid by taking the curl of the equations of motion (9.18),

$$\frac{\partial \boldsymbol{\xi}}{\partial t} + \nabla \wedge (\mathbf{v} \, . \, \nabla \mathbf{v}) = \nu \nabla^2 \boldsymbol{\xi}$$

or

$$\frac{\partial \boldsymbol{\xi}}{\partial t} + \mathbf{v} \, . \, \nabla \boldsymbol{\xi} = \nu \nabla^2 \boldsymbol{\xi} \tag{9.21}$$

where we have used the divergence condition (equation 9.19). This is a simple expression for the time-dependent motion of a fluid since, apart from viscous diffusion, equation 9.21 is a statement that the vorticity is simply advected with the fluid.

The velocity components are defined from the vorticity, by integrating the divergence condition (equation 9.19) to define a vector quantity $\boldsymbol{\psi}$, the *stream function*:

$$\mathbf{v} = \nabla \wedge \boldsymbol{\psi} \tag{9.22}$$

since the divergence of the curl of a vector is zero. We determine the stream function $\boldsymbol{\psi}$ from the vorticity,

$$\begin{aligned} \nabla \wedge (\nabla \wedge \boldsymbol{\psi}) &= \nabla \wedge \mathbf{v} \\ &= \boldsymbol{\xi} \end{aligned} \tag{9.23}$$

and expanding the curl curl operator, we obtain,

$$\nabla^2 \boldsymbol{\psi} - \nabla(\nabla \, . \, \boldsymbol{\psi}) = -\boldsymbol{\xi} \tag{9.24}$$

It is to be noted that the stream function $\boldsymbol{\psi}$ has not been defined from the differential equations (9.22) uniquely, so that we are free to choose the 'gauge' of the stream function. It is most useful to choose the condition:

$$\nabla \, . \, \boldsymbol{\psi} = 0 \tag{9.25}$$

so that the three components of the stream function are separately related to the three components of the vorticity by Poisson's equation:

$$\nabla^2 \boldsymbol{\psi} = -\boldsymbol{\xi} \tag{9.25}$$

We have obtained a simple and complete set of equations to describe the motion of an incompressible fluid: the time-dependent equations for the vorticity (equations 9.21), together with Poisson's equation for the stream function (equation 9.25) from which the velocity field is defined (equation 9.22). In the particularly important two-dimensional case, these equations have considerable simplicity and elegance. We shall assume that the fluid motion takes place in the (x, y) plane $\mathbf{v} = (v_x, v_y, 0)$ and it follows that the vorticity and the stream function are defined with only z-components (Figure 9.1),

$$\boldsymbol{\xi} = (0, 0, \xi)$$

$$\boldsymbol{\psi} = (0, 0, \psi)$$

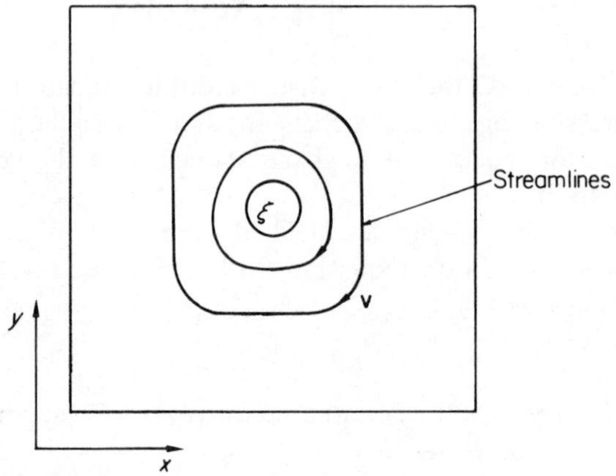

Figure 9.1. The vorticity and stream function for an incompressible fluid. The fluid flows along contours of the stream function

The complete set of equations take the form:

$$\frac{\partial \xi}{\partial t} + \mathbf{v} \,.\, \nabla \xi = \nu \nabla^2 \xi$$

$$\nabla^2 \psi = -\xi$$

$$\mathbf{v} = \nabla \wedge (\psi \mathbf{e}_z) \tag{9.26}$$

where $\mathbf{e}_z$ is a unit vector in the z-direction and the stream function and the vorticity may be treated as pseudo-scalars.

While the vorticity of the fluid is subject to advection and viscous diffusion, the elliptic Poisson's equation for the stream function implies that the transmission of 'information' through the fluid occurs with an infinite speed (Section 3.1). This is a consequence of the effective infinite sound speed which follows from the assumption of incompressibility. The equations remain nonlinear due to the occurrence of the advection term in the vorticity equation.

The set of equations (9.26) may now be differenced on an Eulerian mesh in the conventional manner: for example by the use of the Lax method.

However, to minimize numerical diffusion it is appropriate to employ a scheme of at least second-order accuracy such as the two-step Lax–Wendroff method (Section 3.7). A 'sodium chloride' Eulerian mesh is defined, where each main point has four neighbouring auxiliary points and vice versa (Figure 9.2).

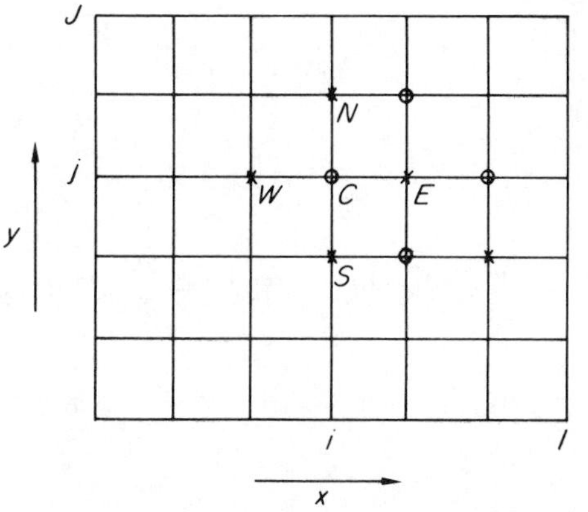

Figure 9.2. The difference mesh used in applying the Lax–Wendroff method to an incompressible fluid. Auxiliary and main points take alternate positions

At each time step n the variables ψ and ξ are defined only on the intermediate main points. Temporary or auxiliary variables are defined at the auxiliary points at the half time step $n + \frac{1}{2}$, by the Lax method:

auxiliary calculation:

$$
\begin{aligned}
v^n_{x i,j+1} &= (\psi^n_{i,j+2} - \psi^n_{i,j})\frac{1}{2\Delta} \\
v^n_{y i,j+1} &= -(\psi^n_{i+1,j+1} - \psi^n_{i-1,j+1})\frac{1}{2\Delta}
\end{aligned}
\tag{9.27}
$$

$$
\begin{aligned}
\xi^{n+\frac{1}{2}}_{i,j+1} &= \tfrac{1}{4}(\xi^n_{i,j} + \xi^n_{i+1,j+1} + \xi^n_{i-1,j+1} + \xi^n_{i,j+2}) \\
&\quad - \frac{\Delta t}{4\Delta} v^n_{x i,j+1}(\xi^n_{i+1,j+1} - \xi^n_{i-1,j+1}) \\
&\quad - \frac{\Delta t}{4\Delta} v^n_{y i,j+1}(\xi^n_{i,j+2} - \xi^n_{i,j})
\end{aligned}
\tag{9.28}
$$

Having obtained the new vorticity, we solve Poisson's equation on the auxiliary mesh to determine auxiliary values of the stream function:

$$(\psi^{n+\frac{1}{2}}_{i,j+1} + \psi^{n+\frac{1}{2}}_{i,j-1} + \psi^{n+\frac{1}{2}}_{i+2,j+1} + \psi^{n+\frac{1}{2}}_{i+2,j-1}) - 4\psi^{n+\frac{1}{2}}_{i+1,j} = -2\Delta^2\xi^{n+\frac{1}{2}}_{i+1,j} \quad (9.29)$$

These auxiliary variables are now used in the main calculation to determine the velocity at the main points at the time step $n + 1$:

main calculation:

$$v^{n+\frac{1}{2}}_{x_{ij}} = (\psi^{n+\frac{1}{2}}_{i,j+1} - \psi^{n+\frac{1}{2}}_{i,j-1})\frac{1}{2\Delta}$$

$$v^{n+\frac{1}{2}}_{y_{ij}} = -(\psi^{n+\frac{1}{2}}_{i+1,j} - \psi^{n+\frac{1}{2}}_{i-1,j})\frac{1}{2\Delta} \quad (9.30)$$

and

$$\xi^{n+1}_{ij} = \xi^{n}_{ij} - \frac{\Delta t}{2\Delta}v^{n+\frac{1}{2}}_{x_{ij}}(\xi^{n+\frac{1}{2}}_{i+1,j} - \xi^{n+\frac{1}{2}}_{i-1,j}) - \frac{\Delta t}{2\Delta}v^{n+\frac{1}{2}}_{y_{ij}}(\xi^{n+\frac{1}{2}}_{i,j+1} - \xi^{n+\frac{1}{2}}_{i,j-1})$$

$$-\frac{\Delta t}{2\Delta^2}\nu(\xi^{n}_{i+1,j+1} + \xi^{n}_{i-1,j+1} + \xi^{n}_{i+1,j-1} + \xi^{n}_{i-1,j-1} - 4\xi^{n}_{i,j}) \quad (9.31)$$

The term in the viscous diffusion of vorticity has been included by an angled first-order explicit method (Section 3.7). Finally, to complete the time step, Poisson's equation is solved (Section 4.4) on the main points to determine the stream function ψ at the time step $n + 1$,

$$(\psi^{n+1}_{i+1,j+1} + \psi^{n+1}_{i-1,j+1} + \psi^{n+1}_{i+1,j-1} + \psi^{n+1}_{i-1,j-1}) - 4\psi^{n+1}_{i,j} = -2\Delta^2\xi^{n+1}_{i,j} \quad (9.32)$$

Several points are to be noted with regard to this method of solution of the incompressible hydrodynamic equations. We have eliminated the explicit dependence on the sound speed, but an explicit method of solution is used for the advection of vorticity. Consequently, stability of the difference scheme is subject to the Courant–Friedrichs–Lewy condition (Chapter 3) where the characteristic velocity here is the advective or centre-of-mass velocity of the fluid,

$$\Delta t \leqslant \frac{2\Delta}{\sqrt{2}|\mathbf{v}|_{\max}} \quad (9.33)$$

$|\mathbf{v}|_{\max}$ is the modulus of the maximum velocity on the mesh.

In addition, the viscous diffusion term (if it exists) has been included by the explicit first-order method, so that there is a limitation on the time step which depends on the kinematic viscosity (Chapter 3):

$$\Delta t \leqslant \frac{\Delta^2}{\nu} \quad (9.34)$$

Usually the viscosity, though important, is small (the Reynolds number is large) and it follows that the time step will not be limited by the diffusion time (equation 9.34) but only by the advection time (equation 9.33). For this reason too, we have included the viscous diffusion term only in the main step of the calculation, so that it is only calculated to first-order accuracy in the time step. By defining the viscous diffusion term as an angled five-point difference, all mesh points in the Lax–Wendroff method are coupled (see Figure 3.14).

The difference method outlined here is accurate for the dominant advective terms to second order in the space step and the time step. It is to be noted that, while the differential system conserves a large number of physical variables and, in particular, the vorticity, the differenced advective term does not conserve the vorticity (Arakawa 1966). At the expense of some complexity for this problem, it is possible to modify the difference scheme so that vorticity is conserved, but all differential variables which are conserved cannot be conserved in the difference method (Section 9.3).

An alternative difference formulation based on the quasi-second-order or Adams–Bashforth method (Section 3.7) avoids the difficulty of a 'sodium-chloride' type mesh (Figure 9.2) and is simple to apply.

(*b*) *The Pressure Method.* In the alternative approach to the solution of the incompressible hydrodynamic equations (9.18, 9.19), we shall use the dependent variables of velocity and pressure directly. An equation for the pressure is obtained by demanding that, if the velocity field is divergence-free at $t = 0$, then it must remain divergence-free at all subsequent times. Therefore, we shall take the divergence of the three acceleration equations (9.18) in the fluid:

$$\begin{aligned} \nabla^2 \bar{p} &= -\boldsymbol{\nabla} . (\mathbf{v} . \boldsymbol{\nabla}\mathbf{v}) \\ &= -(\boldsymbol{\nabla}\mathbf{v}) : (\boldsymbol{\nabla}\mathbf{v}) \end{aligned} \qquad (9.35)$$

where we have used the condition (equation 9.19) on the velocities, $\bar{p}$ is the pressure normalized with respect to the constant density, $\bar{p} = p/\rho$, and a constant coefficient of viscosity has been assumed.

Equation 9.35 is Poisson's equation where the normalized pressure corresponds to the potential function and the 'source' terms on the right-hand side are related to the centrifugal and Coriolis forces due to the rotational motion of the fluid. Again, it is to be noted that the elliptic Poisson's equation represents an infinite speed of propagation of information over the space domain, and is a consequence of the assumption of incompressibility which defines an infinite sound speed. The formulation (equations 9.18, 9.35) is general and equally applicable to the incompressible hydrodynamic equations in two or three space dimensions.

For the sake of simplicity only, we shall express the method of solution for the two-dimensional problem in rectangular Cartesian coordinates (x, y). Defining the velocity in the x-direction as u and the velocity in the y-direction as v, the two-dimensional differential equations are:

$$\frac{\partial u}{\partial t} = -\frac{\partial u^2}{\partial x^2} - \frac{\partial uv}{\partial y^2} - \frac{\partial \bar{p}}{\partial x} + \nu\left\{\frac{\partial^2 u}{\partial x^2} + \frac{\partial^2 u}{\partial y^2}\right\} \tag{9.36}$$

$$\frac{\partial v}{\partial t} = -\frac{\partial v^2}{\partial y} - \frac{\partial uv}{\partial x} - \frac{\partial \bar{p}}{\partial y} + \nu\left\{\frac{\partial^2 v}{\partial x^2} + \frac{\partial^2 v}{\partial y^2}\right\} \tag{9.37}$$

with Poisson's equation for the pressure:

$$\frac{\partial^2 \bar{p}}{\partial x^2} + \frac{\partial^2 \bar{p}}{\partial y^2} = -\left\{\left(\frac{\partial u}{\partial x}\right)^2 + 2\left(\frac{\partial u}{\partial y}\right)\left(\frac{\partial v}{\partial x}\right) + \left(\frac{\partial v}{\partial y}\right)^2\right\} \tag{9.38}$$

It is to be noted that while, in the differential system, if the initial conditions specify a velocity field which has no divergence, the velocity field remains divergence-free for all subsequent times, the same situation does not necessarily apply to a difference solution. A difference scheme must therefore be devised in such a way that, irrespective of truncation errors, the finite difference analogue of the divergence condition (equation 9.19) is strictly satisfied. If this is not achieved, the flow will quickly become compressible.

To satisfy the requirements of the difference solution, we consider an equally spaced Eulerian lattice, where each cell or box is to be regarded as the basic unit (Figure 9.3) in the sense that the flow through each box is to be strictly incompressible. There are three dependent variables $(\bar{p}, u, v)$ and we define each at separate lattice points: the pressure variables are defined only at the centres (x_i, y_j) of each box, the 'horizontal' velocities u in the x-direction are defined only on the 'vertical' (constant x) faces $(x_{i+\frac{1}{2}}, y_j)$ of each box; equally, the 'vertical' velocities v in the y-direction are defined only on the 'horizontal' (constant y) faces $(x_i, y_{j+\frac{1}{2}})$ of each box (Figure 9.3). Consequently, each of the three dependent variables is defined on three separate interlocking point meshes (Figure 9.3). For simplicity we shall use the notation of compass points to define any five neighbouring points for each of the dependent variables (Figure 9.3).

$$\begin{aligned} \bar{p}_C &= \bar{p}(x_i, y_j, t^n) \\ u_{C'} &= u(x_{i+\frac{1}{2}}, y_j, t^n) \\ v_{C''} &= v(x_i, y_{j+\frac{1}{2}}, t^n) \end{aligned} \tag{9.39}$$

The rationale for this mesh formulation is clear, since we may now define the difference analogue of the divergence of the velocities unambiguously:

$$D_C = \frac{1}{\Delta}(u_{C'} - u_{W'}) + \frac{1}{\Delta}(v_{C''} - v_{S''}) \tag{9.40}$$

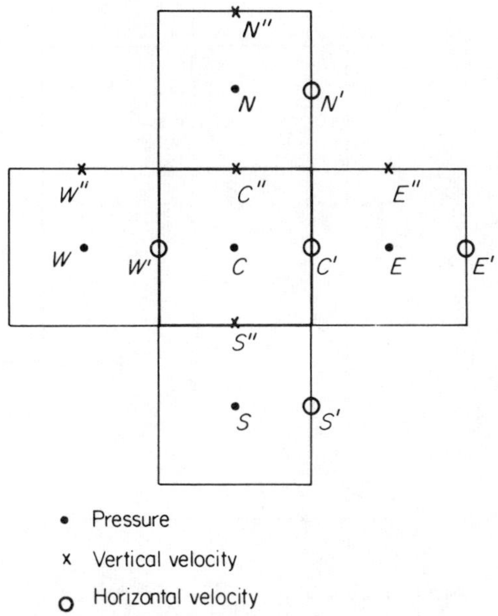

Figure 9.3. The difference mesh used in formulating the equations for an incompressible fluid by the pressure method. The divergence of the flow through each unit cell is made strictly zero

consequently the difference analogue for the incompressible flow of the fluid (equation 9.19) is:

$$D_C = 0 \tag{9.41}$$

for every cell and at all times.

If the dependent variables, defined according to this prescription (equations 9.39), are specified at time t^n, new values for the velocities are obtained over the time step Δt by integrating the Navier–Stokes equations (9.36, 9.37). We use the conservative Lax method (Section 3.8) so that for the velocities in the x-direction (Figure 9.4) new values are obtained at the step $n + 1$:

$$\begin{aligned} u_{C'}^{n+1} &= \frac{1}{4}(u_{N'} + u_{E'} + u_{W'} + u_{S'}) - \frac{\Delta t}{2\Delta}(u_{E'}^2 - u_{W'}^2) \\ &\quad - \frac{\Delta t}{2\Delta}\left\{\frac{v_{E''} + v_{C''}}{2}(u_{N'} + u_{C'}) - \frac{v_{S''} + v_{SE''}}{2}(u_{S'} + u_{C'})\right\} \\ &\quad - \frac{\Delta t}{\Delta}(\bar{p}_E - \bar{p}_C) + \frac{\nu\,\Delta t}{\Delta^2}(u_{N'} + u_{S'} + u_{E'} + u_{W'} - 4u_{C'}) \end{aligned} \tag{9.42}$$

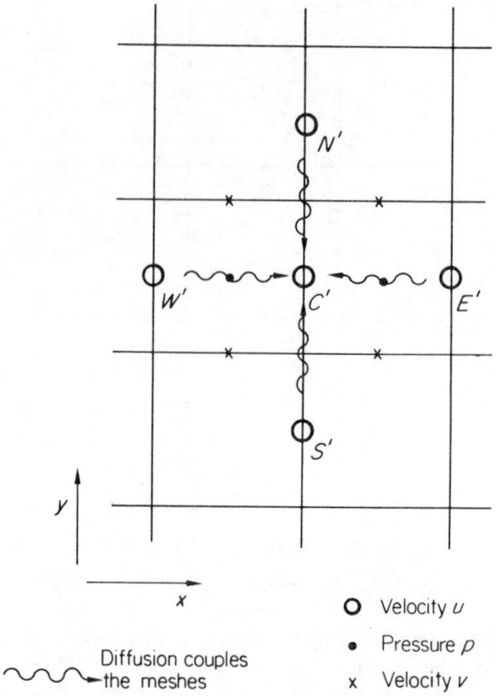

Figure 9.4. The difference mesh on which the horizontal velocity u is defined. The Lax method is used

The first term on the right-hand side is the space average of the dependent variable at the time step n as prescribed in the Lax method and the other terms are evident. It is to be noted, however, that in the nonlinear cross-advective term, $\partial vu/\partial y$, the velocities v are not defined at the same space points as the velocities u, so that a space average has been taken. Correspondingly, we obtain the velocities in the y-direction (Figure 9.5):

$$\begin{aligned} v_{C''}^{n+1} &= \frac{1}{4}(v_{N''} + v_{E''} + v_{S''} + v_{W''}) - \frac{\Delta t}{2\varDelta}(v_{N''}^2 - v_{S''}^2) \\ &- \frac{\Delta t}{2\varDelta}\left\{\frac{u_{N'} + u_{C'}}{2}(v_{E''} + v_{C''}) - \frac{u_{NW'} + u_{W'}}{2}(v_{C''} + v_{W''})\right\} \\ &- \frac{\Delta t}{\varDelta}(\bar{p}_N - \bar{p}_C) + \frac{\nu\,\Delta t}{\varDelta^2}(v_{N''} + v_{E''} + v_{S''} + v_{W''} - 4v_{C''}) \end{aligned} \quad (9.43)$$

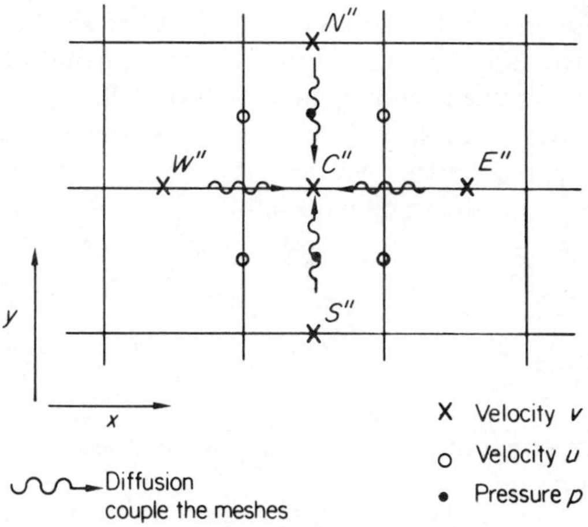

Figure 9.5. The difference mesh on which the vertical velocity v is defined. The Lax method is used

We are now in the position to determine the difference analogue of Poisson's equation from which the pressures at the centres of each cell are to be determined, and we examine the divergence of the velocities at the time t^{n+1} for each cell C (equation 9.40). By using the difference equations (9.42 and 9.43) for the velocity u in the x-direction at the points C' and W' (Figure 9.3) and for the velocity v in the y-direction at the points C'' and S'', we form the sum (equation 9.40) for the divergence:

$$D_C^{n+1} = \frac{1}{4}(D_N + D_S + D_E + D_W) - S_C\frac{\Delta t}{2\Delta^2} - (\bar{p}_N + \bar{p}_S + \bar{p}_E + \bar{p}_W - 4\bar{p}_C)$$

$$\times \frac{\Delta t}{\Delta^2} + \frac{\nu\,\Delta t}{\Delta^2}(D_N + D_S + D_E + D_W - 4D_C) \tag{9.44}$$

where the source function S_C is:

$$S_C = (u_{E'}^2 - u_{C'}^2 - u_{W'}^2 + u_{WW'}^2) + (v_{N''}^2 - v_{C''}^2 - v_{S''}^2 + v_{SS''}^2)$$

$$+ \frac{v_{E''} + v_{C''}}{2}(u_{N'} + u_{C'}) - \frac{v_{S''} + v_{SE''}}{2}(u_{S'} + u_{C'})$$

$$- \frac{u_{NW'} + u_{W''}}{2}(v_{C''} + v_{W''}) + \frac{u_{W'} + u_{SW'}}{2}(v_{S''} + v_{SW''}) \tag{9.45}$$

Since the divergence D at every cell is to be zero, the pressure could be determined at each time level from the velocities in the source function S_C only,

$$\bar{p}_N + \bar{p}_S + \bar{p}_E + \bar{p}_W - 4\bar{p}_C = -S_C \tag{9.46}$$

Such a procedure would be satisfactory in the total absence of errors when obtaining the solution (Chapter 4) to Poisson's equation (9.46). However, round-off errors occur inevitably and certainly, if Poisson's equation (9.46) is solved by an iterative method, the ultimate solutions for the pressure are 'inexact'. In turn, these errors will be reflected in a flow with finite velocity divergence. Therefore, to avoid the accumulation of errors, we solve Poisson's equation as:

$$\bar{p}_N + \bar{p}_E + \bar{p}_S + \bar{p}_W - 4\bar{p}_C = -S_C + (D_N + D_S + D_E + D_W - 4D_C) + \frac{\Delta^2}{\Delta t}(D_N + D_S + D_E + D_W) \tag{9.47}$$

where the divergence of the velocity will remain strictly zero with respect to truncation errors.

The time-dependent equations (9.42, 9.43) and the matrix equation (9.47) define each time level in the difference solution of the incompressible hydrodynamic equations. Advection and viscous diffusion are determined explicity by using the Lax method in conjunction with the explicit first-order method (Chapter 3). It follows that the numerical solutions will be stable, if the time step is limited according to the stability criteria:

$$\Delta t \leqslant \frac{\Delta}{(u_C^2 + v_C^2)^{\frac{1}{2}}\sqrt{2}} \quad \text{(all cells } C\text{)} \tag{9.48}$$

$$\Delta t \leqslant \frac{1}{2}\frac{\Delta^2}{\nu} \tag{9.49}$$

The approach outlined here is certainly more complex than that in which the vorticity and stream function are used (Section 9.2b). However, the method has greater versatility, since it may more readily be applied to three-dimensional problems and, in particular, complex boundary conditions, for example for moving surfaces (Section 9.4), may be incorporated. We shall consider boundary conditions and the application of the method to real, rather than idealized, physical phenomena in Section 9.4. The algorithm described here is in essence due to Harlow and Welch (1965).

3 Incompressible Flow as an Assembly of Vortex Particles

(*a*) *Conservation Laws for Incompressible Fluids.* The fundamental equations for a two-dimensional incompressible fluid are of particular interest both because they describe a wide range of natural phenomena and because they contain in essence the answer to the question of whether turbulent flows can be described by a deterministic model. Certainly, the differential equations determine the state of the fluid uniquely for all subsequent times. However, any computational model must always be finite and will, in general, not describe the short-wavelength phenomena in the fluid: yet in the presence of only a small viscosity many fluids are turbulent and we might question, therefore, how such turbulent short-wavelength phenomena can be described so that the macroscopic motion of the fluid is faithfully represented. This is of importance, for example, with regard to the predictability of the weather (Dutton and Deaven 1972).

It is fruitful, therefore, to gain an alternative perspective of the incompressible fluid equations. In two dimensions and in the absence of viscosity, we may represent the motion of the fluid by a pseudo-scalar vorticity and stream function (equations 9.26):

$$\frac{\partial \xi}{\partial t} + \frac{\partial \psi}{\partial y}\frac{\partial \xi}{\partial x} - \frac{\partial \psi}{\partial x}\frac{\partial \xi}{\partial y} = 0 \tag{9.50}$$

where, as before, the stream function is related to the vorticity through Poisson's equation. This equation may be represented conveniently with a Jacobian transform:

$$\frac{\partial \xi}{\partial t} + \frac{\partial(\xi, \psi)}{\partial(x, y)} = 0$$

or simply in terms of Poisson brackets:

$$\frac{\partial \xi}{\partial t} + (\xi, \psi) = 0 \tag{9.51}$$

Equally, and with no loss of generality, equation 9.50 for the vorticity may be written in either of two conservative forms:

$$\frac{\partial \xi}{\partial t} + \frac{\partial}{\partial x}\left(\xi\frac{\partial \psi}{\partial y}\right) - \frac{\partial}{\partial y}\left(\xi\frac{\partial \psi}{\partial x}\right) = 0$$

$$\frac{\partial \xi}{\partial t} + \frac{\partial}{\partial y}\left(\psi\frac{\partial \xi}{\partial x}\right) - \frac{\partial}{\partial x}\left(\psi\frac{\partial \xi}{\partial y}\right) = 0 \tag{9.52}$$

Each of these equations is a statement of the conservation of vorticity and, on the one hand, if the equations were directly differenced in either of these two forms, the vorticity on the mesh would be conserved. On the other hand, it is clear from equation 9.52 that if we multiply the equation by any power of the vorticity ξ^{p-1}, then any power p of the vorticity will be conserved. In effect, therefore, the incompressible hydrodynamic equations define an infinite set of conserved variables, of which only a few can identically be conserved on the difference mesh.

In addition, it is informative to enquire into the conservation of energy E. In this two-dimensional problem, we shall consider the integral of $\psi\mathbf{v}$ along some contour C, enclosing a surface S in the (x, y) plane, where we obtain by Stokes's theorem:

$$\oint_C \psi\mathbf{v}\,.\,\mathbf{dl} = \iint_S (\text{curl}\;\psi\mathbf{v})\,.\,\mathbf{dS} \tag{9.53}$$

Since ψ and ξ are pseudo-scalars in the sense that they are unidirectional vectors which lie perpendicular to the plane S, we obtain:

$$\begin{aligned}\oint_C \psi\,\mathbf{v}\,.\,\mathbf{dl} &= \iint_S \psi(\text{curl}\;\mathbf{v})\,.\,\mathbf{dS} + \iint_S (\mathbf{v}\,.\,\text{curl}\;\psi)\,\mathrm{d}S \\ &= \iint_S \psi\xi\,\mathrm{d}S + \iint_S \mathbf{v}\,.\,\mathbf{v}\,\mathrm{d}S \end{aligned} \tag{9.54}$$

Thus, if the velocity $\mathbf{v}$ is zero on the contour C, the total energy E of the incompressible fluid is

$$\begin{aligned} E &= \tfrac{1}{2}\rho_0 \iint_S \mathbf{v}\,.\,\mathbf{v}\,\mathrm{d}S \\ &= -\tfrac{1}{2}\rho_0 \iint \psi\xi\,\mathrm{d}S \end{aligned} \tag{9.55}$$

In the first instance, therefore, when devising appropriate finite computational models of incompressible flow, an attempt should be made to conserve as effectively as possible the vorticity of the fluid, powers of vorticity and the kinetic energy E. More important, however, it is apparent from the arguments in this section that the two-dimensional equations for an incompressible fluid can take a Hamiltonian form where the vorticity is related to a density and the stream function is related to a potential.

(*b*) *'Particle-in-cell' Simulation of Incompressible Fluids.* According to the previous section, it is clear that there is a close analogy between, on the one hand, the advective equation for the vorticity with Poisson's equation for the stream function in an incompressible fluid and, on the other hand, the equations for the distribution function and gravitational or electrostatic potential of a Vlasov phase fluid (Chapter 8). As in the Vlasov case, the equations for a two-dimensional incompressible fluid are summarized in Hamiltonian form (Onsager 1949):

$$\frac{\partial\xi}{\partial t} + (\xi, \psi) = 0 \tag{9.56}$$

$$\nabla^2\psi = -\xi \tag{9.57}$$

It is entirely appropriate, therefore, to represent an incompressible fluid in two dimensions as an assembly of 'particle vortices'. A fixed element of vorticity is to be associated with each particle so that the density of particles in the (x, y) space will define the vorticity distribution. Equally, the stream function which corresponds to the potential and to the Hamiltonian (there is no kinetic energy) of the system is to be determined from Poisson's equation. Consequently, we shall simulate an incompressible fluid in two dimensions by a collisionless 'particle-in-cell' model, in a manner closely analogous to an electrostatic or gravitational assembly in a two-dimensional phase space (Christiansen 1971).

The two-dimensional configuration space (x, y) is to be divided into a set $(I \times J)$ of equally spaced Eulerian cells. A sufficiently large number of particle vortices are distributed over the mesh where each particle μ is assigned the canonical coordinates (x_μ, y_μ). By sufficiently large, we mean that the average number of particles in a cell is to be large in order that the fluctuations in the system are minimized and, in analogy to the Vlasov system, the particle motion is, thereby, 'collisionless' (Chapter 6). According to the characteristics of the equation for the vorticity (equation 9.56), the 'equations of motion' for each particle μ are defined as

$$\frac{\mathrm{d}x_\mu}{\mathrm{d}t} = \frac{\partial\psi}{\partial y_\mu}$$

$$\frac{\mathrm{d}y_\mu}{\mathrm{d}t} = -\frac{\partial\psi}{\partial x_\mu} \tag{9.58}$$

In the simplest situation, each particle is assigned a constant vorticity ξ_0, since along the characteristics or trajectory of a particle, the vorticity is constant (equation 9.56) and it follows immediately that the conservation of particles implies the conservation of vorticity and all powers of vorticity (Section 9.3a). In addition, a two-species model may be envisaged, in which positive or negative vortex particles are used. The vorticity of the fluid at any time is simply determined by the density of vortex particles and consequently the stream function ψ is determined from Poisson's equation.

In difference form, the leapfrog scheme is used where all particle coordinates (x_μ, y_μ) are stored in the memory of the computer for two time levels, say t^{n-1} and t^n.

If, for all μ, $1 \leqslant \mu \leqslant N$, cell (i, j) is defined by

$$i = \mathrm{Int}\left(\frac{x_\mu^n}{\Delta}\right), \qquad j = \mathrm{Int}\left(\frac{y_\mu^n}{\Delta}\right) \tag{9.59}$$

then, according to the equations of motion (9.58), particle coordinates are determined at the time step $n + 1$:

$$x_\mu^{n+1} = x_\mu^{n-1} + \frac{\Delta t}{2\Delta}(\psi_{i,j+1}^n - \psi_{i,j-1}^n)$$

$$y_\mu^{n+1} = y_\mu^{n-1} - \frac{\Delta t}{2\Delta}(\psi_{i+1,j} - \psi_{i-1,j}) \tag{9.60}$$

where Δt is the time step, Δ is the space step and the function Int (z) truncates the real number z to an integer.

The particles are summed in each cell (i, j) to determine the fluid vorticity:

$$\zeta_{ij}^{n+1} = \xi_0 \sum_{\mu=1}^{N} \delta\left(i - \text{Int}\left(\frac{x_\mu^{n+1}}{\Delta}\right)\right)\delta\left(j - \text{Int}\left(\frac{y_\mu^{n+1}}{\Delta}\right)\right) \tag{9.61}$$

where

$$\delta(l - m)\begin{cases} = 1 \text{ if } l = m \\ = 0 \text{ if } l \neq m \end{cases}$$

More sophisticated 'area-weighting' methods of summation (Chapter 6) may be employed. To complete the time step, the stream function $\psi_{i,j}^{n+1}$ is determined from the difference form of Poisson's equation (Chapter 4):

$$\psi_{i+1,j}^{n+1} + \psi_{i-1,j}^{n+1} + \psi_{i,j+1}^{n+1} + \psi_{i,j-1}^{n+1} - 4\psi_{i,j}^{n+1} = -\Delta^2\zeta_{i,j}^{n+1} \tag{9.62}$$

It is to be noted that, in this instance and unlike the simulation in phase space of an electrostatic or gravitational assembly (Chapter 6), the two time-dependent coupled equations for the motion of each particle are both dependent on gradients of ψ and, hence, both are dependent on (x_μ, y_μ). Thus both particle coordinates must be stored for two time levels and the use of the leapfrog scheme in this instance introduces a nonphysical computational mode. In order to avoid the two adjacent time levels becoming decoupled (Section 2.6b), the variables at two time levels must be reconstituted, or recoupled, after a certain number (typically 100) of time steps.

(*c*) *Some Fundamental Incompressible Flow Solutions.* In many incompressible flows of interest in the natural world, laminar flow is found to be unstable and turbulence results. Although, in many cases the linear phases (for small perturbations) of flow instabilities have been appreciated for many years, it is the final turbulent nonlinear properties which are of particular interest and which have only recently been studied with the computational models described in this chapter. We shall illustrate some of these phenomena as obtained by an assembly of particle vortices (Christiansen 1971).

Of basic interest is the Kelvin–Helmholtz instability which arises in the absence of viscosity whenever a simple shear exists in the flow of an incom-

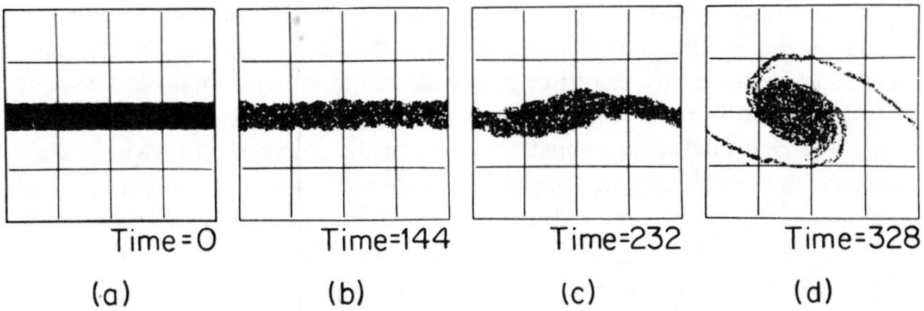

Figure 9.6. The Kelvin–Helmholtz instability. The layer of vortex particles (a) describes a sheared flow. In the nonlinear phase (d) large eddies are formed which tend to mix the two layers of oppositely moving fluid (from Christiansen, J. P., *Vortex: A 2-Dimensional Hydrodynamics Simulation Code*, Culham Laboratory Report, CLM-R 106)

pressible fluid. Such a shear is equivalent to a layer of vorticity (Figure 9.6a) which defines the initial conditions in the particle-in-cell simulation. In Figure 9·6b the linear phase of the Kelvin–Helmholtz instability is illustrated, but this rapidly develops into a nonlinear phase in which a series of large vortices are formed at the boundary of the sheared flow (Figures 9.6c and d). The nonlinear vortices which are formed tend to mix or smooth out the flow, so that from the macroscopic point of view the turbulence has the effect of a viscous diffusion.

An extension of the Kelvin–Helmholtz instability occurs in wakes or whenever an obstruction impedes the flow of a moving fluid: for example when telephone wires obstruct the wind. Behind the obstruction the fluid is stationary while on either side there exists a finite flow. Such an initial configuration is represented by two layers of vorticity of opposite sign (Figure 9.7a). In the linear phase (Figure 9.7b) each layer is separately

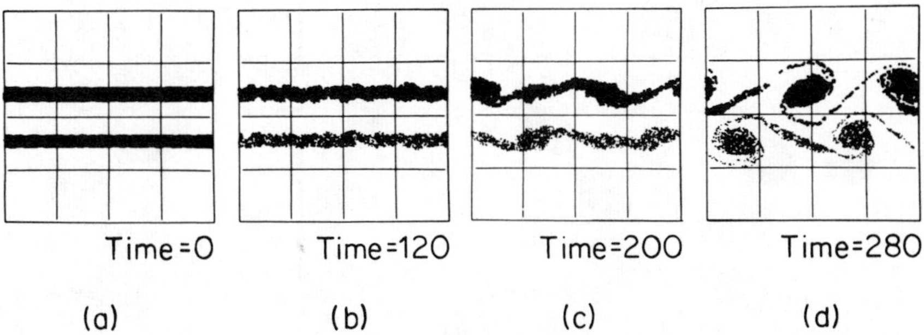

Figure 9.7. The von Karman vortex street. Two layers of vortex particles of opposite sign delineate a wake region of stationary fluid (a). In the nonlinear phase eddies of opposite sign are produced downstream of the obstacle (d) (from Christiansen, J. P., *Vortex: A 2-Dimensional Hydrodynamic Simulation Code*, Culham Laboratory Report, CLM-R 106)

unstable to the Kelvin–Helmholtz instability, but interaction between the layers occurs in the nonlinear regime (Figure 9.7c). Finally, in the nonlinear quasi-steady state, the beautiful structure of the von Karman vortex street (Figure 9.7d) occurs in which a series of large vortices or eddies of opposite sign are produced downstream of the obstacle. It is this series of vortices which produce the 'singing' noise associated with wind through telephone wires and, again, the effect can be observed in the rising smoke from a chimney.

In Figure 9.8, the interaction of two large vortices is illustrated. Vortices of the same sign attract and distort each other and, when sufficiently close, they may coalesce to form one large vortice.

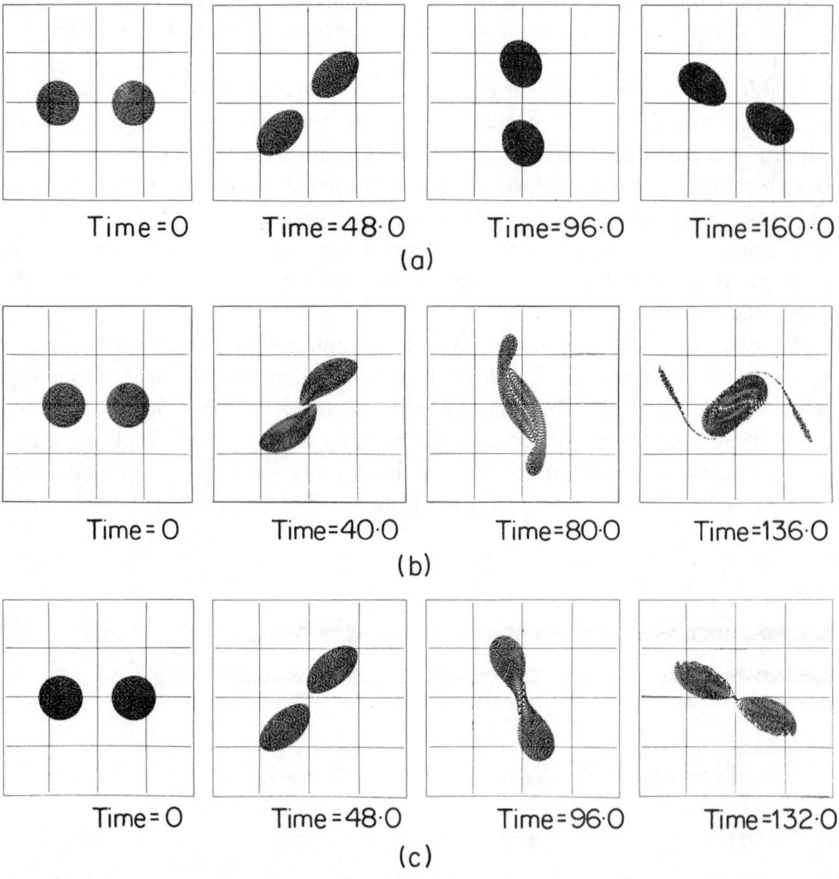

Figure 9.8. The attraction of two finite-sized vortices or eddies of the same sign: (a) rotation about each other when sufficiently far apart; (b) coalescence at sufficient proximity: (c) a critical separation (from Christiansen, J. P., *Vortex: A 2-Dimensional Hydrodynamics Simulation Code*, Culham Laboratory Report, CLM-R 106)

4 The Marker-and-Cell Method for Surfaces and Heavy Fluids: Splashes, Waterfalls and Breaking Waves

In the preceding sections, the appropriate formulation and fundamental methods for solution of incompressible fluids have been discussed. Apart from the theoretical study of ideal incompressible flow, the methods have application in the description of a wide range of applied problems and interesting natural phenomena occurring on the surface of the earth which, because of their nonlinearity and complexity, have defied analysis for many years. To solve, as an initial-value problem, the splash of a liquid drop, the breaking of a wave, or the motion of a waterfall is on the one hand intrinsically satisfying but, perhaps more important, these solutions illustrate the scope and complexity of problems to which these methods can apply.

In applying computational methods for incompressible fluids (Sections 9.2 and 9.3) to 'real' phenomena on the surface of the earth, we need to consider such effects as free surfaces, which mark the boundary of the fluid, and the gravity of the earth. In even the simplest situation of, say, water in a tank, this is a difficult problem since Poisson's equation must be solved in an awkward domain with a moving boundary, while in more complex problems an incompressible fluid of variable density must be described. We may identify two problems in particular. First, how is the position of a moving fluid surface to be defined on an Eulerian mesh in space and second, what are the physical effects of the earth's gravitation on a moving fluid?

On the surface of the earth, the effect of gravity is simply included as a body force in the hydrodynamic equations for an incompressible fluid (equations 9.18 and 9.19):

$$\frac{\partial \mathbf{v}}{\partial t} + \mathbf{v} \,.\, \nabla \mathbf{v} = -\frac{1}{\rho} \nabla p + \mathbf{g} \tag{9.63}$$

$$\nabla \,.\, \mathbf{v} = 0 \tag{9.64}$$

where $\mathbf{g}$ is the constant gravitational acceleration on the surface of the earth and the effect of viscosity is here ignored. Since $\mathbf{g}$ is a constant, predefined acceleration (in no sense self-consistent), it is simply incorporated in any difference formulation, but it has the effect of giving rise to water waves, tides and associated phenomena. We may gain an insight into the properties of a heavy fluid with a free surface by considering shallow water in a channel (Figure 9.9). By shallow water we mean that the typical depth of the water h_0 is very much smaller than the length of waves on the surface of the water and we shall assume that the bottom of the channel ($y = 0$) is in the horizontal plane.

In such a shallow fluid it is reasonable to make the approximation that the vertical velocity v is small, so that we shall replace the Navier–Stokes

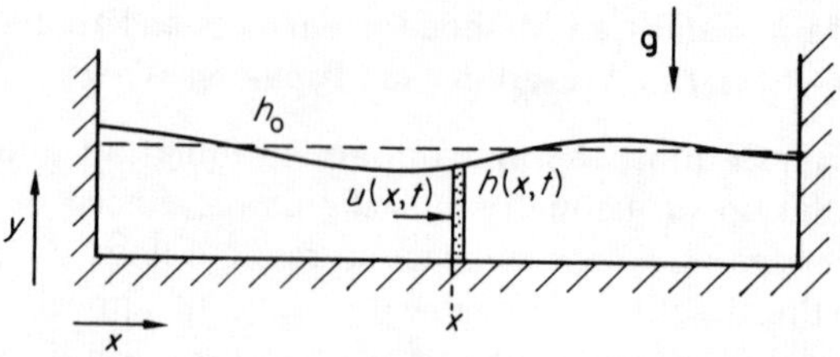

Figure 9.9. Surface waves on a heavy fluid in a shallow tank. Each vertical column of fluid moves rigidly and is described by the height $h(x, t)$ and horizontal velocity $u(x, t)$

equation (9.63) for the vertical component of velocity by the assumption of hydrostatic equilibrium in the vertical direction:

$$\frac{\partial p}{\partial y} = -\rho g. \tag{9.65}$$

By integrating this equation we obtain an expression for the pressure in the fluid:

$$\begin{aligned} p &= -\int_h^y \rho g \, \mathrm{d}y \\ &= -\rho g(y - h) \end{aligned} \tag{9.66}$$

where $h(x, t)$ is the height above the bottom of each point on the surface of the fluid. An equation for the horizontal component of the velocity u is immediately obtained:

$$\frac{\partial u}{\partial t} + u\frac{\partial u}{\partial x} = -g\frac{\partial h}{\partial x} \tag{9.67}$$

where again the inertia associated with the vertical velocity is ignored. An equation for the height of the surface is obtained from the incompressibility condition (equation 9.64):

$$\begin{aligned} v(x, y, t) &= -\int_0^y \frac{\partial u}{\partial x} \mathrm{d}y \\ &= -y\frac{\partial u}{\partial x} \end{aligned} \tag{9.68}$$

where, in accordance with equation 9.67, we have assumed that each column of water moves rigidly so that the horizontal velocity $u(x, t)$ is independent of y. At the surface of the fluid, $y = h$:

$$v = \frac{dh}{dt}$$

thus

$$\frac{\partial h}{\partial t} + u\frac{\partial h}{\partial x} = -h\frac{\partial u}{\partial x} \tag{9.69}$$

or, in conservative form,

$$\frac{\partial h}{\partial t} + \frac{\partial}{\partial x}(hu) = 0$$

Equations 9.67 and 9.69 are the shallow-water equations defining the height of the surface $h(x, t)$ and the horizontal velocity $u(x, t)$. Clearly, the equilibrium solution is a stationary fluid $u = 0$ of constant height $h = h_0$. For a small-amplitude disturbance $h = h_0 + h'$ and $u = u'$ on the surface of the fluid, we may approximate the pair of equations by the linear coupled equations:

$$\frac{\partial u'}{\partial t} + g\frac{\partial h'}{\partial x} = 0$$

$$\frac{\partial h'}{\partial t} + h_0\frac{\partial u'}{\partial x} = 0 \tag{9.70}$$

where we have ignored small second-order terms or, in terms of the second-order equation:

$$\frac{\partial^2 h'}{\partial t^2} - h_0 g\frac{\partial^2 h'}{\partial x^2} = 0 \tag{9.71}$$

Small-amplitude disturbances travel, therefore, as waves with the wave velocity $v_s = \sqrt{(gh_0)}$. For large-amplitude disturbances, equations 9.67 and 9.69 are nonlinear and include both advection and wave phenomena.

The shallow-water equations illustrate the basic effect of surface waves on a heavy fluid but more generally we wish to describe such effects on the full two-dimensional (or three-dimensional) flow of deep fluids and it is appropriate to solve the full two-dimensional equations for an incompressible fluid, including the body gravity force (equations 9.63 and 9.64). These may be resolved according to the methods described in Section 9.2, but the inclusion of the gravitational force adds an additional limitation on the time step due to the effect of surface waves:

$$\Delta t \leqslant \frac{\Delta}{\sqrt{(gh)}} \tag{9.72}$$

where h is the maximum depth of the fluid. The pressure method (Harlow and Welch 1965, Section 9.2c), where the velocity of the fluid is determined directly from the Navier–Stokes equations subject to the pressure obtained from Poisson's equation, is most appropriate to heavy fluids with moving surfaces, since the appropriate complex boundary conditions are more applicable than in the method of vorticity (Section 9.2b).

The problem remains, therefore, to define the position of the moving surface of the dense fluid on the Eulerian space mesh. In the marker-and-cell method (Harlow and Welch 1965), the marker particles are distributed on the Eulerian mesh, not only on the surface but throughout the dense fluid. The marker particles are moved with interpolation from neighbouring cells of the local fluid velocities on the Eulerian mesh, as obtained by the Eulerian difference solution (Section 9.2c). The position of the heavy fluid is defined by the distribution of marker particles in the Eulerian cells, of which, therefore, there are three types: vacuum cells in which there are no marker particles, surface cells, containing one or more marker particles, but adjacent to a vacuum cell, and other fluid cells containing marker particles (Figure 9.10).

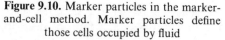

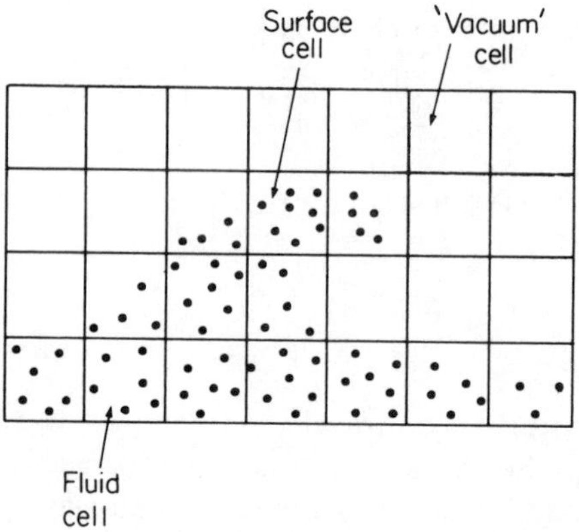

Figure 9.10. Marker particles in the marker-and-cell method. Marker particles define those cells occupied by fluid

The free surface of the fluid is therefore defined by marker particles and in vacuum cells the difference equations are not applied. Boundary conditions for the velocity components are applied along the surface cells of which there are four types (Figure 9.11), and the boundary conditions for the pressure in Poisson's equations are simply the constant applied pressure in the 'vacuum' cells. Simple boundary conditions for the fluid at a rigid wall may generally be taken. If **n** is a unit vector normal to the wall, the condition

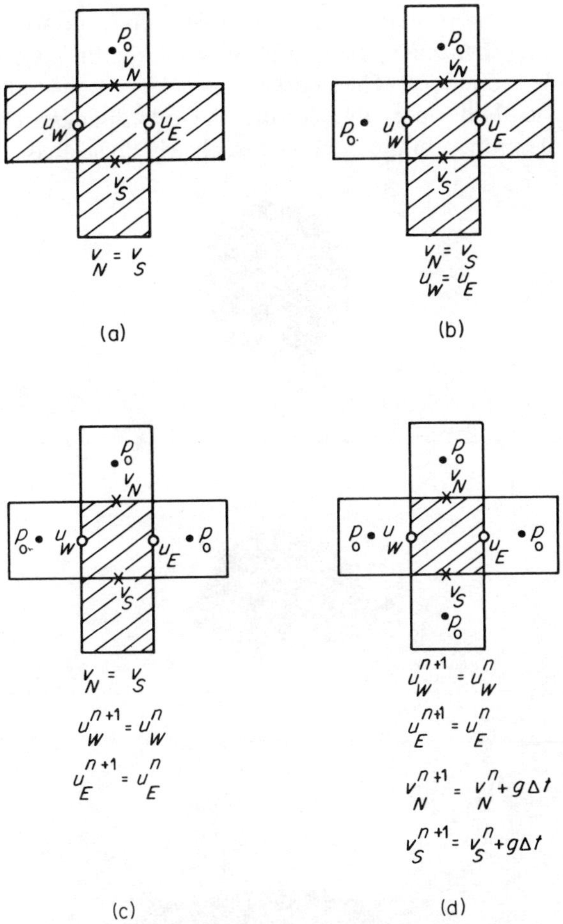

Figure 9.11. The four types of surface cells which occur at the boundary between a moving fluid and the atmosphere. The appropriate boundary conditions for each case are illustrated

on the pressure is

$$\mathbf{n} \cdot \nabla \bar{p} = \mathbf{g} \cdot \mathbf{n}$$

while, for the velocity components, the perpendicular velocity is zero and for the parallel velocity $v_{\parallel}$, either the simple condition of *free slip*, $\partial v_{\parallel}/\partial \mathbf{n} = 0$ can be applied or the condition of *no slip* $v_{\parallel} = 0$ can be applied.

It is to be noted that the 'particles' in the marker-and-cell method are in no sense used to determine the fluid variables self-consistently since the fluid variables are determined by an Eulerian difference method (Section 9.2c).

Their essential purpose is to define the moving free surface of an incompressible fluid in order that appropriate boundary conditions may be applied at the moving boundary. The marker particles are included, however, throughout the fluid since, by plotting the coordinates of the markers graphically, the motion of the fluid is most clearly illustrated.

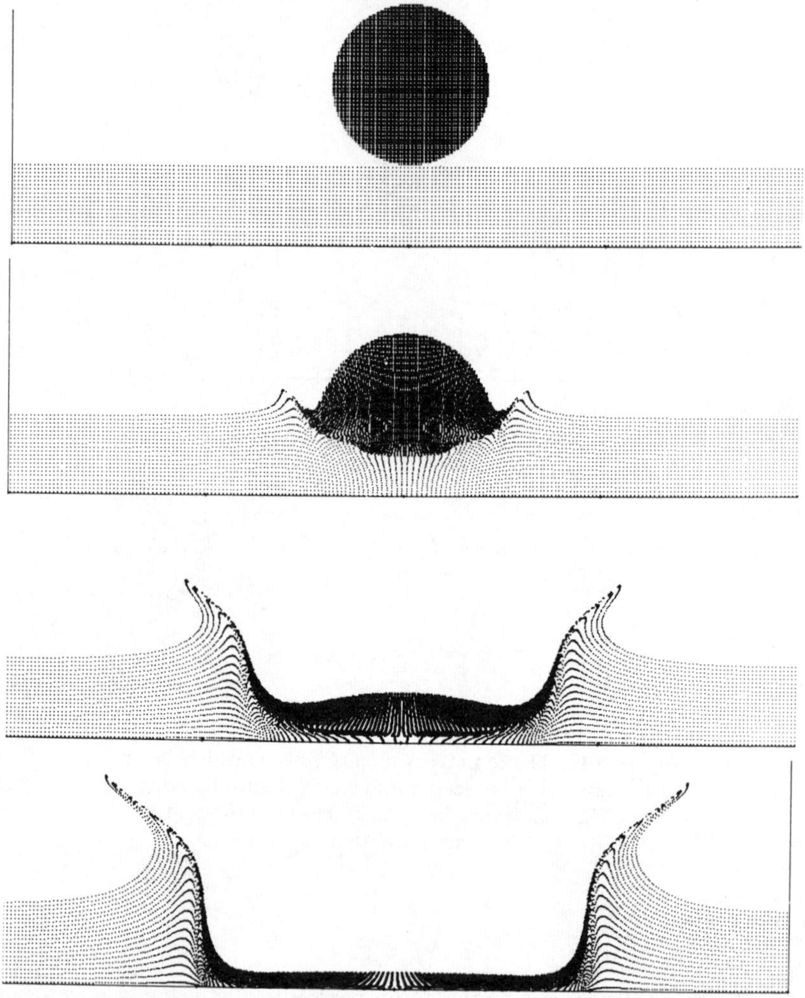

Figure 9.12. The splash of a falling liquid drop into a shallow fluid. The simulation was executed by the marker-and-cell method in two-dimensions (from Harlow, F. H. and Shannon, J. P., *Science*, **157**, pp. 547–550, August, 1967. Copyright 1967 by the American Association for the Advancement of Science). The figure may be compared with the high-speed photograph of the splash of a liquid drop (Figure 9.13) and with the splash of a compressible drop determined by the particle-in-cell method (Figure 6.17)

A wide range of physical phenomena may be studied with the aid of marker-and-cell simulations and a few interesting examples are included in Figures 9.12 to 9.16 (Harlow and Welch 1965, Harlow and Shannon 1967). In Figure 9.12, the splash of a liquid drop into a shallow pool is illustrated, where the characteristic expanding crown of the splash is demonstrated. This can be compared with high-speed photographs (Figure 9.13) of the

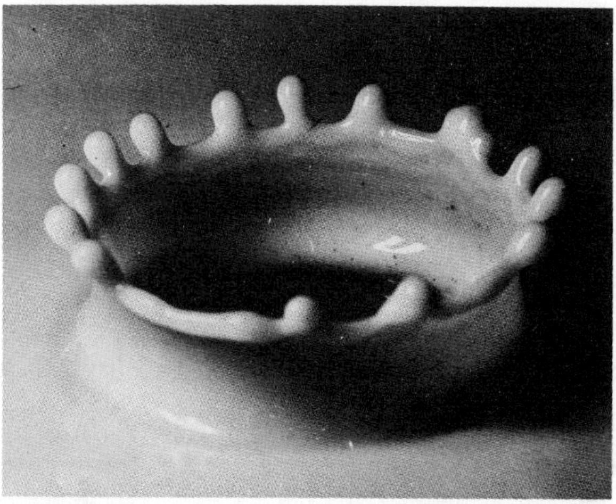

Figure 9.13. High-speed photograph of the splash of a liquid drop (from Harlow, F. H., and Shannon, J. P., 'The Splash of a Liquid Drop', *J. Appl. Phys.*, **38**, No. 10, pp. 3855–3866, 1967). In three-dimensions, the rim of the crown is unstable to Rayleigh–Taylor instabilities

phenomenon (Edgerton and Killian 1954) and the simulation of a splash in a deep pool of water (Figure 9.14), where the features are radically different. The comparison is extremely close, except in so far as the simulation is two-dimensional so that the small three-dimensional effect in the experiment of small droplets occurring azimuthally around the crown are not described. In Figure 9.15 the simulation of the mathematically very complex phenomenon of a wave breaking on the shore is included. The breaking of the wave is analogous to the steepening of a shock wave. The 'surfers' curl is clearly demonstrated and it is to be noted how the marker particles illustrate turbulence as in the surf of the wave and equally laminar flow as in the curl of the wave. The effect of the opening of a sluice gate is shown in Figure 9.16.

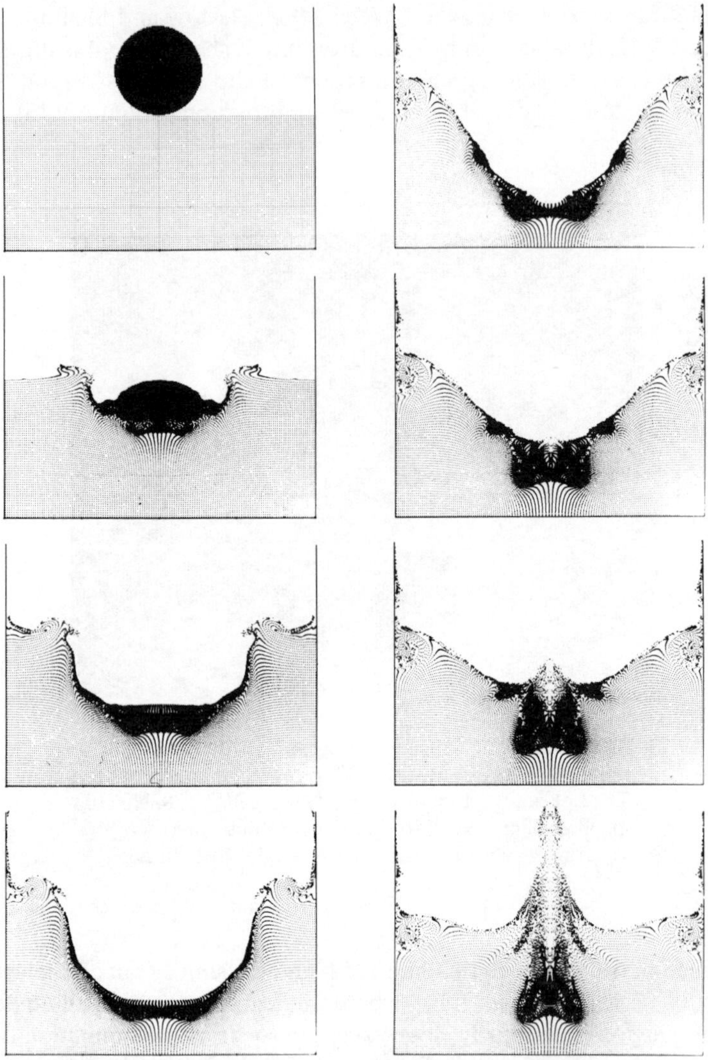

Figure 9.14. The splash of a liquid drop into a deep pool by the marker-and-cell method (from Harlow, F. H., and Shannon, J. P., *Science*, **157**, pp. 547–550, August, 1967. Copyright 1967 by the American Association for the Advancement of Science)

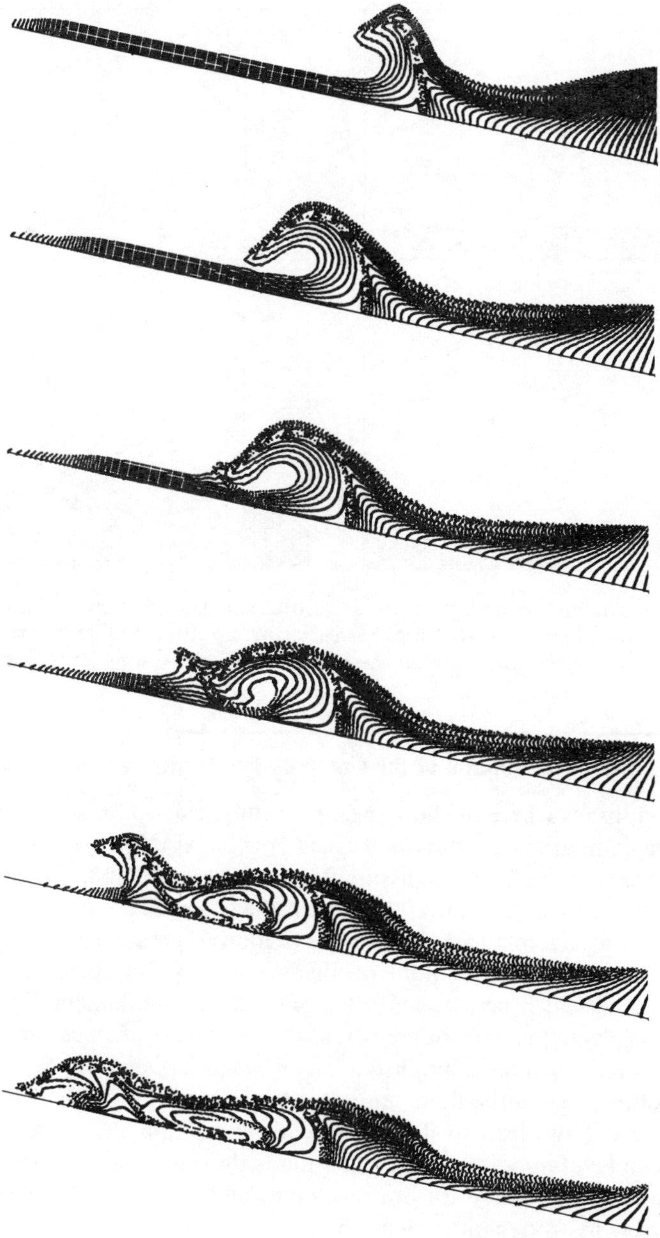

Figure 9.15. The breaking of a wave on the sea-shore as simulated by the marker-and-cell method (from Harlow and Welch 1965). The plotted marker particles illustrate regions of laminar flow and regions of turbulence

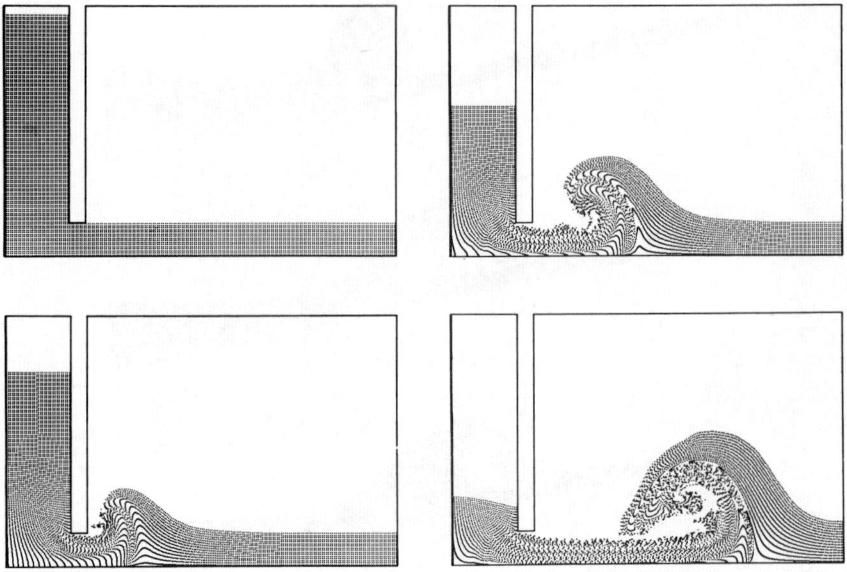

Figure 9.16. The marker-and-cell method used to simulate the flow of water from a sluice gate (from Harlow, F. H., et al., *Science*, **149**, pp. 1092–1093, September, 1965. Copyright 1965 by the American Association of the Advancement of Science)

5 The Difference Solution of the Compressible Hydrodynamic Equations

Under conditions where the flow speeds of a fluid can no longer be regarded as small in comparison to the sound speed or thermal velocity of the particles which constitute the fluid, the assumption of incompressibility is no longer tenable and we must take into full account the effects of compression on the fluid. The compressible hydrodynamic equations (Section 9.1) are therefore appropriate to the general fluid problem and they may describe a great variety of physical processes and phenomena. The fundamental processes are those of advection and sound waves, as described by the basic hyperbolic equations and, when large amplitude disturbances are present, shock waves and discontinuities in the fluid can occur. As in incompressible hydrodynamics, sheared flows lead to fluid instabilities and nonlaminar flow or turbulence can be of importance. In many fluids the diffusion processes of heat conduction and viscosity are of importance and are to be included in the compressible hydrodynamic equations.

The diversity of the physical phenomena described by the compressible hydrodynamic equations is reflected in their intrinsic complexity and it is both because of this diversity and because of the generality of their application, that very many methods have been attempted in their solution. No one single approach in computational physics is most appropriate to all problems in compressible hydrodynamics. We illustrate this point by noting that the equations are to be applied to such differing problems as the atmospheric flow around a re-entrant space vehicle, the motion of the atmosphere, or a one-dimensional shock wave. Each problem or class of problems might require special techniques in the detail of solution and application of boundary conditions.

We have earlier considered one particular approach to the solution by finite methods of compressible hydrodynamics, namely the collision-dominated particle-in-cell method where 'particles' transport the fluid properties of mass, momentum and energy (Section 6.5). The errors there are, however, large and the method has a limited applicability. The fundamental framework in which the compressible hydrodynamic equations are to be solved is the difference method (Chapter 3), but it is to be noted that within the difference method, and depending on the particular problem at hand, consideration must be given as to the choice of various techniques. These techniques include the use of a Lagrangian or an Eulerian mesh, a conservative formulation, an explicit or implicit integration in time and the manner in which the diffusion terms in the equations are to be included in the difference formulation.

In Lagrangian form the compressible hydrodynamic equations (Section 9.1a) take a particularly simple form and in one space dimension it is not difficult to define a Lagrangian mesh, namely a mesh which moves with the fluid and on which the now simpler difference equations are resolved. The difficult problem of advection is therefore reduced and Lagrangian methods provide a general method of solution in one space dimension. However, in the general situation of two or three space dimensions, sheared motion in addition to compressible motion is to be described. Lagrangian methods are then seldom appropriate as a Lagrangian difference mesh is difficult to define and quickly becomes unacceptably distorted.

The most general approach to multidimensional compressible fluid problems is provided by explicit Eulerian difference methods (Sections 3.7 and 3.8). The accuracy in the simulation in describing processes of advection is, however, impaired and, to attempt to minimize this effect, methods of second-order accuracy such as the Lax–Wendroff, leapfrog or quasi-second-order scheme (Section 3.7) are generally to be used. These schemes may be formulated so as to be conservative or nonconservative of appropriate fluid variables (Section 3.6) and from the linear point of view the conservative and nonconservative difference formulations are equivalent. On the one hand, the conservative formulation is generally to be preferred since fundamental integral laws of physics, such as the conservation of mass, momentum and energy, are identically satisfied. It is found too, that the conservative equations tend to minimize spurious nonlinear numerical effects. On the other hand, under particular conditions such as supersonic flow the use of a nonconservative equation for the thermal energy can be advantageous in avoiding errors associated with the large kinetic energy terms.

Where diffusion processes arising from heat conduction and viscosity (Section 9.1) are of importance, they are in essence to be incorporated according to the methods for parabolic equations described in Chapter 3. This is to be achieved, however, consistently with the methods treating advection and compression. In some applications of compressible hydrodynamics the effects of viscosity or heat conduction can be very large and, to avoid the limitations of a small time step in following the flow of the fluid, it is appropriate to consider implicit methods such as the Crank–Nicholson scheme (Section 3.9). This in turn creates a more difficult algebraic problem in that a matrix equation is to be resolved at each time step (Chapter 4). It is to be noted that implicit methods are to be considered too for the effects of pressure, where the flow is subsonic to the extent that flow speeds are small in comparison to the sound speed but not to the extent where the fluid can be regarded as incompressible.

We shall consider the general compressible hydrodynamic problem first in one space dimension and second for the multidimensional case. According to the discussion above, we shall stress in the one-dimensional problem the Lagrangian formulation which may be regarded as the basic approach. Correspondingly, in the multidimensional case we shall illustrate the method of solution by Eulerian explicit methods which provide the most complete solution to the compressible hydrodynamic equations.

(*a*) *Lagrangian Methods for the One-dimensional Problem.* Due to the occurrence of the advective term in each of the hydrodynamic equations, their expression in Lagrangian form is particularly simple (Section 9.1; equations 9.6, 9.7, 9.8):

density:
$$\frac{d\rho}{dt} = -\rho\frac{\partial v}{\partial x} \tag{9.73}$$

momentum:
$$\rho\frac{dv}{dt} = -\frac{\partial p}{\partial x} \tag{9.74}$$

pressure:
$$\frac{d}{dt}\left(\frac{p}{\rho^{\gamma}}\right) = 0 \tag{9.75}$$

x is a Cartesian space variable and $d/dt = \partial/\partial t + v(\partial/\partial x)$ is the Lagrangian time derivative. We have here assumed a particular equation of state, namely that for an ideal gas so that the adiabatic law is appropriate, though the Lagrangian approach to be described may incorporate diffusion terms and a more complex equation of state.

By using Lagrangian coordinates the process of advection is accurately described and equally the difference algorithm is elegantly simple. A Lagrangian mesh is therefore defined so that each mesh point j, to represent the boundary between cells, moves with the local fluid velocity (Figure 9.17):

$$x_j(t') = x_j(t) + \int_t^{t'} v_j(s)\,\mathrm{d}s \tag{9.76}$$

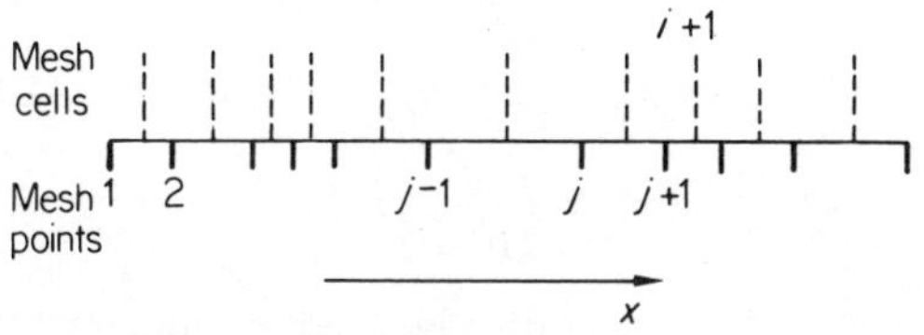

Figure 9.17. A Lagrangian mesh in one dimension. Each cell i is defined by the boundary points j

In practice it is most convenient to define a double mesh. Cell boundaries x_j are defined on the mesh, say $1 \leqslant j \leqslant J$, and equally the fluid velocities v_j, equated now with the cell boundary points x_j, are defined on the mesh j. The intensive properties of density and pressure are defined as cell, rather than point properties, and for this definition, the centres of each cell are defined on a secondary mesh i, $1 < i < I$. Each cell i has width $\Delta_i = x_{j+1} - x_j$. Then the conservation-of-mass equation takes the form:

$$\frac{\mathrm{d}}{\mathrm{d}t}(\rho\Delta) = 0 \tag{9.77}$$

namely the mass contained in each cell is constant (since the cell boundaries move with the fluid velocity). We may now write out the Lagrangian equations very simply. Each point of the fluid x moves with the centre of mass velocity,

$$\frac{\mathrm{d}x}{\mathrm{d}t} = v, \qquad \frac{\mathrm{d}v}{\mathrm{d}t} = -\frac{1}{\rho}\frac{\partial p}{\partial x} \tag{9.78}$$

And an element of the fluid has density ρ and pressure p given by the equations:

$$\frac{\mathrm{d}}{\mathrm{d}t}(\rho\Delta) = 0, \qquad \frac{\mathrm{d}}{\mathrm{d}t}\left(\frac{p}{\rho^{\gamma}}\right) = 0 \tag{9.79}$$

In the integration in time for second-order accuracy, each cell-boundary point x_j is defined at times $n - 1$, while the velocities of the boundary points are defined at the half time step points, $v_j^{n-\frac{1}{2}}$. A time step is executed by first obtaining the new positions of cell boundaries:

$$x_j^n = x_j^{n-1} + v_j^{n-\frac{1}{2}} \Delta t \tag{9.80}$$

Consequently, at time t^n on the mesh i, we may immediately determine the density and pressure in each cell:

$$\rho_i^n = \rho_i^{n-1} \frac{(x_{j+1}^{n-1} - x_j^{n-1})}{(x_{j+1}^n - x_j^n)} \tag{9.81}$$

and

$$p_i^n = \left\{ \frac{\rho_i^n}{\rho_i^{n-1}} \right\}^{\gamma} p_i^{n-1} \tag{9.82}$$

The pressure in each cell i may in turn be used to redetermine the velocity of each cell boundary:

$$v_j^{n+\frac{1}{2}} = v_j^{n-\frac{1}{2}} - \frac{2\,\Delta t}{(\rho_i^n + \rho_{i-1}^n)} \frac{(p_i^n - p_{j-1}^n)}{(x_i^n - x_{i-1}^n)} \tag{9.83}$$

where the centre of each cell x_i^n is specified as the centre point between cell boundaries:

$$x_i^n = \tfrac{1}{2}(x_j^n + x_{j+1}^n)$$

The time loop is then completed and may be repeated for further time steps.

It is to be noted that such a formulation is accurate to second order in both space step and time step. The moving Lagrangian mesh itself illustrates hydrodynamic phenomena where, for example, sound waves will occur on the mesh by successive bunching and opening of cell boundaries (Figure 9.18). There is the advantage too that the effect of the motion of the mesh is

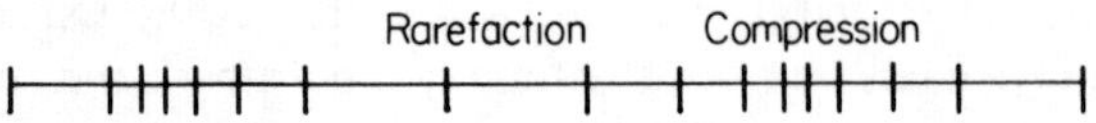

Figure 9.18. Sound waves on a Lagrangian mesh

to concentrate points into regions of interest, such as shocks, so that the spatial resolution and accuracy of the calculation are increased in such regions. Additional terms occurring in the equations, such as those arising from diffusion, are to be included in the usual way.

(*b*) *Eulerian Explicit Methods for the General Multidimensional Problem.* The most general and widely used methods for the solution of multidimensional problems in compressible fluid dynamics are explicit integrations on an

Eulerian mesh (Section 3.7). We shall illustrate the basic features of such an approach by considering the simplest one-dimensional formulation for an ideal gas, where we shall write each term explicitly for the density, velocity and pressure of the fluid (equations 9.6, 9.7, and 9.8):

density:
$$\frac{\partial \rho}{\partial t} = -v\frac{\partial \rho}{\partial x} - \rho\frac{\partial v}{\partial x}$$

velocity:
$$\frac{\partial v}{\partial t} = -v\frac{\partial v}{\partial x} - \frac{1}{\rho}\frac{\partial p}{\partial x}$$

pressure:
$$\frac{\partial p}{\partial t} = -v\frac{\partial p}{\partial x} - \gamma p\frac{\partial v}{\partial x} \tag{9.84}$$

where for the simple example of an ideal gas, the equation of state $\epsilon = p/\rho(\gamma - 1)$ has been assumed. It is to be noted that each of the first terms on the right-hand side of the equations are the advective terms, while the remaining terms each arise from compressibility.

The equations as expressed here are in nonconservative form. If the equations are differenced directly, according to the Lax scheme (equation 3.80) on an Eulerian mesh with points $1 \leqslant j \leqslant J$, a space step Δ, and over the time step $\Delta t = t^{n+1} - t^n$, we obtain:

density:

$$\rho_j^{n+1} = \frac{1}{2}(\rho_{j+1}^n + \rho_{j-1}^n) - v_j^n\frac{\Delta t}{2\Delta}(\rho_{j+1}^n - \rho_{j-1}^n) - \rho_j^n\frac{\Delta t}{2\Delta}(v_{j+1}^n - v_{j-1}^n)$$

velocity:

$$v_j^{n+1} = \frac{1}{2}(v_{j+1}^n + v_{j-1}^n) - v_j^n\frac{\Delta t}{2\Delta}(v_{j+1}^n - v_{j-1}^n) - \frac{\Delta t}{2\Delta}\frac{1}{\rho_j^n}(p_{j+1}^n - p_{j-1}^n)$$

pressure:

$$p_j^{n+1} = \frac{1}{2}(p_{j+1}^n + p_{j-1}^n) - v_j^n\frac{\Delta t}{2\Delta}(p_{j+1}^n - p_{j-1}^n) - \gamma\frac{p_j^n\Delta t}{2\Delta}(v_{j+1}^n - v_{j-1}^n) \tag{9.85}$$

While it has been shown that the Lax method can be stable for a linear advective equation and for a linear wave equation (Chapter 3), the fluid equations (9.85) are more complex and nonlinear. To establish the stability of these nonlinear difference equations, according to the methods of Chapter 3 and for the purposes of the analysis only, we shall *linearize* the equations by making the local approximation that ρ_j^n, v_j^n and p_j^n are constants in the above equations. Certainly, this is a reasonable approach in defining the growth of a small error mode. Consequently, the amplification matrix for a

Fourier mode of wavenumber k is obtained:

$$G(\Delta t, k) = \begin{bmatrix} (\beta - i\alpha v_j^n) & -i\alpha\rho_j^n & 0 \\ 0 & (\beta - i\alpha v_j^n) & -i\alpha(1/\rho_j^n) \\ 0 & i\alpha\gamma p_j^n & (\beta - i\alpha v_j^n) \end{bmatrix} \tag{9.86}$$

where $\beta = \cos(k\Delta)$, $\alpha = \{\Delta t \sin(k\Delta)\}/\Delta$.

The eigenvalues of the amplification matrix are:

$$g_1(\Delta t, k) = \cos(k\Delta) - i\frac{\Delta t}{\Delta} v_j^n \sin(k\Delta)$$

$$g_2(\Delta t, k) = \cos(k\Delta) - i\left\{v_j^n + \sqrt{\left(\frac{\gamma p_j^n}{\rho_j^n}\right)}\right\}\frac{\Delta t}{\Delta}\sin(k\Delta)$$

$$g_3(\Delta t, k) = \cos(k\Delta) - i\left\{v_j^n - \sqrt{\left(\frac{\gamma p_j^n}{\rho_j^n}\right)}\right\}\frac{\Delta t}{\Delta}\sin(k\Delta) \tag{9.87}$$

Clearly, each of these eigenvalues is related to advection, the Doppler-shifted sound wave propagating with the advective velocity and the Doppler-shifted sound wave propagating in the opposite direction to the advective velocity. The eigenvalue of largest modulus will be g_2 or g_3 depending on the sign of v_j^n:

$$|gg|_{max} = 1 - \sin^2(k\Delta)\left[1 - \left\{|v_j^n| + \sqrt{\left(\frac{\gamma p_j^n}{\rho_j^n}\right)}\right\}^2 \frac{(\Delta t)^2}{\Delta^2}\right] \tag{9.88}$$

The von Neumann stability condition (equation 3.47) that the magnitude of the largest eigenvalue be smaller than unity may now be applied and consequently a condition on the time step is obtained:

$$\Delta t \leqslant \frac{\Delta}{|v_j^n| + \sqrt{(\gamma p_j^n/\rho_j^n)}} \tag{9.89}$$

This is the Courant–Friedrichs–Lewy condition for the compressible hydrodynamic equations as differenced by the Lax method. Several points are to be noted with regard to this result and the method by which it was obtained. The two velocities in the denominator are the advective velocity and the sound speed and they sum according to the Doppler effect whereby a sound wave propagates with the sound speed in the frame of the moving fluid. Their sum is the maximum local speed for the propagation of information. Since the original equations are nonlinear, these speeds vary from point to point on the space mesh and consequently the limitation on the time step is a local one to be imposed at every mesh point, j. The Courant–Friedrichs–Lewy condition (equation 9.89) requires that the time step is limited by the maximum speed wherever it occurs on the mesh and at every time step.

It is clearly the case that the differential equations for the density, momentum and energy density of a compressible fluid may be written in a variety of ways (Section 1) and, while these differential forms are exactly equivalent, their analogous difference forms are not equivalent. In Section 3.6 the principle of a conservative difference scheme was introduced, which has the advantage of identically conserving mass, momentum and energy on the difference mesh, but, more important, conservative difference schemes are found to have advantageous numerical properties with regard to nonlinear effects (Arakawa 1966). In general, it is appropriate, therefore, to choose the particular difference scheme which conserves mass, momentum and energy in a compressible fluid (equations 9.1, 9.2, 9.9). In preference to the nonconservative formulation (equations 9.85), we use the conservative difference equations:

density:

$$\rho_j^{n+1} = \tfrac{1}{2}(\rho_{j+1}^n + \rho_{j-1}^n) - \frac{\Delta t}{2\Delta}(\rho_{j+1}^n v_{j+1}^n - \rho_{j-1}^n v_{j-1}^n)$$

momentum:

$$\rho_j^{n+1} v_j^{n+1} = \tfrac{1}{2}(\rho_{j+1}^n v_{j+1}^n + \rho_{j-1}^n v_{j-1}^n) - \frac{\Delta t}{2\Delta}\{\rho_{j+1}^n (v_{j+1}^n)^2 + p_{j+1}^n - \rho_{j-1}^n (v_{j-1}^n)^2 - p_{j-1}^n\}$$

energy:

$$\begin{aligned}\rho_j^{n+1}\{\epsilon_j^{n+1} + \tfrac{1}{2}(v_j^{n+1})^2\} = {} & \tfrac{1}{2}[\rho_{j+1}^n\{\epsilon_{j+1}^n + \tfrac{1}{2}(v_{j+1}^n)^2\} + \rho_{j-1}^n\{\epsilon_{j-1}^n + \tfrac{1}{2}(v_{j-1}^n)^2\}] \\ & - \frac{\Delta t}{2\Delta}[\{\rho_{j+1}^n \epsilon_{j+1}^n + p_{j+1}^n + \tfrac{1}{2}\rho_{j+1}^n (v_{j+1}^n)^2 v_{j+1}^n\} \\ & - \{\rho_{j-1}^n \epsilon_{j-1}^n + p_{j-1}^n + \tfrac{1}{2}\rho_{j-1}^n (v_{j-1}^n)^2\} v_{j-1}^n] \qquad (9.90)\end{aligned}$$

where, for a perfect gas, $p_j^n = (\gamma - 1)\rho_j^n \epsilon_j^n$. With respect to linear effects, these equations are in no manner different from the nonconservative difference equations (9.85) and the stability condition (equation 9.89) is precisely the same.

We have illustrated the difference formulation of the compressible hydrodynamic equations on an Eulerian mesh in terms of the Lax method. The numerical diffusion, which arises as a consequence of the first-order accuracy of this method, is to be avoided in many simulations (Section 3.5), and generally the methods of second-order accuracy (the Lax–Wendroff, leapfrog, quasi-second-order Adams–Bashforth—Section 3.7) are to be used. In three dimensions and in Cartesian coordinates (x, y, z), the compressible hydrodynamic equations are expressed in conservative form (equations

9.1, 9.2, 9.9):

$$\frac{\partial \mathbf{u}}{\partial t} + \frac{\partial \mathbf{F}_x}{\partial x} + \frac{\partial \mathbf{F}_y}{\partial y} + \frac{\partial \mathbf{F}_z}{\partial z} = 0 \tag{9.91}$$

where the dependent variables form the column vector $\mathbf{u} = \{\rho, \rho v_x, \rho v_y, \rho v_z, \frac{1}{2}\rho v^2 + \rho\epsilon\}$. The fluxes are then defined as:

$$\mathbf{F}_x = \begin{bmatrix} \rho v_x \\ \rho v_x^2 + p \\ \rho v_y v_x \\ \rho v_z v_x \\ (\rho\epsilon + \frac{1}{2}\rho v^2 + p)v_x \end{bmatrix} \quad \mathbf{F}_y = \begin{bmatrix} \rho v_y \\ \rho v_x v_y \\ \rho v_y^2 + p \\ \rho v_z v_y \\ (\rho\epsilon + \frac{1}{2}\rho v^2 + p)v_y \end{bmatrix}$$

$$\mathbf{F}_z = \begin{bmatrix} \rho v_z \\ \rho v_x v_z \\ \rho v_y v_z \\ \rho v_z^2 + p \\ (\rho\epsilon + \frac{1}{2}\rho v^2 + p)v_z \end{bmatrix} \tag{9.92}$$

The appropriate algorithms (Sections 3.7 and 3.8) may now be applied directly. For each of the methods the speed to be used in the Courant–Friedrichs–Lewy condition is the maximum Doppler-shifted sound speed:

$$v_c = \sqrt{(v_x^2 + v_y^2 + v_z^2)} + \sqrt{\left(\frac{\gamma p}{\rho}\right)} \tag{9.93}$$

The most severe errors which occur in these methods are those arising from numerical dispersion (Section 3.5), which particularly affect the short wavelengths on the difference mesh and which are only to be avoided at the expense of considerable complexity with schemes of fourth-order accuracy (Roberts and Weiss 1966). In addition discontinuities and shock waves when represented on a difference mesh can give rise to spurious effects and it is frequently necessary to broaden such effects by an artificial diffusion (Section 9.6).

Finally it is to be noted that boundary conditions are to be applied consistently with the numerical scheme employed at a general internal point. While boundary conditions might be well defined in the differential system, there is a degree of arbitrariness introduced in the difference equations. We shall illustrate the application of boundary conditions for the commonly occurring boundaries of an impermeable wall and a geometric singularity such as the axis in cylindrical geometry. At an impermeable wall the perpendicular velocity is zero $v_\perp = 0$ and, correspondingly, the perpendicular fluxes at the wall will be zero. In conservative explicit methods to avoid inconsistencies and spurious numerical boundary effects, it is appropriate to apply the conservative equations in integral form to the boxes or cells immediately adjacent to the boundary (Figure 9.19). For example, in a two-dimensional cylindrical geometry, the boundary cell on the axis is a cylinder of volume $2\pi(\Delta r)^2 \Delta z$ and fluxes are evaluated on the end surfaces of the cylinder, area $\pi(\Delta r)^2$, and on the curved surface of area $2\pi \Delta r \Delta z$. The determination of the dependent variables on the axial boundary at the centre of each cylindrical element is then well defined (Figure 9.19).

$$\frac{du_C}{dt} = \frac{(F_E - F_W)}{2\Delta x} + \frac{2 F_S}{\Delta y}$$

(a)

$$\frac{du_C}{dt} = \frac{(F_E - F_W)}{2\Delta z} - \frac{2F_N}{\Delta r}$$

(b)

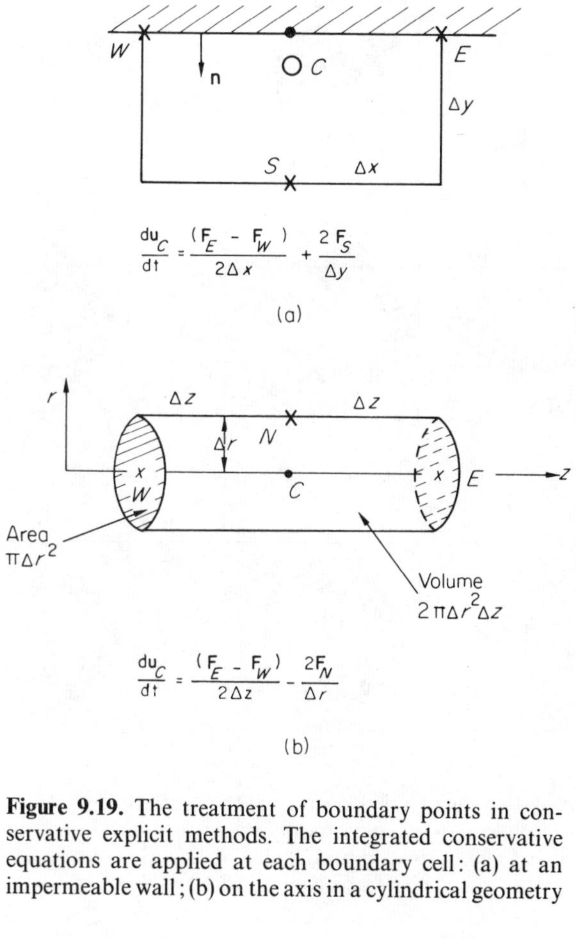

Figure 9.19. The treatment of boundary points in conservative explicit methods. The integrated conservative equations are applied at each boundary cell: (a) at an impermeable wall; (b) on the axis in a cylindrical geometry

Figures 9.20 and 9.21 illustrate the application of a compressible hydrodynamic simulation to the motion of a shock passing over a 30° cone and the subsequent formation of a standing shock equilibrium. We may regard the calculation as a simulation of the re-entry of a space vehicle to the earth's atmosphere. In Figure 9.20, contours of the pressure are drawn and the complex shock structure and diffraction of the shock around the vehicle are illustrated. The steady-state flow density and internal energy density patterns are illustrated in Figure 9.21.

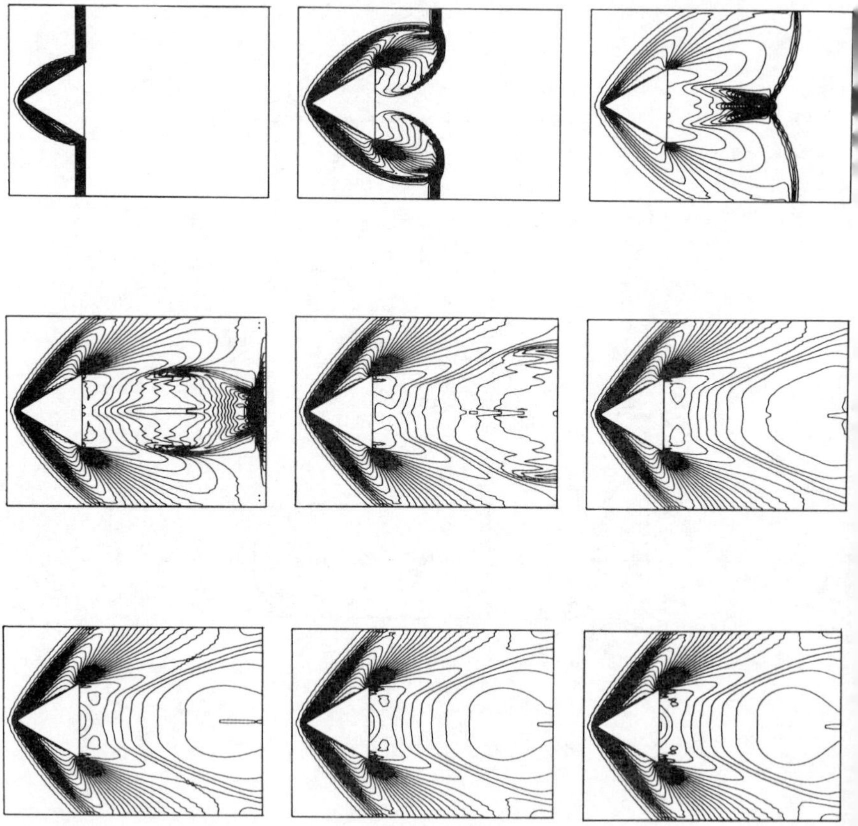

Figure 9.20. The diffraction of a shock wave and the establishment of a steady state as a 30° cone enters a uniform gas supersonically. The figure shows contours of pressures (from Butler 1967). Figure provided by Group T-3, Los Alamos Scientific Laboratory, Los Alamos, U.S.A.

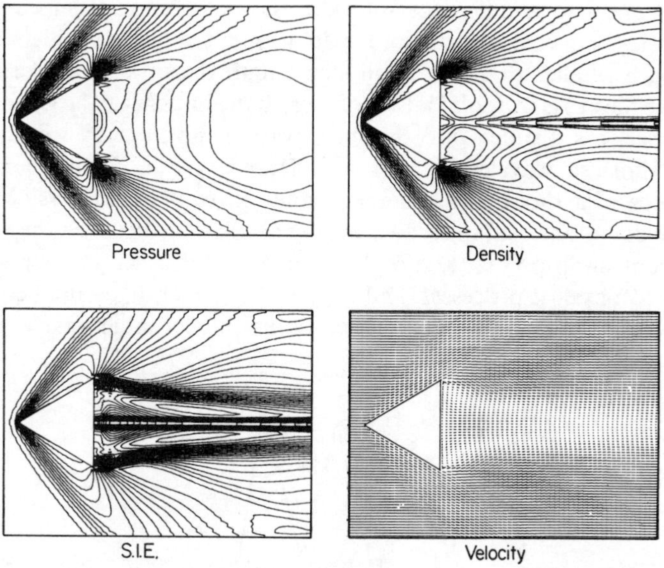

Figure 9.21. The steady-state pressure, density, specific internal energy and velocity patterns produced by the supersonic motion of a 30° cone through a uniform gas. The solutions were obtained by an Eulerian difference method. Figure provided by Group T-3, Los Alamos Scientific Laboratory, Los Alamos, U.S.A.

6 The Treatment of Shocks and Discontinuities

The linear properties of the hydrodynamic equations have previously been considered (Section 9.5; Section 3.1) when waves of small amplitude can propagate independently with the sound speed and with the advective velocity. When, in a compressible fluid, disturbances of large amplitude occur, waves of different wavelength couple or interact and, in particular, energy tends to propagate from the long wavelengths to the short wavelengths. Large-amplitude waves tend to steepen and discontinuities and shocks can occur in a compressible fluid (Courant and Friedrichs 1948). From the point of view of the hyperbolic hydrodynamic equations (no viscosity or heat conduction), changes in the state of the fluid occur abruptly or discontinuously, but in nature the thickness of the shock is defined by a viscous length, which might be very small, associated with the mean free path of the particles in the fluid. Shock waves are entropy-producing so that over the thickness of the shock the kinetic energy of the short wavelengths is transferred into thermal energy.

With regard to a difference solution, the occurrence of shocks or discontinuities in the absence of viscous diffusion is a difficult problem since no wavelength smaller than the mesh step length Δ can be described on the difference mesh. As in a solid lattice of ions, in the presence of large-amplitude waves energy is distilled into the short wavelengths, energy will aggregate in the shortest wavelengths on a difference mesh. Consequently, the occurrence of a shock in a difference solution to the compressible hydrodynamic equations, will induce very large oscillations between the variables at adjacent mesh points (Figure 9.22). Such a numerical effect bears little relation to physical processes, and to the more exact differential equations, and can severely corrupt the numerical solution over large wavelengths.

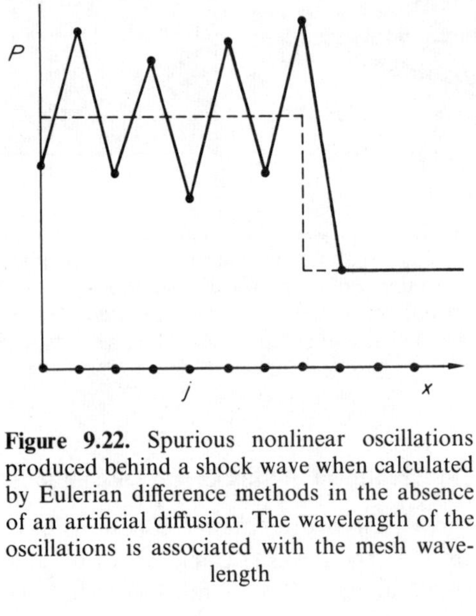

Figure 9.22. Spurious nonlinear oscillations produced behind a shock wave when calculated by Eulerian difference methods in the absence of an artificial diffusion. The wavelength of the oscillations is associated with the mesh wavelength

To avoid the occurrence of these short-wavelength effects in a difference solution, von Neumann and Richtmyer (1950) introduced a purely artificial viscosity to transform the energy of the large-mesh oscillations into thermal energy of the fluid. This has the effect of automatically broadening all shocks on the difference mesh so that any shock thickness is at least several mesh step lengths. In comparison with the physical viscosity in nature (Section 9.1b), an artificial viscosity **W** is therefore to be included in the compressible hydrodynamic equations (equations 9.1, 9.2 and 9.3):

$$\frac{\partial \rho}{\partial t} + \nabla . \rho \mathbf{v} = 0 \tag{9.94}$$

$$\frac{\partial \rho \mathbf{v}}{\partial t} + \nabla . \rho \mathbf{v}\mathbf{v} = -\nabla . (p\mathbf{I} + \mathbf{W}) \tag{9.95}$$

$$\frac{\partial \rho \epsilon}{\partial t} + p\nabla . \mathbf{v} + \nabla . (\rho\epsilon\mathbf{v}) = -\mathbf{W} : \nabla\mathbf{v} \tag{9.96}$$

On the other hand, if in the problem of interest the natural viscosity is large, it is of course not necessary to introduce an additional artificial viscosity.

We shall therefore question what form the artificial viscosity **W** or diffusion should take. The artificial viscosity must have the effect of broadening by diffusion any discontinuities to extend over lengths greater than the mesh step, but this must be achieved without corrupting the solution over long wavelengths in comparison to the mesh interval. Both these requirements can be met by employing an artificial coefficient of viscosity which is wavelength dependent so as to be large for short wavelengths and small for long wavelengths. In one space dimension we may write the acceleration and internal energy equations of the fluid as:

$$\rho \frac{\mathrm{d}v}{\mathrm{d}t} = -\frac{\partial}{\partial x}(p + w) \tag{9.97}$$

$$\rho \frac{\mathrm{d}\epsilon}{\mathrm{d}t} = -p\frac{\partial v}{\partial x} - w\frac{\partial v}{\partial x} \tag{9.98}$$

where w is the artificial diffusion term, which is defined by:

$$w = -c\rho\Delta^2 \left|\frac{\partial v}{\partial x}\right| \frac{\partial v}{\partial x} \tag{9.99}$$

c is a constant and Δ is the mesh step length (von Neumann and Richtmyer 1950, Richtmyer and Morton 1967).

The artificial viscous terms are to be included in the difference scheme according to the explicit first-order method (Section 3.9a). The optimum choice for the magnitude of the constant c depends on the particular difference scheme which is employed to resolve the complete system of equations, and the best value is to be established by numerical experiments. Generally, the constant c will have values in the range:

$$0{\cdot}05 < c < 2{\cdot}0 \tag{9.100}$$

The diffusion coefficient in the artificial viscosity term is:

$$\nu = c\Delta^2 \left|\frac{\partial v}{\partial x}\right| \tag{9.101}$$

so that it is only large in regions of rapidly varying velocity. We may associate a numerical Reynolds number R_N with the artificial viscosity related to the time step Δt and space step Δ:

$$\frac{1}{R_N} = \frac{c\Delta^2\,\Delta t|\delta v|}{\Delta}k^2$$
$$\leqslant c(k\Delta)^3\left(\frac{\Delta t|v|}{\Delta}\right) \tag{9.102}$$

where δv is the difference of fluid velocities between two adjacent mesh points and k is a wavenumber. It is to be noted that each of the factors in the expression for the inverse Reynolds number is smaller or of the order of unity, since k is a wavenumber of the fluid on the mesh ($k\Delta \leqslant 1$) and $\Delta t|v|/\Delta$ must always be taken to be smaller than unity if the Courant–Friedrichs–Lewy condition (equation 9.89) is to be satisfied. It follows that the numerical Reynolds number for long wavelengths is always very small, so that long wavelengths on the mesh are minimally affected by the artificial diffusion.

If the parameter c is large and approaches unity, and since the artificial diffusion term is to be included according to the explicit first-order method, the time step may be limited for linear numerical stability by the artificial coefficient ν (equation 3.53) in conjunction with the Courant–Friedrichs–Lewy condition (equation 9.89):

$$(|v| + v_s)\frac{\Delta t}{\Delta} + \frac{2\,\Delta t\,\nu}{\Delta^2} \leqslant 1$$
$$\Delta t \leqslant \frac{\Delta}{2c|\delta v| + (|v| + v_s)} \tag{9.103}$$

There exists a variety of alternative expressions and methods for treating discontinuities or shocks on a difference mesh (Richtmyer and Morton 1967, Lax and Wendroff 1960). In particular it is possibly advantageous to include the artificial diffusion only during compressions, so that the artificial viscous term w becomes:

$$w = \begin{cases} c\rho\Delta^2\left(\dfrac{\partial v}{\partial x}\right)^2 & \text{if } \dfrac{\partial v}{\partial x} < 0 \\ 0 & \text{if } \dfrac{\partial v}{\partial x} \geqslant 0 \end{cases} \tag{9.104}$$

If alternative methods are used, it is always essential to include the corresponding term in the energy or internal energy equation (9.98) so that the inclusion of an artificial viscosity term is consistent with the conservation of energy. The effect of including an artificial viscosity on the propagation of a shock on a difference mesh is illustrated in Figure 9.23.

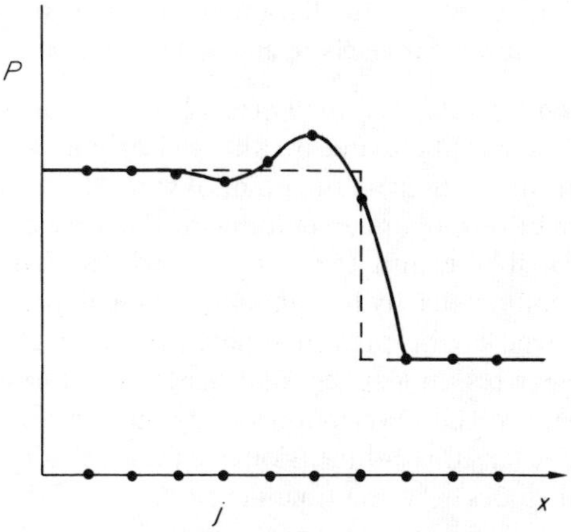

Figure 9.23. A shock wave on a difference mesh when an artificial viscosity is included. The energy in the oscillations produced by the mesh points is transformed into thermal energy

In two space dimensions the artificial viscosity is to be included as a tensor (equations 9.95 and 9.96). The simplest procedure is an extension of the von Neumann artificial viscosity terms to take account of compression along each dimension so that in Cartesian coordinates, with u and v the velocities along the x and y directions respectively, the artificial viscous terms become (Lapidus 1967):

$$\mathbf{W} = -\begin{bmatrix} \rho c \Delta^2 \left|\frac{\partial u}{\partial x}\right| \frac{\partial u}{\partial x} & \rho c \Delta^2 \left|\frac{\partial v}{\partial y}\right| \frac{\partial u}{\partial y} \\ \rho c \Delta^2 \left|\frac{\partial u}{\partial x}\right| \frac{\partial v}{\partial x} & \rho c \Delta^2 \left|\frac{\partial v}{\partial y}\right| \frac{\partial v}{\partial y} \end{bmatrix} \tag{9.105}$$

and the terms are to be included according to the explicit first-order method in analogy to the one-dimensional formulation.

It is to be noted that the procedure of including an artificial viscosity in the difference equations leads to a numerical diffusion which eliminates short wavelengths on the difference mesh. Since the essential approximation of the difference method is a long-wavelength approximation, the inclusion of such an artificial diffusion is entirely appropriate and is to be used in other systems of hyperbolic equations apart from compressible fluid dynamics. On the other hand it is essential to ensure that long wavelengths on the difference mesh are only minimally affected, so that only diffusion terms of third or fourth order (in $k\Delta$) should be used, while numerical diffusion of lower order due to a poor difference scheme is to be avoided (Section 3.5).

7 Hydrostatic Equilibrium in the Simulation of the Atmosphere and World Ocean

(*a*) *Shallow Fluid Equations on the Earth's Surface*. The numerical simulation of the earth's atmosphere and oceans is clearly a major application of computational physics both from the point of view of predicting the weather and from the point of view of understanding the dynamic cycles and structure of the world atmosphere and the world ocean. Richardson (1922) first attempted to formulate the hydrodynamic problem posed by the earth's atmosphere, but even at the present time with the largest computers, the simulation of the earth's atmosphere and oceans is a formidable problem.*

In the first instance the description of the earth's atmosphere and the world ocean are each three-dimensional time-dependent hydrodynamic problems, though to describe the time development of the atmosphere and the ocean with detailed accuracy very complex boundary conditions and thermodynamic processes are required. For the ocean the complex two-dimensional topography of the ocean floor must be included and the effects of salinity, winds and turbulence must be included (Bryan 1969). In the atmosphere, the boundary conditions are simpler though mountain ranges must be included, but the thermodynamic processes of water vapour, rain production, solar radiation etc. can be extremely complicated. These effects occur as local processes but couple to the dynamics of the fluid in providing sources and sinks of energy (Kasahara and Washington 1967). We shall not consider these many interesting and detailed thermodynamic processes nor the detailed inclusion of boundary effects (Leith 1965), but rather we shall outline the essential manner in which the dynamic equations can be formulated. It is the solution of the fluid dynamic problem which is the kernel of atmospheric and ocean models.

The dynamics of the motion of both the atmosphere and ocean are to be described by the Navier–Stokes equations (Section 9.1) in spherical coordinates (Figure 9.24). The problem is simplified considerably, however, by noting that the vertical depth of both the ocean and the atmosphere is small in comparison to horizontal lengths. This permits two essential approximations. First, we shall assume that the radius R and gravitational acceleration g of the earth are constant along the vertical coordinate z, and second we may introduce the shallow fluid approximation that vertical velocities are small in comparison to the horizontal velocities. Consequently it is assumed, both in the ocean and the atmosphere, that the fluid is in vertical hydrostatic equilibrium (equation 9.65) in that the fluid pressure identically balances the weight of the fluid.

* There is some irony in the comment of Richardson in 1922: 'Perhaps some day in the dim future it will be possible to advance the computations faster than the weather advances and at a cost less than the saving to mankind due to the information gained. But that is a dream.'

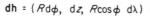

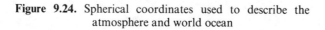

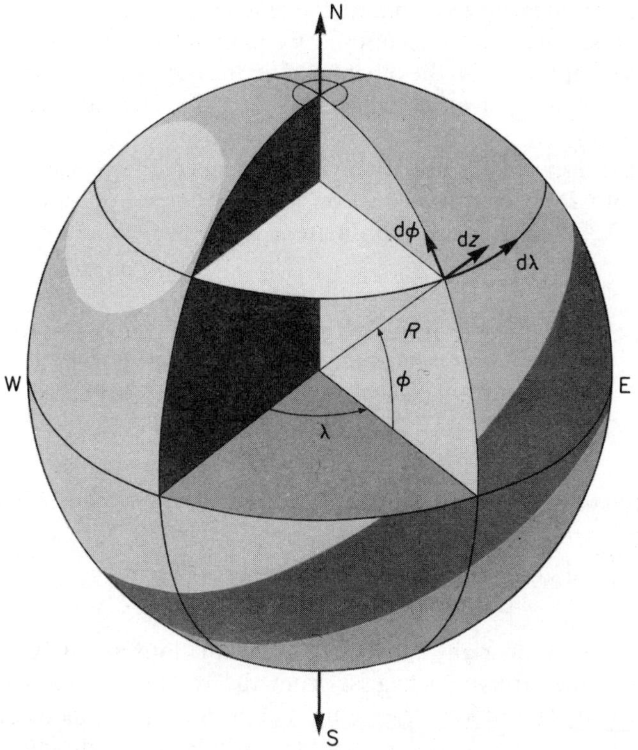

Figure 9.24. Spherical coordinates used to describe the atmosphere and world ocean

We use the coordinates of latitude ϕ, longitude (west to east) λ, and vertical height z. A small element of length dh is described by the vector (Figure 9.24):

$$\mathbf{dh} = (R\, \mathrm{d}\phi, \mathrm{d}z, R \cos \phi\, \mathrm{d}\lambda) \tag{9.106}$$

The three components of fluid velocity are defined as:

$$u = R \cos \phi\, \dot{\lambda}$$

$$v = R\dot{\phi}$$

$$w = \dot{z} \tag{9.107}$$

where

$$w \ll u, v$$

For the atmosphere, although wind speeds are much smaller than the sound speed they are sufficient to compress the tenuous air weakly and to describe such compressible effects, the continuity equation for the density (equation 9.1) must be employed. On the other hand, in the ocean it is assumed that the flow of the fluid is incompressible and that the flow is divergence-free (equation 9.19).

In the first instance, the horizontal components of the velocity of the fluid are to be obtained from the equation for the conservation of momentum (equation 9.2) in spherical coordinates, where we write out the terms explicitly:

$$\frac{\partial}{\partial t}(\rho v) + \frac{1}{R\cos\phi}\frac{\partial}{\partial\phi}\{(\rho vv)\cos\phi\} + \frac{1}{R\cos\phi}\frac{\partial}{\partial\lambda}(\rho uv) + \frac{\partial}{\partial z}(\rho uw)$$
$$= -\rho\left(2\Omega + \frac{u}{R\cos\phi}\right)u\sin\phi - \frac{1}{R}\frac{\partial p}{\partial\phi} + F_\phi \tag{9.108}$$

$$\frac{\partial}{\partial t}(\rho u) + \frac{1}{R\cos\phi}\frac{\partial}{\partial\phi}\{(\rho uv)\cos\phi\} + \frac{1}{R\cos\phi}\frac{\partial}{\partial\lambda}(\rho uu) + \frac{\partial}{\partial z}(\rho uw)$$
$$= \rho\left(2\Omega + \frac{u}{R\cos\phi}\right)v\sin\phi - \frac{1}{R\cos\phi}\frac{\partial p}{\partial\lambda} + F_\lambda \tag{9.109}$$

The first terms on the right-hand side of the equations are the Coriolis and centrifugal force terms which arise from the rotation of the earth, with angular velocity Ω, and from the rotation of the fluid on a spherical surface.* F_ϕ and F_λ are viscous forces to describe turbulence. Assuming a shallow fluid, the vertical velocity is to be determined quite differently, so as always to satisfy vertical hydrostatic equilibrium (equation 9.65):

$$\frac{\partial p}{\partial z} = -\rho g \tag{9.110}$$

Since the vertical mesh interval must be small in comparison to the horizontal mesh intervals, it is essential to determine the vertical velocity (w) implicitly. This may be achieved for both the atmosphere and the ocean by using condition 9.110 to define the vertical velocity at all times as the solution of a boundary-value problem.

For the atmosphere, we supplement these dynamic equations with equations for the internal energy density of the fluid and an advective equation for the local water vapour. In the ocean, the fluid is assumed incompressible, but the effects of temperature and salinity on the buoyancy of the fluid must still be incorporated.

* The form of these terms can easily be seen by considering the effect on the latitudinal velocity (v), of the centrifugal force towards the equator, and the effect on the longitudinal velocity (u), by demanding that angular momentum parallel to the axis of the earth be conserved (see, for example, Haurwitz 1941).

(*b*) *Solution of the Dynamic Equations in the Atmosphere.* In defining the motion of the atmosphere there are three 'force' equations (9.108, 9.109, 9.110) together with the time-dependent equations for density (9.1) and internal energy density (9.3):

$$\frac{\partial \rho}{\partial t} + \nabla \cdot \rho \mathbf{v} = 0 \tag{9.111}$$

$$\frac{\partial}{\partial t}(\rho \epsilon) + p \nabla \cdot \mathbf{v} + \nabla \cdot (\rho \epsilon \mathbf{v}) = Q \tag{9.112}$$

where Q represents additional source terms of internal energy depending on the particular model and typically due to solar radiation, latent heat associated with evaporation and condensation and heating due to eddy diffusion (Leith 1965). In the spherical coordinates used, the divergence of a vector $\mu \mathbf{v}$ takes the form:

$$\nabla \cdot \mu \mathbf{v} = \frac{1}{R \cos \phi} \frac{\partial}{\partial \phi}\{(\mu v) \cos \phi\} + \frac{1}{R \cos \phi} \frac{\partial}{\partial \lambda}(\mu u) + \frac{\partial}{\partial z}(\mu w)$$

An equation of state, usually that for an ideal gas, is used to relate the internal energy density and pressure,

$$\rho \epsilon = \frac{p}{\gamma - 1} \tag{9.113}$$

where γ is the ratio of specific heats. Finally an additional equation might be included to describe the advection of water vapour in the atmosphere.

The time-dependent equations for the density, internal energy density, and horizontal components of momentum are generally to be differenced on a three-dimensional (ϕ, z, λ) Eulerian difference mesh according to explicit second-order difference methods (Section 9.6; Chapter 3). However, the method of solution for the vertical velocity (w) is quite different, and is to be obtained at the completion of each time step (or half time step) consistently with the requirement of vertical hydrostatic equilibrium. To achieve the appropriate equation for w, we shall differentiate the equation for hydrostatic equilibrium (equation 9.110) with respect to the Lagrangian time derivative:

$$\frac{\partial}{\partial z}\left(\frac{\mathrm{d}p}{\mathrm{d}t}\right) + g\left(\frac{\mathrm{d}\rho}{\mathrm{d}t}\right) = 0 \tag{9.114}$$

Therefore, providing that at time $t = 0$ the initial conditions specify the atmosphere as in vertical hydrostatic equilibrium, the pressure and density of the fluid everywhere are to vary according to equation 9.114. This imposes a condition on the vertical velocity at all times and all space points, which we shall obtain by rewriting in Lagrangian form the equations for the density and pressure of the fluid (equations 9.111, 9.112) assuming the equation of state for an ideal gas (equation 9.113):

$$\frac{d\rho}{dt} = -\rho\,\nabla . \mathbf{v} \tag{9.115}$$

$$\frac{dp}{dt} = -\gamma p\,\nabla . \mathbf{v} + (\gamma - 1)Q \tag{9.116}$$

Expressing the divergence of the velocity as the vertical divergence and the horizontal divergence $\nabla . \mathbf{v}_H$,

$$\nabla . \mathbf{v} = \frac{\partial w}{\partial z} + \nabla . \mathbf{v}_H \tag{9.117}$$

we obtain a time-independent equation from equations 9.114, 9.115 and 9.116:

$$-\frac{\partial}{\partial z}\left(\gamma p \frac{\partial w}{\partial z} + \gamma p\,\nabla . \mathbf{v}_H\right) + (\gamma - 1)\frac{\partial Q}{\partial z} - \rho g\left(\frac{\partial w}{\partial z} + \nabla . \mathbf{v}_H\right) = 0$$

Using the hydrostatic equation (9.110) we obtain the more convenient relation:

$$\frac{\partial^2 w}{\partial z^2} + \frac{(\gamma - 1)\rho g}{\gamma p}\frac{\partial w}{\partial z} = \frac{(\gamma - 1)}{\gamma p}\frac{\partial Q}{\partial z} - \frac{\partial}{\partial z}(\nabla . \mathbf{v}_H) - \frac{(\gamma - 1)\rho g}{\gamma p}(\nabla . \mathbf{v}_H) \tag{9.118}$$

This equation may be regarded as a one-dimensional (in z) second-order differential equation in w from which we may determine the vertical velocity along each vertical column of the computation. Differenced on the (ϕ, z, λ) mesh, equation 9.118 becomes a tridiagonal matrix equation for the vertical velocity on each vertical column and it may be resolved according to the simple tridiagonal inversion method (Section 4.3). The usual boundary conditions applied are that the vertical velocity is zero at the top (z_T) and bottom (z_0) of the atmosphere*

$$w(z_0) = 0, \qquad w(z_T) = 0$$

Many types of atmospheric models are used according to the particular application desired. For short-term weather prediction (of the order of days), limited surface areas of the earth are described in order that the mesh intervals may be minimized and corresponding accuracy for short times optimized.

* In many models of the atmosphere the vertical independent variable z is replaced by surfaces of geopotential $d\psi = -g\rho\,dz$, when for example the hydrostatic equation simplifies to $dp/d\psi = 1$. On the other hand a greater complexity is introduced for the boundary conditions at the surface of the earth. The essential procedures used in both models are similar. See for example Leith (1965).

For example, daily prediction of the weather in Europe is usually determined from a model covering the equator to polar latitudes and from the coast of Newfoundland to the Urals. On the other hand, for very long time scales it is essential to describe the total atmosphere and global models are used. In Figure 9.25, examples of solutions from a global model are illustrated (Corby *et al.* 1972).

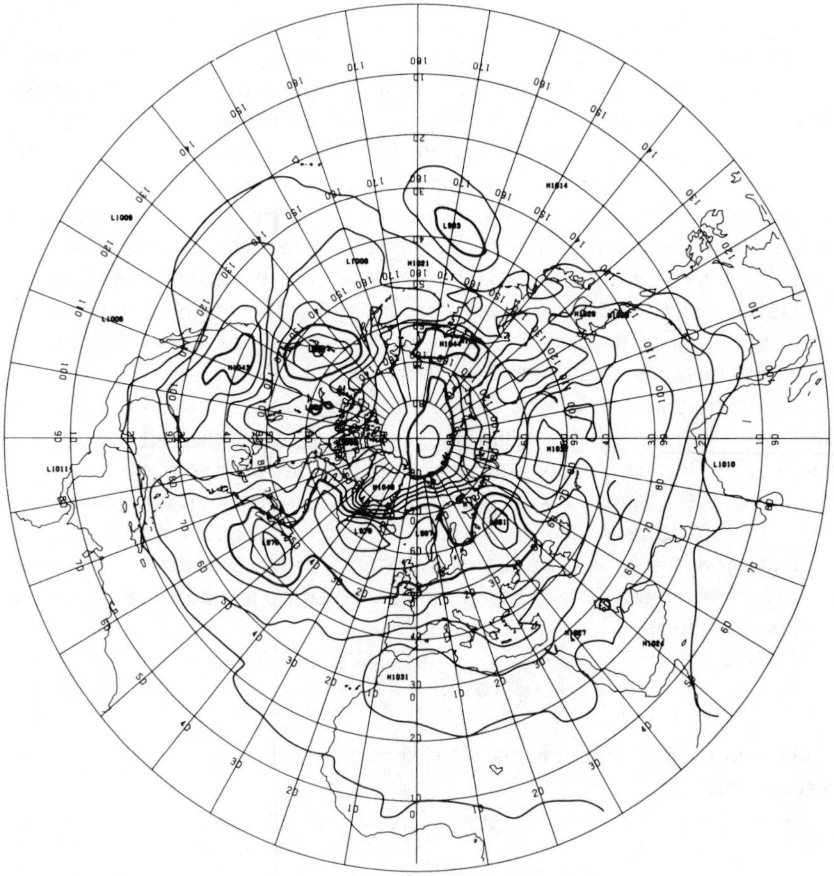

Contour interval = 8·000 millibars

Figure 9.25. Contours of the atmospheric pressure at the surface of the earth in the northern hemisphere. The solutions are obtained from a hemispherical model of the earth's atmosphere and the figure shown corresponds to day 64 of the simulation (Corby *et al*; 1972)

It is to be noted that over the mesh intervals (typically 10–100 miles) involved in atmospheric models, molecular viscosity is negligibly small and fine-scale turbulence may occur. Such short-wavelength phenomena cannot be accurately described in mesh simulations and they gradually destroy the accuracy of a calculation. Heuristic models of eddy or turbulent viscosity are therefore usually employed (Lilly 1967, Deardorff 1971), but this problem will remain a fundamental difficulty until the statistical mechanics of turbulence can be resolved.*

(*c*) *Simulation of the World Ocean.* Models to describe the motion of the oceans are closely allied to those used for the atmosphere. The fundamental difference, however, is that water, even under the extreme pressures at the bottom of the ocean, is incompressible so that the motion of the oceans may be described as an incompressible flow. On the other hand, account must be taken of the buoyancy effect produced by very small changes in the density of water caused by thermal expansion and the saline content of a parcel of water. Hence models must be used in which the flow is strictly divergence-free (equation 9.19) but small changes in density are determined through an equation of state (the Boussinesq approximation). It is essential to eliminate the large pressure terms from the momentum equations, but still retain the effect of buoyancy due to thermal expansion.

In the simplest models of the ocean, the vertical velocity (w) is totally ignored, and a two-dimensional model covering the surface of the earth is used. The flow is simply described by the incompressibility condition (equation 9.19) together with equations for the horizontal components of velocity (equations 9.108, 9.109 where the density is assumed constant ρ_0). The method of solution employing a vorticity and stream function (Section 9.2b) is most appropriate, where Poisson's equation for the stream function ψ_s takes the form:

$$\frac{1}{R^2}\frac{\partial^2}{\partial\phi^2}(\psi_s \cos\phi) + \frac{1}{R^2}\frac{\partial^2\psi_s}{\partial\lambda^2} = -\xi_s \tag{9.119}$$

and the horizontal components of the surface velocity are obtained from the stream function:

$$u = \frac{1}{R\cos\phi}\frac{\partial(\psi_s\cos\phi)}{\partial\phi}, \qquad v = -\frac{1}{R}\frac{\partial\psi_s}{\partial\lambda} \tag{9.120}$$

From the momentum equations we obtain the time-dependent equation:

$$\frac{\partial\xi_s}{\partial t} + \frac{v}{R}\frac{\partial}{\partial\phi}(\xi_s + \Omega') + \frac{u}{R\cos\phi}\frac{\partial}{\partial\lambda}(\xi_s + \Omega') = \frac{1}{R\cos\phi}\left(\frac{\partial F_\phi}{\partial\lambda} - \frac{\partial F_\lambda}{\partial\phi}\right) \tag{9.121}$$

* See, for example, Lundgren (1967).

and

$$\Omega' = \left(2\Omega + \frac{u}{R\cos\phi}\right)\sin\phi$$

Apart from the terms on the right-hand side (due to eddy diffusion and surface winds) this equation is again the advective equation for the total vorticity of the fluid, $\xi_s + \Omega'$, which consists of the rotation of the fluid with respect to the earth and the contribution from the rotation of the earth. The system of equations (9.119, 9.120 and 9.121) may be resolved according to the methods in Section 9.2b.

Bryan (1969) has resolved the full three-dimensional problem where account is taken of the small vertical velocity and the topography of the ocean floor. The procedure employed is to solve the surface equations according to the method above and to determine the deviations at all depths from the surface velocities. If,

$$u' = \frac{\partial u}{\partial z}, \qquad v' = \frac{\partial v}{\partial z}$$

the corrections to the horizontal velocities are obtained from the Navier–Stokes equations:

$$\frac{\partial u'}{\partial t} + \mathbf{v}.\nabla u' + \mathbf{v}'.\nabla u = \left(2\Omega + \frac{u}{R\cos\phi}\right)v'\sin\phi + \frac{u'v\sin\phi}{R\cos\phi} + \frac{g}{R\cos\phi}\frac{\partial\rho}{\partial\lambda} + \frac{\partial F_\lambda}{\partial z} \tag{9.122}$$

$$\frac{\partial v'}{\partial t} + \mathbf{v}.\nabla v' + \mathbf{v}'.\nabla v = \left(2\Omega + \frac{2u}{R\cos\phi}\right)u'\sin\phi + \frac{g}{R}\frac{\partial\rho}{\partial\phi} + \frac{\partial F_\phi}{\partial z} \tag{9.123}$$

The hydrostatic relation (equation 9.110) has been used to replace the pressure but changes occur in the density, which in the simplest situation is related to the temperature by a coefficient of thermal expansion α,

$$\nabla\rho = -\alpha\nabla T \tag{9.124}$$

The vertical velocity is determined along each vertical column to satisfy the condition of a divergence-free flow:

$$\frac{\partial^2 w}{\partial z^2} = -\nabla.\mathbf{v}'_{\mathrm{H}} \tag{9.125}$$

which in difference form may be resolved by the inversion of a tridiagonal matrix equation (Section 4.3). The so-called 'rigid-lid' approximation is used for the surface boundary condition on w, $w_s = 0$, so as to ensure the removal of high-frequency surface waves and tides. The system is supplemented by advective equations for the local temperature and salinity of the fluid.

Figure 9.26 illustrates the surface currents in the world ocean obtained with such a model (Bryan and Cox 1972).

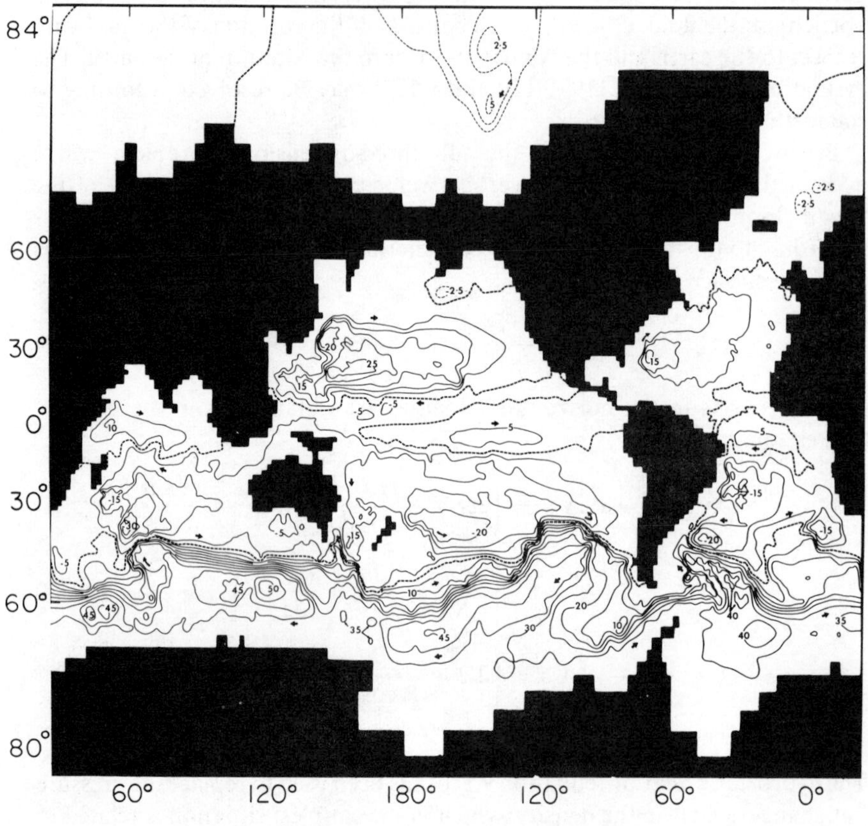

Figure 9.26. Currents on the surface of the world ocean (from Bryan, K., and Cox, M. D., *J. Physical Oceanography*, **2**, pp. 319–335, 1972)

CHAPTER X

Hydrodynamics with Long-Range Forces: Stars, Plasmas, Magneto-Fluids

1 Self-Consistent Fields in Fluid Assemblies

The classical fluids which occur in the laboratory, in the atmosphere and the oceans are assemblies of atoms and molecules which interact with short-range forces. When such assemblies may be regarded as being in local thermodynamic equilibrium, the hydrodynamic equations describe the motion of such an assembly as a fluid and the dynamics of the system are governed by the pressure of the fluid which replaces the short-range forces of the particles (Chapter 9).

The principle of describing an assembly of particles as a fluid is equally appropriate for assemblies which interact with long-range forces, provided that such an assembly may be regarded as being in local thermodynamic equilibrium. In stars, the assembly of nuclear particles may be regarded as a fluid where the dynamics of the fluid is to be described by the gravitational fields produced by the bulk matter in addition to the pressure forces. Similarly in plasmas and conducting fluids, both the thermal pressure and the self-consistent electromagnetic fields of the fluid are to be included in describing the motion of the fluid.

Gravitational hydrodynamics is applicable to the study of stars in describing their structure and their evolution in time as energy is produced by nuclear reactions and lost by radiation and convection and to stellar dynamics as in stellar pulsation, or the catastrophic gravitational collapse of old cold stars. Magneto-hydrodynamics (MHD) has a wide application in astrophysics, geophysics and in laboratory physics. It applies to such diverse problems as solar flares, the structure of the magnetosphere, the earth's dynamo problem and controlled thermonuclear fusion. These problems are strongly nonlinear, since the occurrence of the fluid produces the field. In turn the field interacts as a force on the fluid and, apart from the simplest problems, solutions may only be obtained by computational physics.

The diversity of phenomena to be described by gravitational and magneto-hydrodynamics is extremely broad and for successful models applied to particular problems, many detailed effects and processes are to be included. For example, in many problems it is necessary to include detailed processes of ionization, nuclear reactions, radiation mechanisms, and complex diffusion effects. In addition, and unlike classical hydrodynamics, the theories of gravitational and magneto-hydrodynamics are less well understood, and frequently the equations are less well-defined. We shall here emphasize the fundamental dynamic problem of the interaction of the fluid with self-consistent long-range forces and the manner by which this interaction is to be described in the computational mechanics.

We have considered the hydrodynamic equations for a classical fluid in Chapter 9, where equations for the density, momentum density and internal energy density (9.1, 9.2 and 9.3) define the motion of the fluid. Adopting the simplest point of view, we may include the effect of a long-range force on a fluid as an additional body force **F** in the momentum or acceleration equation of the fluid:

$$\rho\frac{\mathrm{d}\mathbf{v}}{\mathrm{d}t} = -\boldsymbol{\nabla}\,.\,\mathbf{P} + \mathbf{F} \tag{10.1}$$

The pressure has been written as a tensor **P** (with trace $3p$—the scalar pressure) to include the effects of viscosity. The force **F** in both the gravitational and magneto-hydrodynamic case is to be defined self-consistently by the state or properties of the fluid in space and time, $\mathbf{F}(\mathbf{x}, t)$. For a non-relativistic self-gravitating fluid, the body force on the fluid may be defined in terms of a scalar gravitational potential Φ,

$$\mathbf{F} = \rho\mathbf{g} = -\rho\boldsymbol{\nabla}\Phi \tag{10.2}$$

where the potential is defined by the density distribution of the fluid according to Poisson's equation:

$$\nabla^2\Phi = 4\pi G\rho \tag{10.3}$$

G is the gravitational constant.

In addition to the fundamental processes in a classical fluid of advection and of sound waves, the inclusion of the self-consistent gravitational field of the fluid introduces the effect of gravitational oscillations. If L is a characteristic scale length of the fluid, the fundamental frequencies described by the gravitational hydrodynamic equations are the frequency associated with advection ($\omega_{\mathrm{v}} \sim |\mathbf{v}|/L$), the frequency associated with the sound waves ($\omega_{\mathrm{s}} \sim L^{-1}\sqrt{(\gamma p/\rho)}$, where γ is the ratio of specific heats) and the frequency of gravitational oscillations,

$$\omega_{\mathrm{g}} = \sqrt{(4\pi G\rho)} \tag{10.4}$$

These processes might couple together so that, for example, the coupling of sound waves and gravitational oscillations leads to the phenomenon of pulsating stars (Christy 1966). It is these characteristic frequencies which must be followed when integrating the equations explicitly.

In the macroscopic motion of plasmas and conducting fluids, the fluid may be regarded as electrostatically neutral and the fundamental force **F** on the fluid is the magnetic force:

$$\mathbf{F} = \frac{1}{c}\mathbf{j} \wedge \mathbf{B} \tag{10.5}$$

where **B** is the magnetic field permeating the fluid and **j** is the current density in the fluid. The fields are defined by Maxwell's equations. For non-relativistic fluids, the current density carried by the fluid is simply defined by Ampère's law:

$$\mathbf{j} = \frac{c}{4\pi}\nabla \wedge \mathbf{B} \tag{10.6}$$

whence we may re-express the force on the fluid in terms of the magnetic field only:

$$\mathbf{F} = -\nabla\left(\frac{B^2}{8\pi}\right) + \frac{1}{4\pi}(\mathbf{B}\,.\,\nabla)\mathbf{B} \tag{10.7}$$

The first term in this expression has a form analogous to that of the thermal pressure (equations 9.2 and 10.1) and its effect on the fluid is similar, while the second term arises from 'tension' in the field lines. It follows that the fundamental effect of a magnetic field on a conducting fluid is to act as a 'magnetic pressure' on the fluid.

The magnetic field acting on the fluid is to be determined self-consistently, in that the motion of the fluid and the currents carried by the fluid determine the field. The self-consistent magnetic field is defined by Faraday's law for the conservation of magnetic flux (Section 3.1):

$$\frac{1}{c}\frac{\partial \mathbf{B}}{\partial t} + \nabla \wedge \mathbf{E} = \mathbf{0} \tag{10.8}$$

where the fields are subject to the condition,

$$\nabla\,.\,\mathbf{B} = 0 \tag{10.9}$$

The electric field **E** which is used in the conservative equations (10.8) depends upon the particular properties of the conducting fluid to be described. For a wide range of magneto-fluids, the electric field in the frame of the moving fluid (the Lorentz electric field) may be related to the current density according to a simple Ohm's law:

$$\mathbf{E} + \frac{1}{c}\mathbf{v} \wedge \mathbf{B} = \eta\mathbf{j} \tag{10.10}$$

where $\mathbf{v}$ is the velocity of the fluid and η is a resistivity. Inserting this expression for the electric field in Faraday's law (equation 10.8), we obtain a time-dependent equation for the magnetic field of the fluid:

$$\frac{\partial \mathbf{B}}{\partial t} - \nabla \wedge (\mathbf{v} \wedge \mathbf{B}) = -\nabla \wedge \frac{\eta c^2}{4\pi}(\nabla \wedge \mathbf{B}) \tag{10.11}$$

This equation has a form which is similar to the equations for the other properties of a fluid such as density, internal energy or momentum. The second term on the left-hand side (equation 10.11) relates to the advection and compression of the magnetic field, while the term on the right-hand side of the equation is a diffusion term (Section 10.2).

In magneto-hydrodynamics, the characteristic frequencies associated with advection and with sound waves ($\omega_v \sim |\mathbf{v}|/L$, $\omega_s \sim L^{-1}\sqrt{\{\gamma p/\rho\}}$) occur again. But, in addition, the magnetic field acts as a pressure on the fluid (equation 10.7) and consequently Alfven or magnetosonic waves occur with a characteristic frequency $\omega_A \sim L/V_A$, where V_A is the Alfven speed which is related to magnetic pressure (in analogy to sound waves):

$$V_A^2 = \frac{B^2}{4\pi\rho} \tag{10.12}$$

Again it is these characteristic frequencies, for sound waves, advection and Alfven waves, which must be followed in resolving the magneto-hydrodynamic equations explicitly.

In summary, it is stressed that, in contrast to classical hydrodynamic problems, the physical systems to be described here include the effects of field energy. For a self-gravitating fluid, the energy density U of the fluid is:

$$U = \tfrac{1}{2}\rho\Phi + \tfrac{1}{2}\rho v^2 + \rho\epsilon \tag{10.13}$$

where the notation is as before, and each of the terms in this expression relate to gravitational energy density, kinetic energy density and internal energy density respectively. Similarly, in magnetohydrodynamics, the energy density U of the fluid is:

$$U = \frac{B^2}{8\pi} + \frac{1}{2}\rho v^2 + \rho\epsilon \tag{10.14}$$

namely the magnetic energy density, the kinetic energy density and the internal energy density respectively. Consequently, for fluids with long-range forces, field energy and thermal energy are interchangeable and great variations in the fluid parameters can occur. This leads to strongly nonlinear problems and characteristic frequencies (the gravitational frequency or Alfven frequency) can vary very considerably over regions of the problem. For example, in star structure problems, densities between the inner core and outer regions of the star vary by many orders of magnitude, so that the gravitational frequency may be very large in the core of the star, while much smaller in the atmosphere of the star. In explicit calculations, such large variations in the characteristic frequencies can lead to difficulties, if the time step is not to become too small (Section 10.5).

Similarly in the magneto-hydrodynamic case, plasma densities may vary by three or more orders of magnitude and the Alfven speed becomes very large in tenuous plasma ($V_A = B/\sqrt{(4\pi\rho)}$). Again, if the Courant–Friedrichs–Lewy condition on an explicit time step (Chapter 3) is not to be over-restrictive, special models must be introduced for dealing with such low-density vacuum regions.

This fundamental effect of the interchange of field energy and fluid energy gives rise to many similarities in the simulation of gravitational and magneto-hydrodynamic phenomena. In Sections 10.2, 10.3 and 10.4 we shall consider the magneto-hydrodynamic problem and in Section 10.5 the gravitational hydrodynamic problem.

2 The Magneto-hydrodynamic Equations and their Simple Properties

(*a*) *Eulerian Equations and Conservative Properties.* We shall express the closed set of magneto-hydrodynamic equations according to the discussion in Section 1. The equations for the density, momentum density, and internal energy density of a classical fluid (equations 9.1, 9.2 and 9.3) are supplemented by the equations for the conservation of magnetic flux (equation 10.11). In Eulerian coordinates the equations take the form (Jeffrey 1966, Roberts and Potter 1970):

mass:
$$\frac{\partial \rho}{\partial t} = -\nabla \cdot \rho \mathbf{v} \qquad (10.15)$$

momentum:
$$\frac{\partial \rho \mathbf{v}}{\partial t} = -\nabla \cdot \left\{ \rho \mathbf{v}\mathbf{v} + \left(p + \frac{B^2}{8\pi}\right)\mathbf{I} - \frac{\mathbf{B}\mathbf{B}}{4\pi} + \mathbf{V} \qquad (10.16)$$

magnetic flux:
$$\frac{\partial \mathbf{B}}{\partial t} = \nabla \wedge \left(\mathbf{v} \wedge \mathbf{B} - \frac{\eta c^2}{4\pi} \nabla \wedge \mathbf{B}\right) \qquad (10.17)$$

internal energy:
$$\frac{\partial \rho \epsilon}{\partial t} = -p\nabla \cdot \mathbf{v} - \nabla \cdot (\rho \epsilon \mathbf{v} + \mathbf{q}) + \eta j^2 - \mathbf{V} : \nabla \mathbf{v} \qquad (10.18)$$

An 'equation of state' relates the pressure to the internal energy density and mass density of the fluid (Section 9.1a),

$$p = p(\epsilon, \rho) \qquad (10.19)$$

In the simplest magneto-fluids, the heat conduction $\mathbf{q}$ and viscous tensor $\mathbf{V}$

(if finite) may take the same form as the transport fluxes for a classical fluid (Section 9.1b).

Several points are to be noted with regard to these equations. In three space dimensions there are eight time-dependent coupled equations. In the absence of viscosity, resistivity and heat conduction, the equations are hyperbolic (Chapter 3). In accordance with Section 10.1, the effect of the magnetic field has been included under the divergence operator of the momentum equations, in which the magnetic field acts as a pressure together with the effect of tension in the field lines. For the hyperbolic equations, the fundamental processes described are advection, sound waves and Alfven waves.

The effect of viscosity, heat conduction and resistivity is to make the equations parabolic. To illustrate the form of the transport terms in the equations, we assume a simple collisional (Roberts and Potter 1970) ideal fluid with constant transport coefficients of viscosity μ, heat conductivity κ and resistivity η. The viscous terms in the momentum equations, the resistive terms in the magnetic flux equations and the heat conduction term in the internal energy equation then take the form

$$\mu\nabla^2\mathbf{v} - \tfrac{1}{3}\mu\nabla(\nabla\,.\,\mathbf{v}) \tag{10.20}$$

$$\kappa\nabla^2 T \tag{10.21}$$

$$\frac{\eta c^2}{4\pi}\nabla^2\mathbf{B} \tag{10.22}$$

respectively. These terms are each diffusion terms and lead to the diffusion of momentum, internal energy and magnetic flux. Correspondingly, such diffusion processes, reflecting the destruction of field and flow energy, cause heating in the fluid, and the effects of Joule heating (ηj^2) and viscous heating ($-\mathbf{V}\,.\,\nabla\mathbf{v}$) are included in the equation for the internal energy density (equation 10.18).

The properties of the solutions to these equations depend radically on the relative magnitude of the competing processes of, on the one hand, advection, sound waves and Alfven wave ('hyperbolic' processes) and, on the other hand, diffusion by viscosity, resistivity and heat conduction. It is useful, therefore, to establish the relative orders of magnitude of the advective processes and the diffusion processes. This is achieved by defining dimensionless numbers (Reynolds numbers) which are a measure of the dominance of advection in comparison to diffusion. If v is a typical velocity of the fluid which varies over the characteristic scale length L, we define three Reynolds numbers for the viscosity R_ν, heat conductivity R_κ, and resistivity R_η,

$$R_\nu = \frac{vL\rho}{\mu} \tag{10.23}$$

$$R_\kappa = \frac{vL\rho}{\kappa} \tag{10.24}$$

$$R_\eta = \frac{4\pi vL}{\eta c^2} \tag{10.25}$$

If the Reynolds number is large, the effect of the corresponding diffusion is small.

Finally, it is to be noted that each of the equations as expressed for the mass, momentum and magnetic flux of the magneto-fluid (equations 10.15, 10.16 and 10.17) are conservative, so that mass, momentum and magnetic flux in such a fluid are conserved. In contrast, the internal energy of the fluid is not conserved and the corresponding equation (10.18) is non-conservative. We may obtain an alternative equation which is conservative and from which the internal energy may be derived by noting that the total energy density of the fluid is:

$$U = \tfrac{1}{2}\rho v^2 + \rho\epsilon + \frac{B^2}{8\pi}$$

Consequently, by combining the equations (10.15 to 10.17) as the sum,

$$-\frac{1}{2}|\mathbf{v}|^2\frac{\partial \rho}{\partial t} + \mathbf{v}\,.\,\frac{\partial}{\partial t}\rho\mathbf{v} + \frac{\mathbf{B}}{4\pi}\cdot\frac{\partial \mathbf{B}}{\partial t} + \frac{\partial}{\partial t}\rho\epsilon$$

we obtain an equation for the conservation of total energy,

$$\frac{\partial U}{\partial t} = -\boldsymbol{\nabla}\,.\,\mathbf{g} \tag{10.26}$$

The energy flux $\mathbf{g}$ is:

$$\mathbf{g} = \left(\frac{1}{2}\rho v^2 + \rho\epsilon + \frac{B^2}{8\pi} + p\right)\mathbf{v} + (\mathbf{V}\,.\,\mathbf{v}) + \frac{\eta c}{4\pi}\mathbf{j}\wedge\mathbf{B} + \mathbf{q} \tag{10.27}$$

There are seven terms altogether, representing the transport of kinetic, internal and magnetic energy, the work done by material pressure and viscous forces, the Poynting vector associated with resistive diffusion and thermal conduction. The set of equations (10.15, 10.16, 10.17, 10.26) now form a closed set of conservative equations.

(*b*) *Advection and the Lagrangian Formulation.* As in classical hydrodynamics (Chapter 9), all properties of the fluid are transported or advected by the motion of the fluid. The equations, therefore, take a particularly simple form when re-expressed in the frame of the moving fluid, namely in terms of the Lagrangian time derivative,

$$\frac{\mathrm{d}}{\mathrm{d}t} = \frac{\partial}{\partial t} + \mathbf{v} \cdot \nabla$$

Not only is such a formulation informative, but in addition, and certainly for one-dimensional problems, the equations may be resolved on a Lagrangian difference mesh (Section 10.3b).

We pay particular attention to the equations for the magnetic field (10.17) in which, assuming a constant resistivity, we may rewrite the curl operator

$$\frac{\partial \mathbf{B}}{\partial t} + \nabla \cdot (\mathbf{vB} - \mathbf{Bv}) = \frac{\eta c^2}{4\pi} \nabla^2 \mathbf{B} \tag{10.28}$$

The condition that the magnetic field is divergence-free (equation 10.9) has been used. By differentiating the divergence terms, we may now rewrite the Eulerian equations (Section 10.2a) in Lagrangian form:

mass:
$$\frac{\mathrm{d}\rho}{\mathrm{d}t} = -\rho \nabla \cdot \mathbf{v} \tag{10.29}$$

acceleration:
$$\rho \frac{\mathrm{d}\mathbf{v}}{\mathrm{d}t} = -\nabla\left(p + \frac{B^2}{8\pi}\right) + \nabla \cdot \left(\frac{\mathbf{BB}}{4\pi} - \mathsf{V}\right) \tag{10.30}$$

magnetic flux:
$$\frac{\mathrm{d}\mathbf{B}}{\mathrm{d}t} = -\mathbf{B}(\nabla \cdot \mathbf{v}) + (\mathbf{B} \cdot \nabla)\mathbf{v} + \frac{\eta c^2}{4\pi} \nabla^2 \mathbf{B} \tag{10.31}$$

internal energy:
$$\rho \frac{\mathrm{d}\epsilon}{\mathrm{d}t} = -p \nabla \cdot \mathbf{v} + \nabla \cdot \mathbf{q} + \eta j^2 - \mathsf{V} : \nabla \mathbf{v} \tag{10.32}$$

Clearly the magnetic field is advected in a similar manner to the advection of mass, or internal energy. Each of the first terms on the right-hand side of to the equations is associated with the compression of the fluid and gives rise to sound waves and compressional or transverse Alfven waves. The terms $\nabla \cdot \mathbf{BB}$ and $(\mathbf{B} \cdot \nabla)\mathbf{v}$ are associated with the tension in the magnetic field lines and give rise to longitudinal Alfven waves, while the remaining terms in the equations are related to the diffusion processes.

(c) *Electron and Ion Energies in a Plasma.* The most important example of a conducting fluid is the macroscopic description of a plasma and we shall pay particular attention to the magneto-hydrodynamic equations for this case (Spitzer 1962, Roberts and Potter 1970). In a fully ionized plasma, there are two fluid components, namely the electrons and ions and in very dense plasma, there is an equipartition of energy between the electrons and ions. In this instance, the plasma may be treated as a one-component fluid with an ideal-gas equation of state, and the models of Sections (*a*) and (*b*) are appropriate. In more tenuous plasmas, however, the electrons and ions must be treated with separate internal energies and pressures. We define an internal energy per unit fluid mass for the electrons ϵ_e and for the ions ϵ_i, which are each related to their respective pressures through the equation of state for an ideal gas:

$$\epsilon_e = \frac{p_e}{\rho(\gamma - 1)}$$

$$\epsilon_i = \frac{p_i}{\rho(\gamma - 1)} \tag{10.33}$$

To the equations for the mass, momentum and magnetic flux in the fluid, we add two equations for the electron and ion internal energies:

electrons:

$$\frac{\partial}{\partial t}(\rho\epsilon_e) = -p_e \nabla . \mathbf{v} - \nabla . (\rho\epsilon_e \mathbf{v} - \kappa_e \nabla T_e) + \eta j^2 + \frac{\rho(\epsilon_i - \epsilon_e)}{\tau_{eq}} \tag{10.34}$$

ions:

$$\frac{\partial}{\partial t}(\rho\epsilon_i) = -p_i \nabla . \mathbf{v} - \nabla . (\rho\epsilon_i \mathbf{v} - \kappa_i \nabla T_i) - \mathbf{V} : \nabla \mathbf{v} + \frac{\rho(\epsilon_e - \epsilon_i)}{\tau_{eq}} \tag{10.35}$$

The pressure to be used in the momentum equation is the sum of the electron and ion pressures. It is to be noted that separate ion and electron heat conductivities, κ_e, κ_i, have been defined. The Joule heating due to the diffusion of the magnetic field acts only on the electrons (since the electron to ion mass is small), while viscosity is usually only associated with the ions. The remaining terms in the equations describe the equipartition of energy, the rate of which depends on the equipartition time τ_{eq}.

In a plasma the transport coefficients of resistivity, electron heat-conductivity and equipartition time depend on the collision frequency for the electrons when interacting with ions. In general, the coefficients vary rapidly with density and electron temperature so that the diffusion terms become nonlinear. Equally the ion heat-conductivity κ_i and viscosity μ depend on the frequency of ion–ion collisions (Robinson and Bernstein 1963, Spitzer 1962). Simple expressions for these coefficients in a hydrogen-like plasma are included in Table 10.1.

Table 10.1 Transport coefficients for a hydrogen-like plasma

density	$= n$	Boltzmann's constant	$= k$
electron mass	$= m_e$	ion-ion collision frequency	$= \nu_{ii}$
ion mass	$= m_i$	electron-ion collision frequency	$= \nu_{ei}$
electron temperature	$= T_e$	ion temperature	$= T_i$

Resistivity	$\eta = \dfrac{m_e}{ne^2}\nu_{ei}$
Electron heat-conductivity	$\kappa_e = \dfrac{5nk^2 T_e}{m_e \nu_{ei}}$
Ion heat-conductivity	$\kappa_i = \dfrac{5nk^2 T_i}{m_i \nu_{ii}}$
Ion viscosity	$\mu_i = \dfrac{5nk^2 T_i}{m_i \nu_{ii}}$
Equipartition time	$\tau_{eq} = \dfrac{m_i}{2m_e}\dfrac{1}{\nu_{ei}}$

where,

$$\nu_{ei} = \frac{4\pi n e^4 \ln \Lambda_e}{(2m_e)^{\frac{1}{2}} \epsilon_e{}^{3/2}}, \qquad \epsilon_e = \tfrac{3}{2}kT_e$$

$$\nu_{ii} = \frac{4\pi n e^4 \ln \Lambda_i}{(2m_i)^{\frac{1}{2}} \epsilon_i{}^{3/2}}, \qquad \epsilon_i = \tfrac{3}{2}kT_i$$

$$\Lambda_{ei} = \frac{1}{144\pi n}\left(\frac{4\pi n e^2}{kT_{ei}}\right)^{3/2}$$

3 One-Dimensional Methods in Magneto-hydrodynamics

As expressed in Section 10.2, the properties of the magneto-hydrodynamic equations are very similar to the compressible hydrodynamic equations (Section 9.5). This is particularly true in one space dimension, where the vector properties of the magnetic field play no role. It follows that, at least in one space dimension, all the methods appropriate for the hydrodynamic problem are applicable to the magneto-hydrodynamic problem. In particular, we may employ either a Lagrangian or an Eulerian difference mesh and the explicit Lax, Lax–Wendroff, leapfrog and quasi-second-order schemes (Section 3.7) are useful. Again, in one space dimension, the Lagrangian mesh is to be preferred in that the difficult advective term is then accurately treated and the effects of numerical dispersion and diffusion are then minimized.

There are, however, additional difficulties in the magneto-hydrodynamic problem. Particularly in problems arising from controlled thermonuclear fusion research, a near equilibrium or balance can exist between the thermal pressure p and the magnetic pressure $B^2/8\pi$, so that, although very rapid frequencies associated with sound waves and Alfven waves may exist, we are interested in solutions over very long time scales. If an explicit method is employed, a very large, if not prohibitive, number of time steps will be required to satisfy the Courant–Friedrichs–Lewy condition. It is appropriate therefore, to employ an *implicit* method for such problems and we shall formulate such an implicit solution (Hain *et al.* 1960) in Section 10.3b.

Unlike the usual situation in classical hydrodynamics, there is the added difficulty in magneto-hydrodynamics (and indeed in gravitational hydrodynamics) of very large variations in parameters over different regions in the same problem. Thus special methods are required for treating the low-density or 'vacuum' regions (Section 10.3d). Also the transport coefficients can vary accordingly over many orders of magnitude within the same problem so that ideally the diffusion terms, which with variable coefficients are nonlinear, should be treated implicitly (Section 10.3c).

(a) Explicit Methods: Eulerian and Lagrangian Meshes. We shall express the magneto-hydrodynamic equations on an Eulerian Cartesian mesh. We shall ignore the diffusion terms and assume an equation of state for an ideal gas. In one dimension, there is one component of the velocity and we assume one component of the magnetic field. From equations 10.15 through 10.18 we differentiate each term explicitly and obtain the simple differential equations:

density:
$$\frac{\partial \rho}{\partial t} = -v\frac{\partial \rho}{\partial x} - \rho\frac{\partial v}{\partial x}$$

velocity:
$$\frac{\partial v}{\partial t} = -v\frac{\partial v}{\partial x} - \frac{1}{\rho}\frac{\partial p}{\partial x} - \frac{B}{4\pi\rho}\frac{\partial B}{\partial x}$$

magnetic field:
$$\frac{\partial B}{\partial t} = -v\frac{\partial B}{\partial x} - B\frac{\partial v}{\partial x}$$

pressure:
$$\frac{\partial p}{\partial t} = -v\frac{\partial p}{\partial x} - \gamma p\frac{\partial v}{\partial x} \qquad (10.36)$$

This system of equations are in the comparable form of the compressible hydrodynamic equations (Section 9.5b) and, similarly, to illustrate the basic explicit Eulerian method, we difference the equations by the nonconservative Lax scheme (Section 3.7a). On the space mesh, $1 \leqslant j \leqslant J$, new variables are obtained at the time step $n + 1$ according to the algorithms:

density:

$$\rho_j^{n+1} = \frac{1}{2}(\rho_{j+1}^n + \rho_{j-1}^n) - v_j^n \frac{\Delta t}{2\Delta}(\rho_{j+1}^n - \rho_{j-1}^n) - \rho_j^n \frac{\Delta t}{2\Delta}(v_{j+1}^n - v_{j-1}^n)$$

velocity:

$$v_j^{n+1} = \frac{1}{2}(v_{j+1}^n + v_{j-1}^n) - v_j^n \frac{\Delta t}{2\Delta}(v_{j+1}^n - v_{j-1}^n) - \frac{1}{\rho_j^n}\frac{\Delta t}{2\Delta}(p_{j+1}^n - p_{j-1}^n)$$

$$- \frac{B_j^n}{4\pi\rho_j^n}\frac{\Delta t}{2\Delta}(B_{j+1}^n - B_{j-1}^n)$$

magnetic field:

$$B_j^n = \frac{1}{2}(B_{j+1}^n + B_{j-1}^n) - v_j^n \frac{\Delta t}{2\Delta}(B_{j+1}^n - B_{j-1}^n) - B_j^n \frac{\Delta t}{2\Delta}(v_{j+1}^n - v_{j-1}^n)$$

pressure:

$$p_j^{n+1} = \frac{1}{2}(p_{j+1}^n + p_{j-1}^n) - v_j^n \frac{\Delta t}{2\Delta}(p_{j+1}^n - p_{j-1}^n) - \gamma p_j^n \frac{\Delta t}{2\Delta}(v_{j+1}^n - v_{j-1}^n)$$

(10.37)

For the purpose of stability analysis, the equations are linearized by making the local approximation that the coefficients of the space derivatives in the above equations are constant. Again, this is a reasonable approximation when considering the growth of a small error mode. For a Fourier mode in space of wavenumber k, the amplification matrix is

$$\mathbf{G}(\Delta t, k) = \begin{bmatrix} (\beta - i\alpha v_j^n) & -i\alpha\rho_j^n & 0 & 0 \\ 0 & (\beta - i\alpha v_j^n) & \dfrac{-i\alpha B_j^n}{4\pi\rho_j^n} & -i\alpha\dfrac{1}{\rho_j^n} \\ 0 & -i\alpha B_j^n & (\beta - i\alpha v_j^n) & 0 \\ 0 & -i\alpha\gamma p_j^n & & (\beta - i\alpha v_j^n) \end{bmatrix} \qquad (10.38)$$

where $\beta = \cos(k\Delta)$ and $\alpha = \{\Delta t \sin(k\Delta)\}/\Delta$. There are four eigenvalues g of the amplification matrix $\mathbf{G}$:

$$g_{1,2} = \beta - i\alpha v_j^n$$

$$g_{3,4} = \beta - i\alpha\left\{v_j^n \pm \sqrt{\left(\frac{\gamma p_j^n}{\rho_j^n} + \frac{(B_j^n)^2}{4\pi\rho_j^n}\right)}\right\} \qquad (10.39)$$

The eigenvalue of largest modulus is either g_3 or g_4 depending on the sign of the advective velocity v_j^n, so that to satisfy the von Neumann stability condition (equation 3.47), the time step must be limited according to the criterion (the Courant–Friedrichs–Lewy condition):

$$\Delta t \leqslant \frac{\Delta}{|v_j^n| + \sqrt{\left(\dfrac{\gamma p_j^n}{\rho_j^n} + \dfrac{(B_j^n)^2}{4\pi\rho_j^n}\right)}} \tag{10.40}$$

This condition is very similar to that obtained for the compressible hydrodynamic problem (Section 9.5b), but the speed in the denominator is the modular sum of the centre-of-mass velocity and the 'magnetosonic' velocity $v_m = \sqrt{(v_s^2 + v_A^2)}$, where v_s is the sound speed and v_A is the Alfven speed $v_A = B/\sqrt{(4\pi\rho)}$, resulting from the magnetic pressure associated with the field lines. This speed, the Doppler-shifted magnetosonic speed, is the largest speed on the difference mesh and again we are required to ensure that the lattice speed $\Delta/\Delta t$ is smaller. The stability condition is therefore more restrictive than in the hydrodynamic case. Again because the problem is nonlinear, the stability criterion is to be applied at every point, j, on the difference mesh and the time step is therefore limited by the maximum magnetosonic velocity on the mesh. The limitations imposed by a large Alfven speed are here clear, for unlike the situation in hydrodynamics, regions can occur in a magneto-hydrodynamic fluid, where the fluid density may become very small. While the sound speed $v_s = \sqrt{(\gamma p/\rho)} = \sqrt{(\gamma kT/m)}$ (k is Boltzmann's constant and m is the particle mass) depends only on the temperature of the fluid, the Alfven speed is inversely proportional to the square root of the density. To avoid small time steps we are therefore required to employ special methods for the 'vacuum' region (Section 10.3d).

It is to be noted that we have illustrated the explicit Eulerian difference method by the simplest integration procedure. If the difference equations (10.37) are reformulated according to principles of conservation, the linear properties of the difference equations, and in particular the stability criterion (equation 10.40), remain the same. Both from the point of view of nonlinear properties as in shock waves and to satisfy integral physical laws, the conservative difference formulation is to be preferred.

In addition, to minimize the effects of dispersion and diffusion on the difference mesh, second-order conservative methods such as the Lax–Wendroff, leapfrog or quasi-second-order scheme are generally to be used (Chapter 3). In each method, the appropriate Courant–Friedrichs–Lewy condition is satisfied when the Doppler-shifted magnetosonic speed is imposed as the fastest speed on the difference mesh.

In one space dimension and in analogy to the hydrodynamic problem, we may avoid the effects of numerical dispersion and diffusion by the use of a Lagrangian difference mesh. The procedure which may be employed is precisely the same as that used for the hydrodynamic problem (Section 9.5), where a space mesh which moves with the fluid velocity is used. Accordingly we define the boundaries (mesh points j) of cells at integer time steps:

$$x_j^{n+1} = x_j^n + v_j^{n+\frac{1}{2}}\,\Delta t \tag{10.41}$$

and correspondingly the width of each cell (i) is defined by its boundaries:

$$\Delta_i = x_{j+1}^n - x_j^n \tag{10.42}$$

The basic set of magneto-hydrodynamic equations (10.36) integrated on such a Lagrangian mesh takes a particularly simple form:

$$\begin{aligned}
&\frac{\mathrm{d}}{\mathrm{d}t}(\rho\Delta) = 0\\
&\frac{\mathrm{d}}{\mathrm{d}t}(B\Delta) = 0\\
&\frac{\mathrm{d}}{\mathrm{d}t}\left(\frac{p}{\rho^{\gamma}}\right) = 0\\
&\frac{\mathrm{d}v}{\mathrm{d}t} = -\frac{1}{\rho}\frac{\partial}{\partial x}\left(p + \frac{B^2}{8\pi}\right)
\end{aligned} \tag{10.43}$$

The first three of these equations in the absence of diffusion express the conservation of mass, the conservation of magnetic flux and the identical satisfaction of the adiabatic law respectively, in each cell Δ. Velocities are defined at the half time steps and on the boundary points j between cells according to the method of Section 9.5.

Again, in analogy with the hydrodynamic problem, shock waves and discontinuities can occur in the magneto-hydrodynamic problem and indeed shock waves associated with both the thermal pressure and magnetic pressure occur. In general, therefore, it is necessary to broaden such discontinuities on the difference mesh by the use of an artificial viscosity, where precisely the same algorithms as in the hydrodynamic problem are used (Section 9.6).

(*b*) *Nonlinear Diffusion Terms*. We have noted that due to the possibility of interchange between magnetic energy and fluid thermal and kinetic energy, the variables in a given magneto-hydrodynamic problem may change by several orders of magnitude in space and time. In addition the diffusion coefficients of resistivity, heat conductivities and viscosity are in general rapidly varying functions of the dependent variables (Table 10.1), so that the diffusion terms in the magneto-hydrodynamic problem are nonlinear and difficult to treat. Furthermore, the corresponding Reynolds numbers (Section 10.1) may become small compared to unity in some regions of the fluid. It is frequently necessary, therefore, to solve the diffusion terms implicitly.

On either an Eulerian or a Lagrangian difference mesh, which is defined by the method of resolving the advective and compressible flow terms, we are required to include a nonlinear diffusion term:

$$\frac{\partial u}{\partial t} - \frac{\partial}{\partial x}\kappa\frac{\partial u}{\partial x} = S \tag{10.44}$$

where S is the remaining terms associated with the hyperbolic equations and $\kappa = \kappa(u)$. For the internal energy $(u = \epsilon)$ or temperature equation, for example,

$$\kappa \propto \epsilon^{\frac{5}{2}}$$

To resolve this term independently of a stability criterion, we shall apply the Crank–Nicholson implicit method which is unconditionally stable (Section 3.9b). During each time step, however, we are required to iterate over the nonlinear diffusion coefficient κ. If the superscript p refers to the pth iteration step during the time interval $\Delta t = t^{n+1} - t^n$, we obtain successively improved values for the dependent variable u:

$$u^{p+1} - \frac{\partial}{\partial x}\frac{\bar{\kappa}\,\Delta t}{2}\frac{\partial u^{p+1}}{\partial x} = u^n + \frac{\partial}{\partial x}\frac{\bar{\kappa}\,\Delta t}{2}\frac{\partial u^n}{\partial x} + \int \mathrm{d}t\, S \tag{10.45}$$

Values of the variable u at the previous iteration step p are used to define the time average for the diffusion coefficient:

$$\bar{\kappa} = \tfrac{1}{2}\kappa(u^p) + \tfrac{1}{2}\kappa(u^n) \tag{10.46}$$

The terms on the right-hand side of equation 10.45 are always explicitly known and it follows that on the difference mesh, u^{p+1} is determined as the solution to a tridiagonal matrix equation (Section 4.3). The new values of the dependent variable so obtained serve to define an improved value of the diffusion coefficient (equation 10.46). If, after many iterative steps, convergence is achieved,

$$\lim_{p\to\infty} u^{p+1} = u^p$$

then u^{p+1} is the required solution at the time step $n + 1$:

$$u^{n+1} = \lim_{p\to\infty} u^{p+1} \tag{10.47}$$

since equation 10.45 is then the Crank–Nicholson algorithm for the diffusion terms.

For many problems only two or three iterations are required. This implicit method is particularly successful because of the ease with which a tridiagonal matrix equation may be inverted.

(c) *The Hain Implicit Method.* In many magneto-hydrodynamic problems, the limitation on the time step imposed by the Courant–Friedrichs–Lewy condition (equation 10.40) is severe, and in particular, the Alfven speed can lead to very small time steps. As an example, in controlled thermonuclear fusion plasmas a near-equilibrium can exist between the plasma pressure and magnetic pressure of the field so that, while the plasma does not move significantly, magnetosonic waves of large frequency still propagate.

In the Hain method (Hain *et al.* 1960), an implicit method is used for resolving the terms associated with Alfven waves and sound waves. The approach is similar to that of the implicit second-order method for ordinary differential equations (Sections 2.4 and 2.6d). We shall first consider any simple pair of coupled first-order partial differential equations which describe waves:

$$\frac{\partial v}{\partial t} = -\beta \frac{\partial B}{\partial x} \tag{10.48}$$

$$\frac{\partial B}{\partial t} = -\beta \frac{\partial v}{\partial x} \tag{10.49}$$

where β is related to a wave velocity. If this pair of equations is differenced in time according to the second-order implicit method, we obtain at step $n + 1$:

$$v^{n+1} = v^n - \frac{\Delta t \beta}{2}\left(\frac{\partial B^n}{\partial x} + \frac{\partial B^{n+1}}{\partial x}\right) \tag{10.50}$$

$$B^{n+1} = B^n - \frac{\Delta t \beta}{2}\left(\frac{\partial v^n}{\partial x} + \frac{\partial v^{n+1}}{\partial x}\right) \tag{10.51}$$

We are required to solve for B^{n+1} and v^{n+1} in space, and equation (10.51) may be used to substitute for B^{n+1} in equation 10.50:

$$v^{n+1} - \frac{\Delta t \beta}{2}\frac{\partial}{\partial x}\left(\frac{\Delta t \beta}{2}\frac{\partial v^{n+1}}{\partial x}\right) = v^n - \Delta t \beta \frac{\partial B^n}{\partial x} + \frac{\Delta t \beta}{2}\frac{\partial}{\partial x}\left(\frac{\Delta t \beta}{2}\frac{\partial v^n}{\partial x}\right) \tag{10.52}$$

The variables on the right-hand side of this equation are known from the previous time step, while the second term on the left-hand side is a second-order spatial differential operator, similar to the diffusion operator, acting on the unknown variable v^{n+1}. When equation 10.52 is applied on a difference mesh in space, either Lagrangian or Eulerian, the left-hand side becomes a tridiagonal matrix acting on the variable v^{n+1} defined at every space point. Again, we may simply invert the tridiagonal matrix equation (Section 4.3) to obtain the required solutions v^{n+1} at all space points. The complementary variable B^{n+1} is now simply determined from equation 10.51.

The corresponding amplification matrix has eigenvalues of modulus unity for all time steps Δt and there is therefore no limitation on the time step.

In application to the magneto-hydrodynamic problem it is sufficient to consider a restricted set of the equations 10.36 in Lagrangian form where we pay particular attention to the nonlinear magnetic pressure term:

$$\frac{d\rho}{dt} = -\bar{\rho}\frac{\partial v}{\partial x}$$

$$\frac{dB}{dt} = -\bar{B}\frac{\partial v}{\partial x}$$

$$\bar{\rho}\frac{dv}{dt} = -\frac{\bar{B}}{4\pi}\frac{\partial B}{\partial x} \tag{10.53}$$

We shall iterate around the nonlinear factors in a manner entirely analogous to the procedure adopted for the nonlinear diffusion equation, and the bar in equations 10.53 has been introduced to illustrate the variables held constant during an iteration step p to $p + 1$. Applying the implicit method above, the fluid velocity is to be determined at each iteration according to the pseudo-diffusion equation:

$$\bar{\rho}\frac{dv}{dt} - \frac{\Delta t\bar{B}}{8\pi}\frac{\partial}{\partial x}\bar{B}\frac{\partial v}{\partial x} = -\frac{\bar{B}}{4\pi}\frac{\partial B^n}{\partial x} \tag{10.54}$$

which is resolved by the Crank–Nicholson method according to the algorithm (10.52). The magnetic field and density are subsequently determined:

$$\alpha = \frac{\Delta t}{4}\left(\frac{\partial v^n}{\partial x} + \frac{\partial v^{p+1}}{\partial x}\right)$$

$$B^{p+1} = \frac{(1-\alpha)}{(1+\alpha)}B^n, \qquad \rho^{p+1} = \left(\frac{1-\alpha}{1+\alpha}\right)\rho^n \tag{10.55}$$

Typically only two or three iterations are required.

Sound waves and the associated pressure term in the acceleration equations are treated in the same way. On the other hand, since the advective terms will in general still be treated explicitly, the associated stability criterion for the advective velocity will remain:

$$\Delta t \leqslant \frac{\Delta}{|v|_{\max}} \tag{10.56}$$

This condition ensures that the parameter α in equations 10.55 remains small.

Hain *et al.* (1960) have applied this method in resolving the magneto-hydrodynamic equations in cylindrical geometry.

(*d*) *The 'Vacuum' Region.* The magneto-hydrodynamic problem is quite different from the hydrodynamic problem in the effect of the interchange of magnetic energy with fluid energy. Frequently, moving regions in the domain of interest occur where the magnetic energy is large and the fluid density becomes extremely small. This situation occurs in astrophysical problems such as solar flares and it occurs in laboratory experiments in the study of controlled thermonuclear fusion. In such regions, the Alfven speed may become very large and certainly in an explicit solution, the time step would become unacceptably small. It is necessary, in general, to avoid such a restriction on the time step by employing one of a number of approximations.

In the simplest approximation, we assume that such a vacuum region cannot support a current, so that it is assumed that the 'vacuum' acts as an insulator. The electromagnetic effects may then be decoupled from the fluid equations, so that where the density falls below some small arbitrarily chosen minimum density, $\rho_{\min}$, the fluid is treated as a simple hydrodynamic fluid. The field in the vacuum region is determined according to Ampère's law and, as an example, in Cartesian coordinates the field is a constant:

$$j = \frac{c}{4\pi}\frac{\partial B}{\partial x} = 0$$

therefore

$$B = B_0 \tag{10.57}$$

The constant B_0 is determined according to the boundary conditions which might be coupled to an external circuit and from which the current flowing is determined self-consistently (Roberts and Potter 1970). An upper limit is imposed on the Alfven speed and the time step Δt is maintained at an acceptable magnitude.

On the other hand, if the magnetic field in the vacuum region is of varying direction, large force-free currents may flow parallel to the magnetic field so that the assumption that the vacuum region is an insulator is untenable. An alternative model has been used (Hain *et al.* 1960) in which the vacuum region is avoided by arbitrarily introducing a small fluid density at the boundary and the density is always maintained above some minimum value

ρ_{min}. Such a prescription does clearly not conserve mass, but if the minimum value of the density ρ_{min} is chosen to be sufficiently small, the effect of non-conservation may be small.

4 Multidimensional Magneto-hydrodynamics

It is particularly in two and three space dimensions that the study of magneto-hydrodynamics is of greatest interest. This is because the vector and topological properties of the magnetic field can be demonstrated only in several space dimensions and in both astrophysical problems and in laboratory plasma physics the topological properties of the magnetic field are of paramount importance. Examples of interest which illustrate this feature are solar flares and toroidal magnetic field configurations for the containment of thermonuclear plasmas.

In developing computational models to describe particular multi-dimensional magneto-hydrodynamic problems, three dimensionless numbers which relate to the hyperbolic equations are of particular importance:

$$\beta = \frac{p}{B^2/8\pi} \tag{10.58}$$

$$M_s^2 = \frac{\rho v^2}{\gamma p} \tag{10.59}$$

$$M_A^2 = \frac{\frac{1}{2}\rho v^2}{B^2/8\pi} \tag{10.60}$$

M_s and M_A are the sound and Alfven Mach numbers respectively, and the order of magnitude of these numbers compared to unity establishes the relative importance of the thermal energy, the magnetic energy and the flow or kinetic energy. When the sound Mach number is small, the fluid may be regarded as incompressible in three dimensions and the methods of Chapter 9 for an incompressible fluid may be extended. When the Alfven Mach number is small (but not the sound Mach number), or the 'beta' of the fluid is small, the flow of the fluid is subject to a quasi-incompressibility condition in two dimensions only—namely perpendicular to the magnetic field. This case, where the beta of the plasma is small but finite, is particularly important in the study of controlled thermonuclear fusion.

In the general situation where the fluid energy and the magnetic energy can interact strongly (the beta of the magnetofluid is of the order of unity), explicit conservative methods are applicable (Section 10.3a). These methods are not appropriate to the low-beta problem, where large Alfven frequencies can lead to limitingly small time steps and where the effects of numerical dispersion and diffusion in an explicit determination of the magnetic fields can produce severe and unacceptable errors. In addition, the vector properties of the magnetic field lead to anisotropic phenomena such as torsional waves, which are 'piped' along magnetic field lines, and diffusion processes which may be large along a magnetic field line but small in the directions perpendicular to a magnetic field line (Robinson and Bernstein 1963). If such effects are important, an Eulerian difference is not effective and resort must be made to special techniques such as the use of magnetic field lines as coordinates.

(*a*) *Explicit Conservative Methods.* In three space dimensions and in Cartesian coordinates, the magneto-hydrodynamic equations for a one-component compressible fluid may be written in conservative form:

$$\frac{\partial \mathbf{u}}{\partial t} + \frac{\partial \mathbf{F}_x}{\partial x} + \frac{\partial \mathbf{F}_y}{\partial y} + \frac{\partial \mathbf{F}_z}{\partial z} = 0 \tag{10.61}$$

There are nine dependent variables,

$$\mathbf{u} = \left\{\rho, \rho v_x, \rho v_y, \rho v_z, B_x, B_y, B_z, \left(\frac{1}{2}\rho v^2 + \frac{B^2}{8\pi} + \rho\epsilon\right)\right\} \tag{10.62}$$

for the density, three components of momentum density, three components of magnetic field, and total energy density. For the hyperbolic equations (when the Reynolds numbers are large), the fluxes are defined by the column vectors $\mathbf{F}_x$, $\mathbf{F}_y$ and $\mathbf{F}_z$, for which it is sufficient to define the x-component:

$$\mathbf{F}_x = \begin{pmatrix} \rho v_x \\ \rho v_x v_x + p + \dfrac{B_y^2}{8\pi} + \dfrac{B_z^2}{8\pi} \\ \rho v_x v_y - \dfrac{B_x B_y}{4\pi} \\ \rho v_x v_z - \dfrac{B_x B_z}{4\pi} \\ v_x B_y - B_x v_y \\ v_x B_z - v_z B_x \\ \left(\dfrac{1}{2}\rho v^2 + \rho\epsilon + p + \dfrac{B^2}{8\pi}\right) v_x \end{pmatrix} \tag{10.63}$$

$\mathbf{F}_y$ and $\mathbf{F}_z$ have an analogous form. An equation of state relates the internal energy density to the pressure and density of the fluid.

In this form, the magneto-hydrodynamic equations may be resolved on an Eulerian difference mesh to which the explicit conservative methods of Sections 3.7 and 3.8 are applied. To minimize the effects of numerical diffusion, methods of second-order accuracy such as the Lax–Wendroff, leapfrog, and quasi-second-order Adams–Bashforth method are to be preferred. In addition, providing the Reynolds numbers are larger than unity, the diffusion terms may be included explicitly on the same difference mesh (Section 3.9). As in the hydrodynamic problem, shocks and discontinuities can occur and the artificial viscosity terms are to be included (Section 9.6).

The general principle of the explicit conservative method for the magneto-hydrodynamic equations is illustrated here, but it is to be appreciated that there are particular difficulties for particular problems. We shall raise several questions about this method of solution which may or may not be important depending on the problem at hand. In the first instance we have defined three components of the magnetic field, which are, however, not independent since,

$$\nabla . \mathbf{B} = 0 \tag{10.64}$$

If the initial conditions specify a magnetic field which is divergence-free, then the differential equations ensure that the magnetic field is divergence-free for subsequent times. On the other hand, the difference equations do not ensure this condition. Errors occur, which yield a magnetic field which increasingly has a finite divergence. This effect may be avoided either by the geometry of the problem, or by defining a magnetic vector potential (Section 10.4b).

A second source of errors may occur from defining the temperature or internal energy ϵ of the fluid from the equation for the total energy density U. Although such an equation is conservative, it is evident that in regions of large magnetic energy, any errors which arise in the remaining fluid variables are automatically distilled into the variable for the internal energy. It is frequently preferable, therefore, to employ the equation for the internal energy density directly (equation 10.18). This is particularly true in a plasma simulation when separate equations for the electron and ion internal energies are specified. These equations may be differenced with consistency on the same Eulerian difference mesh as the remaining conservative equations.

Finally it is to be noted that, as in one dimension, low-density or vacuum regions can occur and a vacuum approximation must be employed to avoid the effect of very large Alfven speeds (Section 10.4b). Either the vacuum region may be treated as an insulator, or fluid must arbitrarily be added to maintain the density above some small minimum value. The problem is, however, more severe than in one dimension since the vacuum region now consists of a two- or three-dimensional domain with an awkward moving boundary.

An example of such a two-dimensional simulation in cylindrical coordinates is indicated in Figures 10.1 and 10.2. The simulation model (Roberts and Potter 1970, Potter 1971) has been applied to a laboratory pinch device in which thermonuclear densities (10^{19} particles cm^{-3}) and temperatures ($T \sim 1{\cdot}5$ kev) are produced. The particular feature of this problem which simplifies the nature of the simulation considerably is that the magnetic field is only in the azimuthal direction ($\mathbf{B} = (0, B_\theta, 0)$), so that the magnetic field is always perpendicular to the plane of the calculation. Consequently the field may be treated as a pseudoscalar and the condition that the magnetic field be divergence-free is automatically satisfied. The 'vacuum' region of large magnetic field and low density can clearly be seen.

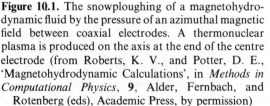

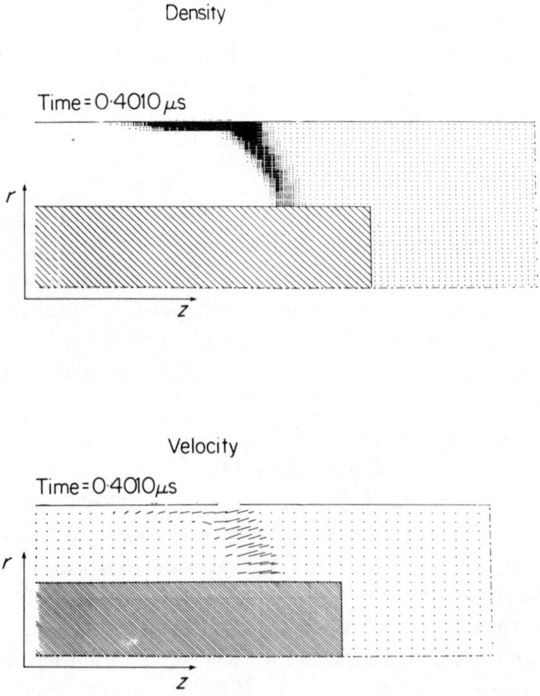

Figure 10.1. The snowploughing of a magnetohydrodynamic fluid by the pressure of an azimuthal magnetic field between coaxial electrodes. A thermonuclear plasma is produced on the axis at the end of the centre electrode (from Roberts, K. V., and Potter, D. E., 'Magnetohydrodynamic Calculations', in *Methods in Computational Physics*, **9**, Alder, Fernbach, and Rotenberg (eds), Academic Press, by permission)

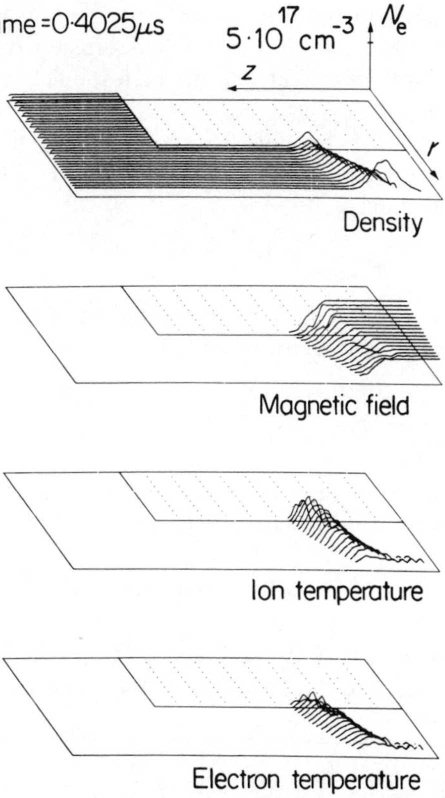

Figure 10.2. The density, azimuthal magnetic field and electron and ion temperatures in a two-dimensional magnetohydrodynamic simulation of the plasma focus (from Roberts, K. V., and Potter, D. E., 'Magnetohydrodynamic Calculations', in *Methods in Computational Physics*, **9**, Alder, Fernbach, and Rotenberg (eds), Academic Press, by permission)

(*b*) *The Magnetic Vector Potential.* In the foregoing sections of this chapter, the magnetic fields have been treated from the point of view of the conservation of magnetic flux as specified by Faraday's law. The magnetic field equations have therefore been determined in a manner entirely analogous to the fluid variables of density, momentum and total energy density.

Nevertheless, the magnetic field components are not independent and must satisfy a divergence-free condition,

$$\nabla . \mathbf{B} = 0$$

In one space dimension there is no difficulty, but in two or three space dimensions when Faraday's law is applied directly (equation 10.17) the difference scheme will lead to errors in which the field components do not satisfy this condition.

It is frequently useful therefore to determine the magnetic fields from time-dependent equations for the magnetic vector potential **A**. We may integrate equation 10.64:

$$\mathbf{B} = \nabla \wedge \mathbf{A} \tag{10.65}$$

and equally Faraday's law may be integrated to yield the time-dependent equation for the magnetic vector potential:

$$\frac{1}{c}\frac{\partial \mathbf{A}}{\partial t} + \mathbf{E} = -\nabla \Phi \tag{10.66}$$

where Φ is the scalar potential. This equation is coupled to the motion of the fluid according to the appropriate Ohm's law of the fluid (equation 10.10):

$$\frac{\partial \mathbf{A}}{\partial t} - \mathbf{v} \wedge (\nabla \wedge \mathbf{A}) = -c\eta\mathbf{j} - c\nabla\Phi \tag{10.67}$$

In turn, the current density is specified by Ampère's law (equation 10.6), and is therefore related to the magnetic vector potential (equation 10.65) by Poisson's equation:

$$\nabla \wedge (\nabla \wedge \mathbf{A}) = \frac{4\pi\mathbf{j}}{c}$$

$$\nabla^2 \mathbf{A} = -\frac{4\pi\mathbf{j}}{c} \tag{10.68}$$

The Coulomb 'gauge', $\nabla \cdot \mathbf{A} = 0$, has been assumed.

The usefulness of this procedure depends on the geometry of the problem at hand. We shall illustrate the application of such an approach to the two-dimensional (x–y) Cartesian plane. The magnetic fields in the x–y plane may then be specified by a perpendicular component of the magnetic vector potential (A_z) which can be treated as a pseudo-scalar. No gradient of the potential exists in the axial direction so that the equation for the vector potential (equation 10.67) takes the simple form:

$$\frac{\partial A_z}{\partial t} + \mathbf{v} \cdot \nabla A_z = \frac{\eta c^2}{4\pi}\nabla^2 A_z \tag{10.69}$$

It is evident that in the absence of resistive diffusion, the magnetic vector potential is simply advected by the motion of the fluid, and we may resolve this equation by simple second-order explicit methods (Section 3.7) in the conventional way. The magnetic field components are determined as simple derivatives:

$$B_x = \frac{\partial A_z}{\partial y}, \qquad B_y = -\frac{\partial A_z}{\partial x} \tag{10.70}$$

which ensures the *identical* satisfaction of the divergence-free condition.

If, in such a two-dimensional problem, all three components of the magnetic field occur, the third component of the magnetic field (B_z) is specified in the usual way according to Faraday's law:

$$\frac{\partial B_z}{\partial t} + \nabla \cdot (\mathbf{v}B_z - \mathbf{B}v_z) = \nabla \cdot \frac{\eta c^2}{4\pi} \nabla B_z \tag{10.71}$$

The three components of the magnetic field are defined by the two dependent variables B_z, A_z.

It is useful too, in this geometry, to consider the low-density or 'vacuum' regions where the Alfven speed becomes very large (Section 10.3d). To avoid the restriction of a limitingly small time step as imposed by the Courant–Friedrichs–Lewy condition due to the large Alfven speed in such regions (equations 10.40), we have introduced the approximation of treating the vacuum region as an insulator thereby avoiding very fast Alfven waves. This implies that in the vacuum region B_z is a constant which is readily specified according to boundary conditions. In the cylindrical (r–z) plane, the corresponding vacuum field for the azimuthal component of the magnetic field is:

$$B_\Theta = \frac{2I}{rc}$$

where I is a total current flowing in an external circuit.

On the other hand, the two components of the magnetic field in the plane of the problem are to be determined from the magnetic vector potential (equation 10.70) which in the vacuum region for the (x, y) plane satisfies Laplace's equation:

$$\nabla^2 A_z = 0 \tag{10.72}$$

This equation must be resolved at every time step in an awkwardly shaped domain with a moving boundary (Figure 10.3). Because of the shape of the domain, exact methods of solution are inapplicable and the methods of successive over-relaxation or Chebyshev iteration (Chapter 4) are to be used.

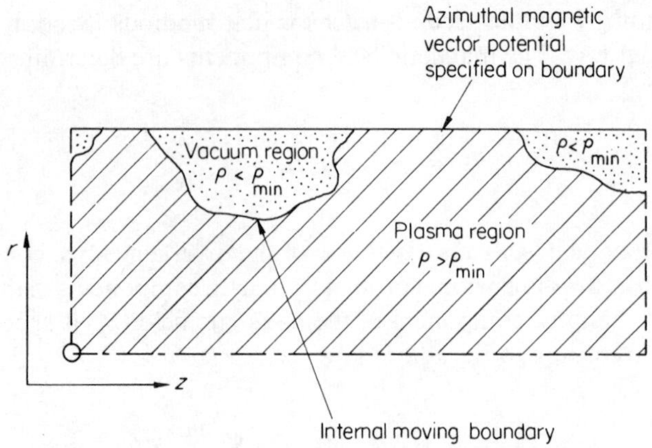

Figure 10.3. The vacuum region in magnetohydrodynamic simulations. In the vacuum region magnetic energy replaces fluid energy. Laplace's equation must be solved in the vacuum region

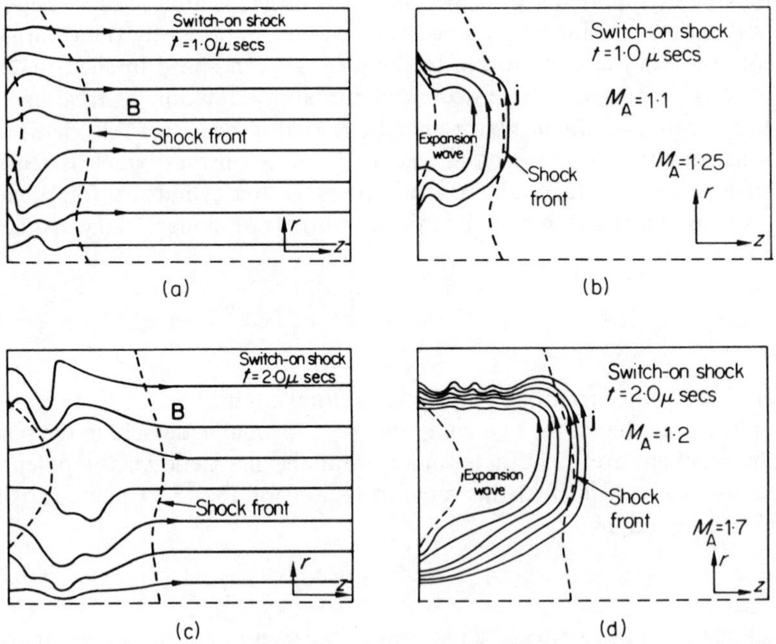

Figure 10.4. The simulation of a switch-on shock in two dimensions by an explicit conservative method: (a) the magnetic fields in the $(r - z)$ plane; (b) the currents in the $(r - z)$ plane associated with the shock which switch on the third component of the magnetic field (from Watkins 1972)

Such an approach to the solution of the two-dimensional magneto-hydrodynamic problem has been applied by Freeman and Lane (1968) and Watkins (1972). An example of such a simulation which illustrates the complex time evolution of a three-dimensional magnetic field is illustrated in Figure (10.4). The figure illustrates the anisotropic motion of a 'switch-on' shock parallel to an initial magnetic field, where the shock switches on a third component of the magnetic field. The inverse process of a 'switch-off' shock is believed to be of importance in the dissipation of magnetic energy in solar flares.

(*c*) *Concluding Remarks on Special Techniques in Magneto-hydrodynamics.* The complexity of the magneto-hydrodynamic equations and the range of phenomena which they describe imply that no one method of solution can be applied to describe all regimes of interest. The problem of the simulation of multidimensional magneto-hydrodynamics is a partially unsolved problem, though some special methods have been developed to deal with particular aspects.

An alternative to the difference method in the solution of the magneto-hydrodynamic equations for supersonic problems is the particle-in-cell collision-dominated method (Section 6.5), which may be applied in a similar manner as to the hydrodynamic problem (Butler *et al.* 1969). The PIC method applies to the case where the diffusion processes are negligible and where the beta of the fluid is of the order of unity.

Of particular importance in the study of magneto-hydrodynamics is the regime of a low-beta fluid. This is the problem of greatest interest in controlled thermonuclear fusion physics, but equally it is the most difficult problem to study for a number of reasons (Roberts and Potter 1970). If the beta of the fluid is small, Alfven waves will propagate with very high frequencies so that to avoid small time steps an implicit method is required. While this may be achieved in one space dimension due to the simplicity of the inversion of tridiagonal matrices, much slower iterative methods are available in two and three space dimensions (Chapter 4). An approach has been made to the implicit problem by using the alternating-direction implicit method (Section 4.6f) (Lindemuth 1971).

In multidimensional plasmas there is also the effect of anisotropy due to the vector properties of a strong magnetic field. In low-density plasmas processes of diffusion and waves can occur preferentially along a magnetic field (Robinson and Bernstein 1963, Kaufman 1960) and it is clear that difficulties arise in the simulation of an anisotropic fluid on an anisotropic Eulerian lattice or mesh. Providing the fields are not distorted or perturbed very much, it is evident that a difference scheme can be devised in which the magnetic field lines are used as the coordinates of the problem. Winsor *et al.* (1970) have applied this method for resolving the hydrodynamic equations on a specified unchanging magnetic field configuration (limit of zero beta).

5 Gravitational Hydrodynamics

The main application of gravitational hydrodynamics is in the study of the structure and properties of stars (Schwarzschild 1958). There are many problems associated with the macroscopic study of stars, though we may identify three major areas of interest: the evolution of stars as energy is produced by nuclear reactions and lost through radiation; the structure of stellar atmospheres which may be regarded as a time-independent problem in planar geometry (Mihalas 1967); and the dynamical properties of stars such as stellar pulsation (Christy 1966) or the gravitational collapse of old cold stars when dynamics employing special relativity or general relativity might be required (May and White 1967).

We shall here be concerned with the classical (nonrelativistic) formulation of the gravitational fluid problem and in particular with the nonlinear interaction of the field and pressure of the fluid on the dynamics of the star. The nature of the problem is very similar to the magneto-hydrodynamic problem, though on the one hand, many problems are simpler since they may be treated in one space dimension assuming spherical symmetry of the star. On the other hand in application to particular problems, many detailed *local* processes of ionization phenomena, of nuclear reactions, and of non-equilibrium radiation effects might have to be described.

(*a*) *The Dynamic Equations with Radiation.* According to the discussion in Section 10.1, in the nonrelativistic problem, we may include the effect of the self-gravitational field of a star as a body force in the hydrodynamic equation for the acceleration or momentum change of the fluid (equations 10.1, 10.2). In three space dimensions the Eulerian hydrodynamic equations for the density, momentum density and internal energy density (9.1, 9.2, 9.3) of a self-gravitating fluid take the form:

mass:
$$\frac{\partial \rho}{\partial t} = -\mathbf{\nabla} \cdot \rho \mathbf{v} \qquad (10.73)$$

momentum:
$$\frac{\partial \rho \mathbf{v}}{\partial t} = -\mathbf{\nabla} \cdot (\rho \mathbf{v} \mathbf{v}) - \mathbf{\nabla} p - \rho \mathbf{\nabla} \Phi \qquad (10.74)$$

internal energy:
$$\frac{\partial}{\partial t} \rho \epsilon = -p \mathbf{\nabla} \cdot \mathbf{v} - \mathbf{\nabla} \cdot (\rho \epsilon \mathbf{v} + \mathbf{q}) + S \qquad (10.75)$$

where the gravitational potential is defined by Poisson's equation:

$$\nabla^2 \Phi = 4\pi G \rho \qquad (10.76)$$

$\mathbf{q}$ is an energy conduction or diffusion term and S is a source term to describe the production of energy either by chemical or ionization processes, or by fusion and fission in nuclear processes.

The fundamental phenomena associated with the hyperbolic equations of advection, sound waves and gravitational oscillations have previously been stressed (Section 10.1). But in addition to these effects, the most important property to be included within these equations is that of radiation. There are several ways of treating the problem of the radiation field. In the most complete approach to the problem, we might regard any small volume element of the star as a neutrally-charged assembly of electrons, of nuclei (or ions) and of photons and consequently separate equations for the internal energy density of each species could be developed. However, since in general the density of particles is extremely large, we shall regard each species of the assembly as being in local thermodynamic equilibrium with every other species. Under this approximation, the internal energy density ϵ (equation 10.75) is the total internal energy density in the sense of being the sum over all species. It is related to the pressure and density of the fluid by a possibly complex equation of state:*

$$\epsilon = \epsilon(p, \rho) \tag{10.77}$$

This assumption, of local thermodynamic equilibrium between species, permits a particularly simple description of the radiation field which is included directly in equations 10.74 and 10.75. Thus the total pressure on the fluid is the sum of the partial pressures of the radiation p_r, the electrons p_e and the ions p_i:

$$p = p_r + p_e + p_i \tag{10.78}$$

According to statistical mechanics, the radiant power emitted by a black body is:

$$W = \sigma T^4 \tag{10.79}$$

where T is the equilibrium temperature. This is the Stefan–Boltzmann law where σ is Stefan's constant. Correspondingly the radiation pressure is (Eyring *et al.* 1964):

$$p_r = \frac{4}{3}\frac{\sigma T^4}{c} \tag{10.80}$$

where c is the velocity of light.

Equally, the effect of the transport of radiative energy is to be included in the internal energy equation (10.75). With the assumption of thermodynamic equilibrium between the radiation field and the fluid, the transport of radiative

* For example, see Kippenhahn *et al.* (1967).

energy may be included in the energy diffusion term $\mathbf{q}$, since the radiative energy flux is:

$$\mathbf{q} = \frac{\partial}{\partial \tau}(p_r c) \tag{10.81}$$

where $\partial\tau$ is defined as an optical depth:

$$\partial \tau = -\kappa\rho\,\mathbf{dr} \tag{10.82}$$

κ is the opacity of the fluid material. Using equations 10.80, 10.81 and 10.82, the effect of the flux of radiation-energy occurs as a diffusion term in the internal energy equation:

$$\mathbf{q} = -\alpha \nabla T$$

with the diffusion coefficient:

$$\alpha = \frac{16}{3}\frac{\sigma T^3}{\kappa\rho} \tag{10.83}$$

These equations (10.73 to 10.76) with the effect of radiation pressure and radiative diffusion are the classical partial differential equations for the study of the dynamics of stars. Nuclear reactions and chemical reactions occur as local effects and may, therefore, be included by a set of ordinary differential equations at each local point. They affect the dynamics through the source S in the internal energy equation and in specifying the equation of state of the fluid and the opacity (Cox and Stewart 1965). The rates of nuclear reactions depend on the density, the temperature and the relative abundance of each nuclear component (Reeves 1965).

(*b*) *The Lagrangian Mass Coordinate.* A simulation in one space dimension is appropriate for many problems in the structure and internal dynamics of stars. Such an assumption relying on spherical symmetry ignores the effect of convective processes which might be important in producing turbulent mixing of fluid variables. Nevertheless, the simplicity introduced by employing one radial space variable is considerable, since a Lagrangian mesh may be used and implicit methods are easily soluble in one space dimension.

We shall define r as a dependent variable describing a radial shell in the star:

$$\frac{\mathrm{d}r}{\mathrm{d}t} = v \tag{10.84}$$

where v is the radial centre-of-mass velocity. It is convenient to define the independent variable as the mass within the shell of radius r:

$$M(r) = \int_0^r 4\pi s^2 \rho(s)\,\mathrm{d}s \tag{10.85}$$

From the continuity equation (10.73) we obtain in the Lagrangian frame:

$$\frac{\mathrm{d}M}{\mathrm{d}t} = 0 \tag{10.86}$$

Clearly the mass contained within each spherical Lagrangian shell is a constant.

It follows that in this simple geometry, Poisson's equation (10.76) may be integrated immediately to define the local gravitational acceleration g:

$$g = -\frac{\partial \Phi}{\partial r}$$

$$= -\frac{GM}{r^2} \tag{10.87}$$

We shall rewrite the momentum equation (10.74) as an acceleration equation in the Lagrangian coordinate, and using the result (equation 10.87) for the gravitational force we obtain:

$$\frac{\mathrm{d}v}{\mathrm{d}t} = -\frac{GM}{r^2} - \frac{1}{\rho}\frac{\partial p}{\partial r} \tag{10.88}$$

Since M is to be used as the independent space variable, the acceleration equation may be rewritten as

$$\frac{\mathrm{d}v}{\mathrm{d}t} = -\frac{GM}{r^2} - 4\pi r^2 \frac{\partial p}{\partial M} \tag{10.89}$$

The final partial differential equation for the internal energy density (equation 10.75) may be rewritten in Lagrangian form:

$$\rho\frac{\mathrm{d}\epsilon}{\mathrm{d}t} = -p\nabla . \mathbf{v} - \nabla . \mathbf{q} + S$$

Again assuming spherical symmetry and defining the space variable as the mass coordinate M, we obtain:

$$\frac{\mathrm{d}\epsilon}{\mathrm{d}t} = -p\frac{\partial}{\partial M}(4\pi r^2 v) - \frac{\partial}{\partial M}(4\pi r^2 q) + S/\rho \tag{10.90}$$

It is to be noted that an alternative formulation (apart from nuclear reactions) may be used to replace the specific internal energy equation. The specific energy density of the fluid is

$$u = \frac{1}{2}v^2 - \frac{GM}{r} + \epsilon \tag{10.91}$$

and, by multiplying the acceleration equation (10.89) and adding it to the internal energy equation (10.90) a conservative equation is obtained. In stellar problems, however, large variations occur between the gravitational and thermal energy in different regions of the star and, in particular, there exist regions of low density on the surface of the star where the calculation of the temperature from the total energy equation would yield large errors from the small thermal energy term. Hence it is most appropriate to use the equation for the internal energy directly.

The stellar problem has here been reduced to three first-order (in time) one-dimensional (in the 'space' variable M) equations, for the radius $r(M, t)$, velocity $v(M, t)$ and internal energy $\epsilon(M, t)$ (equations 10.84, 10.89. 10.90). The equations are essentially hyperbolic, describing advection, sound waves and gravitational oscillations, though the inclusion of the radiative diffusion term makes the energy equation parabolic with a diffusion decay rate:

$$\omega_R = \frac{\Delta^2}{4\sigma T^3/3\kappa\rho} \tag{10.92}$$

In problems of stellar evolution occurring over a very long time scale, the inertia of the fluid may be neglected and the assumption of hydrostatic equilibrium employed. The acceleration equation (10.89) may be replaced by:

$$\frac{GM}{r^2} + 4\pi r^2 \frac{\partial p}{\partial M} = 0 \tag{10.93}$$

This assumption has the effect of removing the rapid frequencies associated with sound waves and gravitational oscillations.

(c) *An Explicit Difference Solution for Stellar Pulsation.* The formulation of the gravitational hydrodynamic problem in one dimension is very similar to the magneto-hydrodynamic problem and each of the methods employed for the latter (Section 10.3) are equally appropriate for the gravitational case. Either implicit or explicit methods may be used for the integration of the equations in time.

We shall illustrate the solution of the above equations by the explicit Lagrangian method as applied for the equivalent hydrodynamic problem (Section 9.5a). The Lagrangian independent variable M is discretized by spherical shell boundaries j, which define mesh cells $j - \frac{1}{2}$ (Figure 9.17):

$$M_j = M_{j-1} + \Delta M_{j-\frac{1}{2}} \tag{10.94}$$

The dependent variables of the radius for each mass point and velocity are defined at the mesh boundaries r_j^n, $v_j^{n-\frac{1}{2}}$, with the velocities defined at the half time steps $t^{n-\frac{1}{2}}$. All other variables (the intensive variables) may be defined as cell quantities at $j - \frac{1}{2}$, and at integer time steps. Thus given the radius and pressure (from the equation of state) at integer times and cell points, the acceleration equation (10.87) is integrated with a time-centred, space-centred scheme,

$$v_j^{n+\frac{1}{2}} = v_j^{n-\frac{1}{2}} - \Delta t \frac{GM_j}{(r_j^n)^2} - \frac{\Delta t\, 4\pi (r_j^n)^2}{\frac{1}{2}(\Delta M_{j+\frac{1}{2}} + \Delta M_{j-\frac{1}{2}})}(p_{j+\frac{1}{2}}^n - p_{j-\frac{1}{2}}^n) \tag{10.95}$$

and it follows that the new radial positions of the cell boundaries at integer times may be determined:

$$r_j^{n+1} = r_j^n + v_j^{n+\frac{1}{2}}\, \Delta t \tag{10.96}$$

The local density of each cell is determined:

$$\rho_{j-\frac{1}{2}}^{n+1} = \frac{M_j - M_{j-1}}{\frac{4}{3}\pi\{(r_j^{n+1})^3 - (r_{j-1}^{n+1})^3\}} \tag{10.97}$$

It remains to solve the equation for the specific internal energy density (equation 10.90). The essential difficulty here is the nonlinear radiation diffusion term **q** (equation 10.83). Since the radiative diffusion coefficient α is a strong function of temperature and density, it is essential to solve the diffusion equation implicitly in order that the time step Δt is not limited by the diffusion decay time. We may express the internal energy equation in the form:

$$\frac{\mathrm{d}}{\mathrm{d}t}(\bar{\epsilon} T) + \bar{p} T \frac{\partial}{\partial M}(4\pi r^2 v) - \frac{\partial}{\partial M}\left(4\pi r^2 \bar{\alpha} \frac{\partial T}{\partial M}\right) = \frac{S}{\rho} \tag{10.98}$$

where the internal energy density is related to the temperature and pressure according to an equation of state and we have written

$$\epsilon = \bar{\epsilon} T$$

$$p = \bar{p} T$$

The second term in the equation is the adiabatic compression term where the rate of change of each volume element is known from the change in density (equation 10.97)

$$\frac{\partial}{\partial M}(4\pi r^2 v) = -\frac{1}{\rho}\frac{\mathrm{d}\rho}{\mathrm{d}t}$$

The problem of resolving equation 10.98 implicitly is therefore entirely analogous to the nonlinear diffusion equation considered for the magnetohydrodynamic problem (Section 10.3b), and it may be solved in the same way by the Crank–Nicholson method. The notation of a bar in equation 10.98 has been used to illustrate the variables which are held constant with respect to the temperature during an iteration step (Section 10.3b). At each iteration a tridiagonal matrix equation coupling the temperature at cell points is resolved. When convergence is achieved the time step is complete.

Because the gravitational frequency is an increasing function of the density and the time step must be chosen smaller than the period of this frequency, the inner boundary is not taken at $r = 0$, but over a rigid radiating small sphere at the centre of the star. Christy (1966) has applied such a model to the study of pulsating stars. It is found in the simulations that for a wide range of stellar parameters, nonlinear steady-state oscillations exist, of which the typical amplitude is 10 per cent of the equilibrium configuration. The results are in essential agreement with observation.

Bibliography

Alder, B. J., and Wainwright, T. E. (1959). 'Studies in molecular dynamics I: general method', *J. Chem. Phys.*, **31**, 459.

Alder, B. J., and Wainwright, T. E. (1960). 'Studies in molecular dynamics II: behaviour of a small number of elastic spheres', *J. Chem. Phys.*, **33**, 1439.

Alder, B. J., and Wainwright, T. E. (1962). 'Phase transition in elastic disks', *Phys. Rev.*, **127**(2), 359.

Amsden, A. A. (1966). *The Particle-in-Cell Method for the Calculation of the Dynamics of Compressible Fluids* (Report LA-3466). Los Alamos, New Mexico: Los Alamos Sci. Lab.

Arakawa, A. (1966). 'Computational design for long-term numerical integration of the equations of fluid motion: two-dimensional incompressible flow', *J. of Comp. Phys.*, **1**, 119.

Berk, H. L., and Roberts, K. V. (1970). 'The waterbag model.' In Alder, B., Fernbach, S., and Rotenberg, M. (editors), *Methods in Computational Physics*, Volume 9, *Plasma Physics*. New York/London: Academic Press. p. 87.

Birdsall, C. K., and Fuss, D. (1969). 'Clouds-in-clouds, clouds-in-cells physics for many-body plasma simulation', *J. Comp. Phys.*, **3**, 494.

Birdsall, C. K., Langdon, A. B., and Okudu, H. (1970). 'Finite-size particle physics applied to plasma simulation.' In Alder, B., Fernbach, S., and Rotenberg, M. (editors), *Methods in Computational Physics*, Volume 9, *Plasma Physics*. New York/London: Academic Press. p. 241.

Birkhoff, G., and Mac Lane, S. (1965). *A Survey of Modern Algebra*, Third edition. New York: Macmillan.

Bohm, D., and Pines, D. (1950). 'Screening of electronic interactions in a metal', *Phys. Rev.*, **80**, 903.

Bohm, D., and Pines, D. (1953). 'A collective description of electron interactions III: Coulomb interactions in a degenerate electron gas', *Phys. Rev.*, **92**, 609.

Boris, J., and Roberts, K. V. (1969). 'The optimization of particle calculations in two and three dimensions', *J. Comp. Phys.*, **4**, 552.

Brust, D. (1968). 'The pseudopotential method and the single-particle electronic excitation spectrum of crystals.' In Alder, B., Fernbach, S., and Rotenberg, M. (editors), *Methods of Computational Physics*, Volume 8, *Energy Bands of Solids*. New York/London: Academic Press. p. 33.

Bryan, K. (1965). 'Non-linear effects in the theory of wind-driven ocean circulation.' In Alder, B., Fernbach, S., and Rotenberg, M. (editors), *Methods in Computational Physics*, Volume 4, *Applications in Hydrodynamics*. New York/London: Academic Press. p. 29.

Bryan, K. (1969). 'A numerical method for the study of the circulation of the world ocean', *J. Comp. Phys.*, **4**, 347.

Bryan, K., and Cox, M. D. (1972). 'The circulation of the world ocean: a numerical study. Part I', *J. Physical Oceanography*, **2**, 319.

Buneman, O. (1959). 'Dissipation of currents in ionized media', *Phys. Rev.*, **115**, 503.

Buneman, O. (1969). *Report SUIPR 294*. Stanford, California: Inst. for Plasma Res., Stanford Univ.

Butler, T. D. (1967). 'Numerical solutions of the hypersonic sharp-leading edge problem', *Phys. Fluids*, **10**, 1205.

Butler, T. D., Henins, I., Jahoda, F., Marshall, J., and Morse, R. L. (1969). 'Co-axial snowplow discharge', *Phys. Fluids*, **12**, 1904.

Chapman, S., and Cowling, T. G. (1953). *The Mathematical Theory of Non-Uniform Gases*. London/New York: Cambridge University Press.

Christiansen, J. (1971). *Vortex: A 2-Dimensional Hydrodynamics Simulation Code* (Report CLM-R 106). Abingdon: UKAEA, Culham Laboratory.

Christy, R. F. (1966). 'A study of pulsation in RR lyrae models', *Astrophys. J.*, **144**, 108.

Christy, R. F. (1967). 'Computational methods in stellar pulsation.' In Alder, B., Fernbach, S., and Rotenberg, M. (editors), *Methods in Computational Physics*, Volume 7, *Astrophysics*. New York/London: Academic Press. p. 191.

Cohen, M. H., and Heine, V. (1961). 'Cancellation of kinetic and potential energy in atoms, molecules and solids', *Phys. Rev.*, **122**, 1821.

Cooley, J. W., and Tukey, J. W. (1965). 'An algorithm for the machine calculation of complex Fourier series', *Math. Comp.*, **19**, 297.

Corby, G. A., Gilchrist, A., and Newson, R. L. (1972). 'A general circulation model of the atmosphere suitable for long-period integrations', *Quar. J. of the Meteorological Soc.*, **98**, 809.

Courant, R., and Friedrichs, K. O. (1948). *Supersonic Flow and Shock Waves*. New York: Interscience.

Courant, R., Friedrichs, K. O., and Lewy, H. (1928). 'Über die partiellen Differenzengleichungen der mathematischen Physik', *Math. Ann.*, **100**, 32.

Courant, R., and Hilbert, D. (1953). *Methods of Mathematical Physics*, Volume I. New York: Interscience.

Courant, R., and Hilbert, D. (1962). *Methods of Mathematical Physics*, Volume II. New York: Interscience.

Cox, A. N., and Stewart, J. N. (1965). 'Radiative and conductive opacities for eleven astrophysical mixtures', *Astrophys. J. Suppl.*, **11**, 22.

Crank, J., and Nicholson, P. (1947). 'A practical method for numerical integration of solutions of partial differential equations of heat-conductive type', *Proc. Camb. Philos. Soc.*, **43**, 50.

Dawson, J. (1962). 'One-dimensional plasma model', *Phys. Fluids*, **5**, 445.

Dawson, J. (1970). 'The electrostatic sheet model for a plasma and its modification to finite size particles.' In Alder, B., Fernbach, S., and Rotenberg, M. (editors), *Methods in Computational Physics*, Volume 9, *Plasma Physics*. New York/London: Academic Press. p. 1.

Deardorff, J. W. (1971). 'On the magnitude of the subgrid scale eddy coefficient', *J. Comp. Phys.*, **7**, 120.

Dekker, A. J. (1957). *Solid-state Physics*. London: Macmillan.

Dicke, R. H., and Wittke, J. P. (1960). *Introduction to Quantum Mechanics*. Reading, Massachusetts/London: Addison-Wesley.

Dufort, E. C., and Frankel, S. P. (1953). 'Stability conditions in the numerical treatment of parabolic differential equations', *Math. Tables and Other Aids to Comp.*, **7**, 135.

Duhem, P. (1954). *The Aim and Structure of Physical Theory*. Translated by Philip L. Wiener. Princeton, New Jersey: Princeton University Press.

Dutton, J. A., and Deaven, D. G. (1972). 'Some observed properties of atmospheric turbulence.' In Rosenblatt, M., and van Atta, C. (editors), *Statistical Models and Turbulence*. Berlin/New York: Springer-Verlag.

Edgerton, H. E., and Killian, J. R., Jnr. (1954). In *Flash*. Boston, Massachusetts: Charles, T. Branford.

Eyring, H., Henderson, D., Stover, B. J., and Eyring, E. M. (1964). *Statistical Mechanics and Dynamics*. New York/London: Wiley.

Faddeev, D. K., and Faddeeva, V. N. (1963). *Computational Methods of Linear Algebra.* Translated from the Russian by R. C. Williams. San Francisco/London: Freeman.

Fernbach, S., and Taub (editors) (1971). *Computers and their Role in the Physical Sciences.* New York/London: Academic Press.

Fox, L. (1964). *An Introduction to Numerical Linear Algebra.* Oxford: Clarendon Press.

Fox, L., and Mayers, D. F. (1968). *Computing Methods for Scientists and Engineers.* Oxford: Clarendon Press.

Freeman, J. R., and Lane, F. O. (1968). *Initial Results from a Two-Dimensional Lax–Wendroff Hydromagnetic Code.* (Report LA-3990. Los Alamos Scientific Laboratory, Los Alamos, New Mexico, paper C7.)

Goldstein, H. (1962). *Classical Mechanics.* Reading, Massachusetts/London: Addison-Wesley.

Golub, G. S., and Varga, R. S. (1961). 'Chebyshev semi-iterative methods, successive over-relaxation methods, iterative methods and second-order Richardson iterative methods, Part II', *Numerische Math.*, **3**, 157.

Hain, K., Hain, G. Roberts, K. V., Roberts, S. J., and Koppendorfer, W. (1960). 'Fully ionized pinch collapse', *Z. Naturforsch.*, **15a**, 1039.

Harlow, F. H. (1964). 'The particle-in-cell computing method for fluid dynamics.' In Alder, B., Fernbach, S., and Rotenberg, M. (editors), *Methods in Computational Physics*, Volume 3, *Fundamental Methods in Hydrodynamics.* New York/London: Academic Press. p. 319.

Harlow, F. H., and Shannon, J. P. (1967). 'The splash of a liquid drop', *J. Appl. Phys.*, **38**, 3855.

Harlow, F. H., and Welch, J. E. (1965). 'Numerical calculation of time-dependent viscous incompressible flow', *Phys. Fluids*, **8**, 2182.

Hartree, D. R. (1957). *The Calculation of Atomic Structures.* New York: Wiley/London: Chapman and Hall.

Haurwitz, B. (1941). *Dynamic Meteorology.* New York: McGraw-Hill.

Herring, C. (1940). 'A new method for calculating wave functions in crystals', *Phys. Rev.*, **57**, 1169.

Hockney, R. W. (1970). 'The potential calculation and some applications.' In Alder, B., Fernbach, S., and Rotenberg, M. (editors), *Methods in Computational Physics*, Volume 9, *Plasma Physics.* New York/London: Academic Press. p. 135.

Hockney, R. W. (1971a). 'Measurements of collision and heating times in a two-dimensional thermal computer plasma', *J. Comp. Phys.*, **8**, 19.

Hockney, R. W. (1971b): 'Self-consistent electron motion through the triode substrate'. Unpublished work.

Hockney, R. W., and Hohl, F. (1969). 'Effects of velocity dispersion on the evolution of a disk of stars', *Astron. J.*, **74**, 1102.

Hohl, F. (1970). *Dynamical Evolution of Disk Galaxies.* (Report NASA-TR, R-343.) Springfield, Virginia: National Tech. Inform. Service.

Hohl, F. (1971). *Computer Experiments on the Structure and Dynamics of Spiral Galaxies.* (Report NASA-TN D-6630.) Springfield Virginia: National Tech. Inform. Service.

Jackson, J. D. (1963). *Classical Electrodynamics.* New York/London: Wiley.

Jeffrey, A. (1966). *Magnetohydrodynamics.* New York: Wiley.

Kasahara, A., and Washington, W. M. (1967). 'NCAR global general circulation model of the atmosphere', *Monthly Weather Rev.*, **95**(7), 389.

Kaufman, A. N. (1960). 'Plasma viscosity in a magnetic field', *Phys. Fluids*, **3**, 610.

Kippenhahn, R., Weigert, A., and Hofmeister, E. (1967). 'Methods for calculating stellar evolution.' In Alder, B., Fernbach, S., and Rotenberg, M. (editors), *Methods in Computational Physics*, Volume 7, *Astrophysics*. New York/London: Academic Press. p. 129.

Landau, L. D., and Lifshitz, E. M. (1959). *Fluid Mechanics*. Translated from the Russian by J. B. Sykes and W. H. Reid. London: Pergamon.

Lapidus, A. (1967). 'A detached shock calculation by second-order finite differences', *J. Comp. Phys.*, **2**, 154.

Lax, P. D. (1954). 'Weak solutions of non-linear hyperbolic equations and their numerical computation', *Comm. Pure Appl. Math.*, **7**, 135.

Lax, P. D., and Wendroff, B. (1960). 'Systems of conservation laws', *Comm. Pure Appl. Math.*, **13**, 217.

Leith, C. E. (1965). 'Numerical simulations of the earth's atmosphere'. In Alder, B., Fernbach, S., and Rotenberg, M. (editors), *Methods in Computational Physics*. Volume 4, *Applications in Hydrodynamics*. New York/London: Academic Press. p. 1.

Lilly, D. K. (1967). 'The representation of small-scale turbulence in numerical simulation experiments.' In *Proceedings of the IBM Scientific Computing Symposium on Environmental Sciences* (IBM Form No. 320-1951).

Lindemuth, I. R. (1971). *The Alternating-Direction-Implicit Numerical Solution of Time-Dependent, Two-Dimensional Two-Fluid Magnetohydrodynamic Equations*. (Report TID-4500, UC. 34.) Livermore, California: Univ. of Calif. Lawrence Rad. Lab.

Lundgren, T. S. (1967). 'Distribution functions in the statistical theory of turbulence', *Phys. Fluids*, **10**, 969.

May, M. M., and White, R. H. (1967). 'Stellar dynamics and gravitational collapse.' In Alder, B., Fernbach, S., and Rotenberg, M. (editors), *Methods in Computational Physics*, Volume 7, *Astrophysics*. New York/London: Academic Press. p. 219.

Matthews, P. T. (1968). *Introduction to Quantum Mechanics*. New York/London: McGraw-Hill.

Metropolis, N., Rosenbluth, A. W., Rosenbluth, M. N., Teller, A. H., and Teller, E. (1953). 'Equation of state calculations by fast computing machines', *J. Chem. Phys.*, **21**(6), 1087.

Mihalas, D. (1967). The calculation of model stellar atmospheres.' In Alder, B., Fernbach, S., and Rotenberg, M. (editors), *Methods in Computational Physics*, Volume 7, *Astrophysics*, New York/London: Academic Press. p. 1.

Miller, R. H., Prendergast, K. H., and Quirk, W. J. (1970). 'Numerical experiments on spiral structures', *Astrophys. J.*, **161**, 903.

Morse, R. L. (1970). 'Multidimensional plasma simulation by the particle-in-cell method.' In Alder, B., Fernbach, S., and Rotenberg, M. (editors), *Methods in Computational Physics*, Volume 9, *Plasma Physics*. New York/London: Academic Press. p. 213.

Morse, R. L., and Nielson, C. W. (1969). 'Numerical simulation of warm two-beam plasma', *Phys. Fluids*, **12**, 2418.

Onsager, L. (1949). 'Statistical hydrodynamics', *Nuovo Cimento*, *Suppl.* 6, 279.

Oort, J. H. (1965). 'Stellar dynamics.' In Blaauw, A., and Schmidt, M. (editors), *Galactic Structure*. Chicago, Illinois: University of Chicago Press. p. 455.

Peaceman, D. W., and Rachford, H. H., Jnr. (1955). 'The numerical solution of parabolic and elliptic differential equations', *J. Soc. Indus. Appl. Math.*, **3**, 28.

Philips, G. M., and Taylor, P. J. (1969). *Computers*. London: Methuen.

Potter, D. E. (1971). 'Numerical studies of the plasma focus', *Phys. Fluids*, **14**, 1911.

Reeves, H. (1965). 'Stellar energy sources.' In Aller, L. H., and McLaughlin, D. B. (editors), *Stellar Structure*. Chicago/London: University of Chicago Press. p. 113.

Richardson, L. F. (1922). *Weather Prediction by Numerical Process*. London: Cambridge University Press.

Richtmyer, R. D. (1962). *A Survey of Difference Methods for Non-Steady Fluid Dynamics*. (Report NCAR-TN 63-2.) Boulder, Colorado: Natl. Center for Atmos. Res.

Richtmyer, R. D., and Morton, K. W. (1967). *Difference Methods for Initial-Value Problems*. Second edition. New York/London: Interscience.

Roberts, K. V., and Potter, D. E. (1970). 'Magnetohydrodynamic calculations.' In Alder, B., Fernbach, S., and Rotenberg, M. (editors), *Methods in Computational Physics*, Volume 9, *Plasma Physics*. New York/London: Academic Press. p. 339.

Roberts, K. V., and Weiss, N. O. (1966). 'Convective difference schemes', *Math. Comp.*, **20**, 272.

Robinson, B. B., and Bernstein, I. B. (1963). 'A variational description of transport phenomena in a plasma', *Annals Phys.*, **18**, 110.

Roothan, C. C. J. (1951). 'New developments in molecular orbital theory', *Revs. Modern Phys.*, **23**, 69.

Roothan, C. C. J., and Bagus, P. S. (1963). 'Atomic self-consistent field calculations by the expansion method.' In Alder, B., Fernbach, S., and Rotenberg, M. (editors), *Methods in Computational Physics*, Volume 2, *Quantum Mechanics*. New York/London: Academic Press. p. 47.

Schwarzschild, M. (1958). *Structure and Evolution of the Stars*. Princeton, New Jersey: Princeton University Press.

Slater, J. C. (1953). 'A generalized self-consistent field method', *Phys. Rev.*, **91**, 528.

Slater, J. C. (1963). *Quantum Theory of Molecules and Solids*, Volume I, *Electronic Structure of Molecules*. New York/London: McGraw-Hill.

Slater, J. C. (1965). *Quantum Theory of Molecules and Solids*, Volume II, *Symmetry and Energy Bands in Crystals*. New York/London: McGraw-Hill.

Slater, J. C. (1968). 'Exchange in spin-polarized energy bands', *Phys. Rev.*, **165**, 658.

Spitzer, L. (1962). *Physics of Fully Ionized Gases*. Second edition. New York: Wiley.

Tolman, R. C. (1967). *The Principles of Statistical Mechanics*. London: Oxford University Press.

Varga, R. S. (1962). *Matrix Iterative Analysis*. Englewood Cliffs, New Jersey: Prentice-Hall.

von Neumann, J., and Richtmyer, R. D. (1950). 'A method for the numerical calculation of hydrodynamic shocks', *J. Appl. Phys.*, **21**, 232.

Watkins, M. L. (1972). 'Magnetohydrodynamic simulation of the switch-on shock in two-dimensions'. Unpublished work.

Wilkinson, J. H. (1964). *The Algebraic Eigenvalue Problem*. Oxford: Clarendon Press.

Winsor, N. K., Johnson, J. L., and Dawson, J. M. (1970). 'Model for toroidal plasma containment with flow', *J. Comp. Phys.*, **6**, 430.

Young, D. M. (1962). 'The numerical solution of elliptic and parabolic differential equations.' In Todd, J. (editor), *A Survey of Numerical Analysis*. New York: McGraw-Hill.

Author Index

Subject Index